CAM DESIGN
AND MANUFACTURE

MECHANICAL ENGINEERING

A Series of Textbooks and Reference Books

EDITORS

L. L. FAULKNER

Columbus Division
Battelle Memorial Institute

and

Department of Mechanical Engineering
The Ohio State University
Columbus, Ohio

S. B. MENKES

Department of Mechanical Engineering
The City College of the
City University of New York
New York, New York

1. Spring Designer's Handbook, *by Harold Carlson*
2. Computer-Aided Graphics and Design, *by Daniel L. Ryan*
3. Lubrication Fundamentals, *by J. George Wills*
4. Solar Engineering for Domestic Buildings, *by William A. Himmelman*
5. Applied Engineering Mechanics: Statics and Dynamics, *by G. Boothroyd and C. Poli*
6. Centrifugal Pump Clinic, *by Igor J. Karassik*
7. Computer-Aided Kinetics for Machine Design, *by Daniel L. Ryan*
8. Plastics Products Design Handbook, Part A: Materials and Components; Part B: Processes and Design for Processes, *edited by Edward Miller*
9. Turbomachinery: Basic Theory and Applications, *by Earl Logan, Jr.*
10. Vibrations of Shells and Plates, *by Werner Soedel*
11. Flat and Corrugated Diaphragm Design Handbook, *by Mario Di Giovanni*
12. Practical Stress Analysis in Engineering Design, *by Alexander Blake*
13. An Introduction to the Design and Behavior of Bolted Joints, *by John H. Bickford*
14. Optimal Engineering Design: Principles and Applications, *by James N. Siddall*
15. Spring Manufacturing Handbook, *by Harold Carlson·*
16. Industrial Noise Control: Fundamentals and Applications, *edited by Lewis H. Bell*
17. Gears and Their Vibration: A Basic Approach to Understanding Gear Noise, *by J. Derek Smith*

18. Chains for Power Transmission and Material Handling: Design and Applications Handbook, *by the American Chain Association*

19. Corrosion and Corrosion Protection Handbook, *edited by Philip A. Schweitzer*

20. Gear Drive Systems: Design and Application, *by Peter Lynwander*

21. Controlling In-Plant Airborne Contaminants: Systems Design and Calculations, *by John D. Constance*

22. CAD/CAM Systems Planning and Implementation, *by Charles S. Knox*

23. Probabilistic Engineering Design: Principles and Applications, *by James N. Siddall*

24. Traction Drives: Selection and Application, *by Frederick W. Heilich III and Eugene E. Shube*

25. Finite Element Methods: An Introduction, *by Ronald L. Huston and Chris E. Passerello*

26. Mechanical Fastening of Plastics: An Engineering Handbook, *by Brayton Lincoln, Kenneth J. Gomes, and James F. Braden*

27. Lubrication in Practice, Second Edition, *edited by W. S. Robertson*

28. Principles of Automated Drafting, *by Daniel L. Ryan*

29. Practical Seal Design, *edited by Leonard J. Martini*

30. Engineering Documentation for CAD/CAM Applications, *by Charles S. Knox*

31. Design Dimensioning with Computer Graphics Applications, *by Jerome C. Lange*

32. Mechanism Analysis: Simplified Graphical and Analytical Techniques, *by Lyndon O. Barton*

33. CAD/CAM Systems: Justification, Implementation, Productivity Measurement, *by Edward J. Preston, George W. Crawford, and Mark E. Coticchia*

34. Steam Plant Calculations Manual, *by V. Ganapathy*

35. Design Assurance for Engineers and Managers, *by John A. Burgess*

36. Heat Transfer Fluids and Systems for Process and Energy Applications, *by Jasbir Singh*

37. Potential Flows: Computer Graphic Solutions, *by Robert H. Kirchhoff*

38. Computer-Aided Graphics and Design, Second Edition, *by Daniel L. Ryan*

39. Electronically Controlled Proportional Valves: Selection and Application, *by Michael J. Tonyan, edited by Tobi Goldoftas*

40. Pressure Gauge Handbook, *by AMETEK, U.S. Gauge Division, edited by Philip W. Harland*

41. Fabric Filtration for Combustion Sources: Fundamentals and Basic Technology, *by R. P. Donovan*

42. Design of Mechanical Joints, *by Alexander Blake*

43. CAD/CAM Dictionary, *by Edward J. Preston, George W. Crawford, and Mark E. Coticchia*

44. Machinery Adhesives for Locking, Retaining, and Sealing, *by Girard S. Haviland*

45. Couplings and Joints: Design, Selection, and Application, *by Jon R. Mancuso*

46. Shaft Alignment Handbook, *by John Piotrowski*

47. BASIC Programs for Steam Plant Engineers: Boilers, Combustion, Fluid Flow, and Heat Transfer, *by V. Ganapathy*

48. Solving Mechanical Design Problems with Computer Graphics, *by Jerome C. Lange*

49. Plastics Gearing: Selection and Application, *by Clifford E. Adams*

50. Clutches and Brakes: Design and Selection, *by William C. Orthwein*

51. Transducers in Mechanical and Electronic Design, *by Harry L. Trietley*

52. Metallurgical Applications of Shock-Wave and High-Strain-Rate Phenomena, *edited by Lawrence E. Murr, Karl P. Staudhammer, and Marc A. Meyers*

53. Magnesium Products Design, *by Robert S. Busk*

54. How to Integrate CAD/CAM Systems: Management and Technology, *by William D. Engelke*

55. Cam Design and Manufacture, Second Edition; with cam design software for the IBM PC and compatibles, disk included, *by Preben W. Jensen*

ADDITIONAL VOLUMES IN PREPARATION

Fundamentals of Robotics, *by David D. Ardayfio*

Belt Selection and Application for Engineers, *edited by Wallace D. Erickson*

Solid-State AC Motor Controls: Selection and Application, *by Sylvester Campbell*

Mechanical Engineering Software

Spring Design with an IBM PC, *by Al Dietrich*

Mechanical Design Failure Analysis: With Failure Analysis System Software for the IBM PC, *by David G. Ullman*

CAM DESIGN
AND MANUFACTURE

Second Edition

With Cam Design Software for the IBM PC and Compatibles

Disk Included

Preben W. Jensen

School of Physics, Engineering and Technology
Mankato State University
Mankato, Minnesota

CRC Press
Taylor & Francis Group
Boca Raton   London   New York

CRC Press is an imprint of the
Taylor & Francis Group, an **informa** business

First published 1987 by Marcel Dekker

Published 2019 by CRC Press
Taylor & Francis Group
6000 Broken Sound Parkway NW, Suite 300
Boca Raton, FL 33487-2742

ISBN 13: 978-0-367-45150-9 (pbk)
ISBN 13: 978-0-8247-7512-4 (hbk)

Visit the Taylor & Francis Web site at
http://www.taylorandfrancis.com

and the CRC Press Web site at
http://www.crcpress.com

Library of Congress Cataloging-in-Publication Data

Jensen, Preben W., [date]
  Cam design and manufacture.

  (Mechanical engineering; 55)
  "With cam design software for the IBM PC and
compatibles, disk included."
  Bibliography: p.
  Includes index.
  1. Cams—Design and construction—Data processing.
2. Computer-aided design. 3. IBM Personal Computer—
Programming. I. Title. II. Series.
TJ206.J48 1987   621.8′38   86-32776
ISBN 0-8247-7512-0

Preface

This book was written to give the practicing engineer a sound grasp of the methods of solving the problems connected with cams—their design, application, and manufacture. The above goal is as valid for this new edition as it was for the first edition. Since that time the most important change has been the improvement of numerically controlled machine tools (NC-machines) and the availability of computers in general. Therefore the emphasis on graphical and analytical methods has shifted toward the latter; but for a design engineer who has put his creative thoughts into metal it is of the utmost importance that he can visualize the problems; therefore the graphical approach has not been neglected. Major changes occur in Chapters 5 and 12 where analytical expressions that can be programmed on a home computer have been developed.

The plan of the book is as follows: In Chapter 1 the basic types of cam and follower systems are described and illustrated. In Chapter 2 the construction and use of displacement diagrams are explained and formulas are given for the displacement, velocity, and acceleration curves for various types of cam motion.

In Chapter 3 displacement diagram synthesis is explained and methods of combining various curves to obtain a desired motion are given. Chapter 4 outlines the methods of determining cam profiles graphically when different types of followers are used and formulas for determining the cam profile using both rectangular and polar coordinates are given.

Chapter 5 takes up the importance of pressure angle, and the procedure for proportioning a cam with respect to pressure angle limitations is explained. The importance of avoiding too small a radius of curvature is also discussed.

In Chapter 6 the advantages of circular cams and methods of proportioning these are outlined. Chapter 7 continues with a discussion of circular-arc and straight-line cams, which have advantages particularly from the standpoint of ease of production.

Chapter 8 considers the important factors of forces, contact stresses, and materials. This is followed in Chapter 9 by a discussion of various methods of cam manufacture.

When cams rotate at high speed, the factors of elasticity and backlash must be taken into consideration if the desired accuracy of motion is to be obtained. One way of doing this is to use the polydyne method of cam design. In Chapter 10 this method is described in detail and the effect of various members of the cam train are determined.

In Chapter 11 the use of various types of mechanisms with cams to offset the disadvantages of the latter is illustrated with various examples.

In Chapter 12, formulas to determine velocities, accelerations, and forces in linkages have been developed. This chapter together with Chapter 8 should enable the reader to find forces in complex mechanisms other than cam mechanisms.

Chapter 13 includes six computer programs that allow the user to determine minimum cam size for given maximum pressure angles for eight different cam displacement diagrams and calculate the maximum compressive stress by both rise and return for translating as well as swinging roller followers.

My thanks to Mr. A. F. Abou-Ghaledum from Cleveland State University, who in no time linked the six programs together to make them user-friendly.

Twenty-eight nomograms are included in order to facilitate computations.

The bibliography at the end of the book lists more than 1800 titles in English and German. The list in the first edition comprised "only" approximately 500 titles. The list is in alphabetical order and is referenced in eleven groups to facilitate its use.

An author usually puts a little of his heart and convictions into his book and I am happy to express that the first edition was well accepted.

Preben W. Jensen

Contents

Preface   iii

Using the Cam Design Software   vii

Nomenclature   ix

List of Nomograms   xiii

1. Cam and Follower Systems   1

2. Displacement Diagrams   8

3. Displacement Diagram Synthesis   31

4. Cam Profile Determination   46

5. Pressure Angle and Radius of Curvature   75

6. Circular Cams   106

7. Circular-arc and Straight-line Cams   126

8. Forces, Contact Stresses, and Materials   141

9. Methods of Cam Manufacture   159

10. Dynamics of Cam Mechanisms   170

11. Cam Mechanisms   188

12. Velocities, Accelerations, and Dynamic Forces in Linkages and Cam Mechanisms   206

13. Computer Programs for Analysis and Synthesis of Cam Mechanisms   252

Bibliography   321

Index   425

Using the Cam Design Software

The following should help the new user in using the cam programs included on the supplied disk. There are a number of requirements needed before these programs and the disk can be used. These requirements are concerned with the type of system you are using, and the operating system used.

The programs were developed on an IBM-PC running on DOS 2.0, and using BASICA. However, since the programs do not use any special IBM formats and system calls they could run virtually on any compatible running MS-BASIC, GW-BASIC, or similar basic editors.

The following procedure is recommended for proper use of these programs:

1. Switch your computer ON, insert a DOS disk or a disk containing a BASIC Editor into drive A.
2. When you receive the A> prompt, type in the command to run basic (*i.e.,* A>BASICA).
3. Remove the disk from drive A and insert the cam programs disk in A.
4. Type the following command: RUN "MAIN"

This should get you into the MAIN program, which will then prompt you with the master MENU. At this point refer to Chapter 13.

Nomenclature

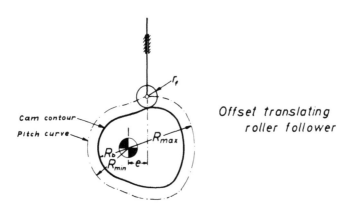

Cam contour

Pitch curve

R_{max}

R_b

R_{min}

e

r_f

Offset translating
roller follower

Fixed center
of rotation

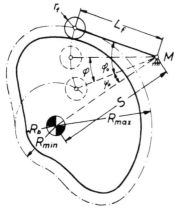

r_f

L_f

M

φ

φ_0

ψ_0

S

R_{max}

R_b

R_{min}

Swinging roller follower

$$a = \frac{dv}{dt} = \frac{d^2y}{dt^2} = \text{acceleration of follower, in/sec}^2$$

b = thickness of contacting cam and follower, in.

d_s = shaft diameter, in.

e = offset or eccentricity, in.

F_n = normal load, lbs.

g = gravitational constant = 386 in/sec^2 = 32.16 ft./sec^2

h = maximum displacement of follower, in.

L_f = length of oscillating follower arm, in.

N = cam speed, rpm

$$p = \frac{da}{dt} = \text{pulse of follower, in/sec}^3$$

r = radius to trace point, in.

r_f = roller radius, in.

S = distance between cam and oscillating follower centers, in.

t = time for cam to rotate angle θ, sec

T = time for cam to rotate angle β, sec

T_1 = time for cam to rotate angle β_1, sec

T_2 = time for cam to rotate angle β_2, sec

$$v = \frac{dy}{dt} = \text{velocity of follower, in/sec}$$

y = displacement of follower, in.

$$y' = \frac{dy}{dt} = \text{follower velocity, in/sec.}$$

$$y'' = \frac{d^2y}{dt^2} = \text{follower acceleration, in/sec}^2$$

$$y''' = \frac{d^3y}{dt^3} = \text{follower pulse or jerk, in/sec}^3$$

R_{max} = maximum radius of cam (to center of roller), in.

R_{min} = minimum radius of cam (to center of roller), in.

R_b = radius of base circle (to actual cam shape), in.

R_g = radius of cutter or grinder, in.

S_c = pressure, lb./in.^{2}

α = pressure angle, degrees

α_1 = pressure angle by rise, degrees

α_2 = pressure angle by return, degrees

α_{m1} = max. pressure angle by rise, degrees

α_{m2} = max. pressure angle by rise, degrees

β = cam angle rotation for total rise h, degrees
β_1 = cam angle rotation for total rise, degrees
β_2 = cam angle rotation for total return, degrees
θ = cam angle rotation for follower displacement y, degrees
μ = coefficient of friction
μ = transmission angle, degrees
μ_c, μ_f = Poisson's ratio for cam and follower, respectively
R_c = radius of curvature of cam, in.
ϕ = angle of oscillating follower movement for cam angle θ, degrees
ϕ_0 = total angle of oscillating follower movement, degrees
ω = cam angular velocity, rad/sec

List of Nomograms

Figure

5-5	Pressure angle—constant velocity motion	80
5-6	Pressure angle—parabolic motion	81
5-7	Pressure angle—simple harmonic motion	82
5-8	Pressure angle—cycloidal motion	83
5-9	Pressure angle—3-4-5 polynomial	84
5-10	Pressure angle—modified trapezoidal acceleration	85
5-19	Pressure angle—parabolic, cycloidal and simple harmonic motions	91
5-23	Minimum radius of curvature—parabolic motion, translating follower	97
5-24	Minimum radius of curvature—simple harmonic motion, translating follower	98
5-25	Minimum radius of curvature—cycloidal motion, translating follower	99
5-26	Minimum radius of curvature—double-harmonic motion, translating follower	100
6-9	Best μ_{min}	112
6-10	$\dfrac{a}{d}$ for best μ_{min}	113
6-11	$\dfrac{b}{d}$ for best μ_{min}	114
6-12	$\dfrac{c}{d}$ for best μ_{min}	115

6-13   Best μ_{min} ($\phi_c = 180°$)   116
6-20   Proportioning of a slider-crank, μ_{min} 122

6-21   Proportioning of a slider-crank, $\dfrac{a}{s}$ 122

6-22   Proportioning of a slider-crank, $\dfrac{b}{s}$ 123

6-23   Proportioning of a slider-crank, λ 123
6-24   Proportioning of a slider-crank for given ϕ_c, best μ_{min} 124
8-3   Acceleration—translating follower; parabolic, cycloidal,
            and simple harmonic motions ($50 \leqslant N \leqslant 500$)   144
8-4   Acceleration—translating follower; parabolic, cycloidal,
            and simple harmonic motions ($500 \leqslant N \leqslant 5000$)   145
8-9   Hertz's pressure for convex surface   155
8-10   Hertz's pressure for concave surface   156
8-11   Hertz's pressure for plane surface   157
12-15   Acceleration—inverted crossed slide-crank
            ($60 \leqslant N \leqslant 600$)   244
12-16   Acceleration—inverted crossed slide-crank
            ($600 \leqslant N \leqslant 6000$)   245

CAM DESIGN
AND MANUFACTURE

CHAPTER 1

Cam and Follower Systems

Cams are used in a wide variety of machines; such as packaging machines, can-making machinery, wire-forming machines, engines, computing mechanisms, and mechanical and electronic computers. One important reason why cam mechanisms are preferred over other types is that the use of cams makes it possible to obtain an unlimited variety of motions and when certain basic requirements are followed, cams perform satisfactorily year after year.

Cams are used to transform a rotary motion into a translating or oscillating motion. In certain cases they are also used to transform a translating or oscillating motion in to a different translating or oscillating motion.

The requirements which are imposed on cams vary from machine to machine because the requirements depend not only on the speed of the cam, but also on the kind of machine in which they are being used. In certain kinds of wrapping machines, for example, the forces imposed on the material to be wrapped should be kept as low as possible, but it doesn't matter if these forces are applied suddenly, whereas in other machines it is very important for the proper performance of the machinery that the variation of forces is smooth and gradual. The basic limiting requirements are: kind of time-displacement diagram, pressure angle, radius of curvature, and finally, the contact pressure between follower and cam. These requirements will be discussed in subsequent chapters.

The most commonly used cam is the plate cam which is cut out of a piece of flat metal or plate. Dependent on the kind of follower, various types of following systems are often employed. A *radial translating roller follower* is shown in Fig. 1-1a and is so called because the center line of the follower-stem passes through the center of the cam shaft. An *offset trans-*

lating roller follower is shown in Fig. 1-1b; here the center line of the follower-stem does not pass through the cam shaft center.

Figure 1-1c shows a *swinging roller follower* which is preferred over the translating follower because a much higher pressure angle can be allowed, and hence the overall proportions of the mechanism can be reduced. The question is often raised as to whether the rotation of the cam should be away from or toward pivot point M; in the case of Fig. 1-1c this would mean respectively CCW (counterclockwise) and CW (clockwise) rotation of the cam. There is a slight advantage in letting the cam rotate away from the pivot point, but in most cases the advantage is insignificant.

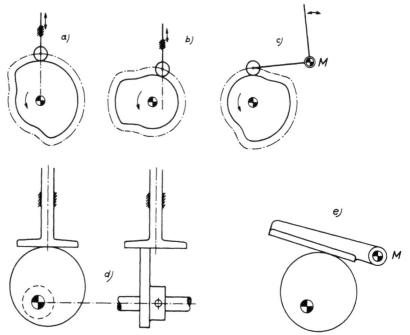

Fig. 1-1. (a) Radial translating roller follower.   (b) Offset translating roller follower.
(c) Swinging roller follower.   (d) Flat-faced translating follower.   (e) Flat-faced
swinging follower.

Figure 1-1d shows a translating flat-faced follower. The flat does not necessarily have to be perpendicular to the follower-stem, although it usually is. Sometimes the center line of the follower-stem is offset as shown in the right-hand view of Fig. 1-1d. This arrangement will tend to distribute and reduce wear on the flat because the friction force created between the cam and the follower will tend to rotate the follower around its axis. It should be noticed that whether the center line of the follower-stem passes through the cam shaft center or not has no effect on the cam profile. In certain kinds of textile machines, the arrangement shown in Fig. 1-1e with a swinging flat-faced follower is used.

In the systems so far shown, the follower is kept in contact with the cam with the help of gravity forces. Obviously, this is only possible in the case of low-speed cams. For moderate- and high-speed cams other means must be employed. The use of springs is one obvious solution to keep cam and follower in contact with each other.

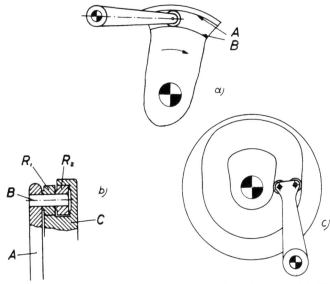

Fig. 1-2. (a) Closed-track cam. (b) Double closed-track cam. (c) Closed-track cam with two rollers.

Other possibilities are shown in Figs. 1-2a, b, and c. In Fig. 1-2a part of a closed-track cam is shown. When the roller follower is driven upward, the roller contacts B, and when driven downward the roller contacts A. A certain clearance is therefore necessary to permit the roller to roll. However, this clearance should be kept as small as possible, because the larger the clearance, the larger the impact will be when the roller changes contact from one side of the track to the other. This change takes place when the motion of the roller changes from acceleration to deceleration. If this change is made gradually, as with a cycloidal motion, the impact is greatly reduced as compared to the case of parabolic motion where the acceleration changes suddenly.

It is therefore natural to try to reduce the clearance to a minimum and it can be done with the help of the arrangement shown in Fig. 1-2b. The follower arm A carries the two rollers R_1 and R_2 on the same pin B. R_1 rolls on the inner side of the track of cam C and R_2 rolls on the outer track.

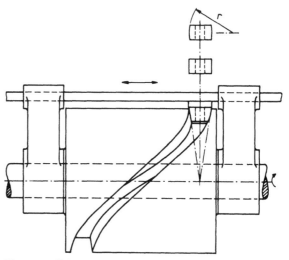

Fig. 1-3a. Cylindrical cam with translating roller follower.

Because of its cost, however, this arrangement is seldom used. In Fig. 1-2c two rollers are placed on different pins. There seems to be no advantage over that of Fig. 1-2b. The track shown in Fig. 1-2c is also more difficult to machine than that of Fig. 1-2b.

A cylindrical cam with translating roller follower, Fig. 1-3a, has the characteristic that the direction of motion of the follower is parallel to the cam shaft. Theoretically, the roller has to be cone-shaped with its apex at the cam shaft center, as shown in Fig. 1-3a, in order that the roller may roll without sliding. However, a cylindrical roller or one with a slightly spherical shape will operate satisfactorily in most cases.

Figure 1-3b shows a similar arrangement but two rollers are used instead of one. The advantage as compared with Fig. 1-3a, where one roller

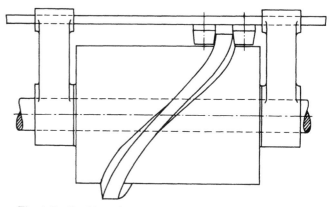

Fig. 1-3b. Double-end cam with translating roller follower.

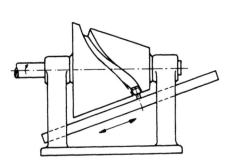

Fig. 1-4(a). Conical closed-track cam with translating roller follower

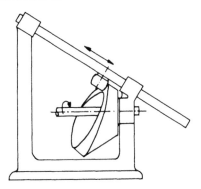

Fig. 1-4(b). Conical open-track cam with translating roller follower

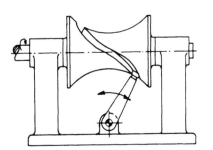

Fig. 1-5. Globoidal closed-track cam with swinging roller follower

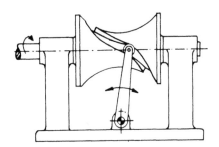

Fig. 1-6. Globoidal closed-track cam with swinging roller follower

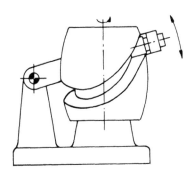

Fig. 1-7. Globoidal closed-track cam with swinging roller follower

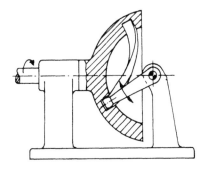

Fig. 1-8. Spherical closed-track cam with swinging roller follower

Figs. 1-4 to 1-8. Special types of cams.

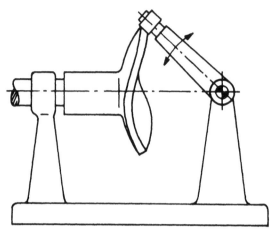

Fig. 1-9. Spherical open-track cam with swinging roller follower.

changes contact from one side of the track to the other whenever acceleration is changed in direction, is that with both rollers in contact with the track at the same time there will be no backlash.

A conical cam, Fig. 1-4a, has much the same characteristics as a cylindrical one and is used in cases when the direction of the output motion is parallel to an element of the base cone. The conical cam in Fig. 1-4a has a closed track and that in Fig. 1-4b an open track.

The cam and follower systems shown in Figs. 1-5 to 1-10 have to be cut by special devices. Because of the cost of making them, they are

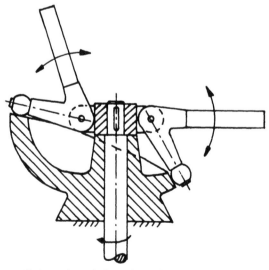

Fig. 1-10. Kinematic inversion of the spherical cam; cam is stationary and swinging
roller follower rotates around cam.

seldom employed and then only for rather small mechanisms. Figures 1-5, 1-6, and 1-7 show globidal cams with swinging roller followers; the only difference is in how the roller is placed relative to the input and output shafts. Figure 1-8 shows a spherical cam with swinging roller follower and closed-track, and in Fig. 1-9 the mechanism is shown with an open track.

Figure 1-10 is a kinematic inversion of the spherical cam; the cam is stationary and the swinging roller follower rotates. This kind of mechanism is used in agricultural machinery.

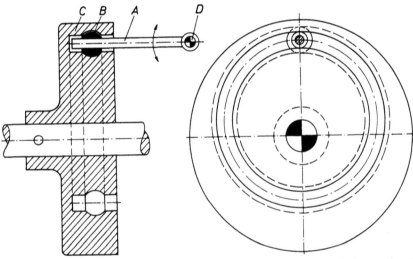

Fig. 1-11. Ball runs in groove of closed-track cam, imparting an oscillating motion to output member A.

A very special kind of cam and follower system is shown in Fig. 1-11. The cam C is a closed-track cam and the track is formed so that a ball B can be guided by either side of the track dependent upon the direction of motion of arm A. A round arm A has a sliding fit in the hole of the ball and is fastened to shaft D, which is the output shaft. This mechanism is used in a sewing machine and the cam rotates at 2000 to 3000 rpm. The advantage of this mechanism becomes clearly obvious when compared with Fig. 1-9. To cut the cam in Fig. 1-9 would require that the milling cutter be moved exactly the same way relative to the spherical cam as the roller follower, and this requires a special set-up. However, the cam in Fig. 1-11 can be cut with a milling cutter which has the form of the groove, exactly as if it were a plate cam.

Displacement Diagrams

In Chapter 1 many varieties of cams and followers are illustrated and in all of them the cam rotates at a constant angular velocity and the follower moves in a manner prescribed by the functional requirements of the machine.

The simplest follower motion is a constant velocity rise followed by a similar return with a dwell in between. A simple graph called a displacement diagram illustrates this sequence of events.

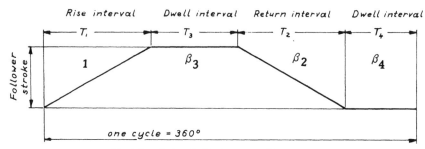

Fig. 2-1.   A simple displacement diagram.

Such a diagram is shown in Fig. 2-1. Here, one cycle is taken to mean one complete revolution of the cam; that is, one cycle represents 360 degrees. From this it follows that the horizontal distances, T_1, T_2, T_3, T_4 is expressed in seconds and β_1, β_2, β_3, β_4 in degrees of rotation. Degrees are mostly used. The vertical distances represent the motion of the follower dependent on time. From Fig. 2-2 it can be seen that in place of a rotating cam it is possible to achieve the identical follower action by means of a translating

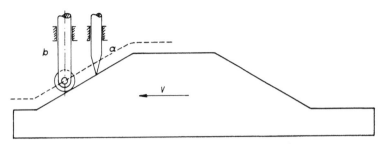

Fig. 2-2.   Translating cam made from diagram in Fig. 2-1.

member, which has a profile the same as that of the displacement diagram, translating with a constant velocity v. This device is often used in machinery. To cause the follower to move as indicated by Fig. 2-1, but by means of a rotating member instead of a slider, we merely "wrap" the time displacement diagram around a circular disk as shown in Fig. 2-3. (This will be explained in more detail in Chapter 4.) Thus, the diagram in Fig. 2-1 represents the follower movement of either Figs. 2-2 or 2-3.

Cam Followers

It is important to study the effect of the kind of contact between the cam surface and the follower. For example, in Fig. 2-2, shown at (a) is a follower having a pointed end which makes line contact with the cam. It is clear that if continuous contact is maintained, the follower stem will have truly the

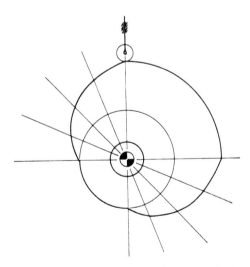

Fig. 2-3.   Rotating cam made from diagram in Fig. 2-1.

motion prescribed by the time-displacement diagram. If, however, a roller is used as shown at (b), the follower stem cannot possibly have exactly the movement prescribed because of the sharp corners of the cam at the start and finish of each rise and return. The movement of the roller follower will be as shown by the dashed line in Fig. 2-2. Note the curved path at the end of the rise.

It is obvious from Fig. 2-2 that since a constant slope is used, this has the effect of moving the follower with constant velocity. Now constant velocity is a desirable form of motion for a cam follower provided that the acceleration from rest to the constant velocity value is moderate. Theoretically, an instantaneous jump from zero velocity to any value of velocity results in an infinite acceleration and since mass is always involved in machines, this theoretically results in an infinitely large force. Actually, an instantaneous change in velocity is impossible due to flexure of the machine parts and other factors. Nevertheless, any shock effect is serious and must be kept to a minimum. For this reason the rise and fall portions of a cam displacement diagram are of vital importance and need to be studied in considerable detail.

It is fortunate that in many design problems the required operation merely requires a particular machine member to be at a given point at a given time. How it gets to this point is not specified. The cam designer is therefore at liberty to choose an approach which gives the lowest shock values possible so as to reduce wear and tear on the cam assembly. It is also fortunate that what is best for the cam is almost always best for the machine and its product.

Types of Cam Displacement Curves

A wide variety of cam curves are available for moving the follower and these will be thoroughly analyzed. In the following sections only the rise portions of the total time-displacement diagram are studied. The return portions can be analyzed in a similar manner. However, formulas for the return portions are added for convenience. Complex cams are frequently employed which may involve a number of rise-dwell-return intervals in which the rise and return aspects are quite different. To analyze the action of a cam it is necessary to study its time-displacement and associated velocity and acceleration curves. The latter are based on the first and second time-derivatives of the equation describing the time-displacement curve:

$$y = \text{displacement} = f(t) = f(\theta)$$

$$\frac{dy}{dt} = \text{velocity} \qquad \frac{d^2y}{dt^2} = \text{acceleration}$$

A variety of displacement curves will now be briefly discussed. The equations for return motion are based upon the point A in Fig. 2-4 and the following figures being at the top of the diagram and point B at the bottom. The displacement curves will be more thoroughly analyzed in subsequent chapters. In the group called polynomial curves the significance of numerical prefixes is taken up in detail in Chapter 10.

y = displacement of follower, in.
h = maximum displacement of follower, in.
t = time for cam to rotate through angle θ, sec.
T_1 = time for total rise, sec
T_2 = time for total return, sec
θ = cam angle rotation for follower displacement y, degrees
β_1 = cam angle for total rise, degrees
β_2 = cam angle for total return, degrees
v = velocity of follower, in./sec
a = follower acceleration, in./sec^2
t_x = a function of t

In all the following formulas for y, θ and β_1, β_2 are used but can be replaced with t and T_1, T_2, respectively. This will facilitate calculation of velocities and accelerations when θ and β are known.

Constant-velocity Motion (Fig. 2-4)

$$\left. \begin{array}{l} y = h\,\dfrac{\theta}{\beta_1} \\[2ex] v = \dfrac{h}{T_1} \\[2ex] a = 0 \end{array} \right\} \quad \begin{array}{l} 0 < \theta < \beta_1 \\ \text{(rise)} \end{array} \qquad (2.1a)$$

except at $\theta = 0$ and $\theta = \beta_1$ where the acceleration is theoretically infinite

$$\left. \begin{array}{l} y = h\left(1 - \dfrac{\theta}{\beta_2}\right) \\[2ex] v = -\dfrac{h}{T_2} \\[2ex] a = 0 \end{array} \right\} \quad \begin{array}{l} 0 < \theta < \beta_2 \\ \text{(return)} \end{array} \qquad (2.1b)$$

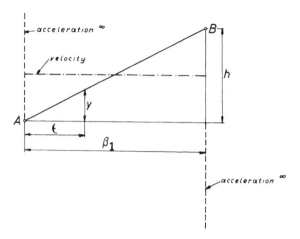

Fig. 2-4. Cam displacement, velocity, and acceleration curves for constant velocity motion.

This motion and its disadvantages were discussed earlier in the chapter. This curve is in general only to be used as a composite curve.

The angle for rise β_1 and the corresponding time T_1 are related by the formula

$$T_1 = \frac{60}{N} \frac{\beta_1}{360}$$

For return the formula becomes

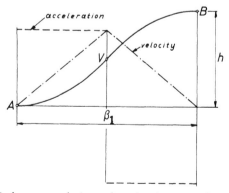

Fig. 2-5. Cam displacement, velocity, and acceleration curves for parabolic motion.

$$T_2 = \frac{60}{N} \frac{\beta_2}{360}$$

where N is the rotational speed of the cam in RPM (revolutions pr. minute).

In the unaltered form shown it is rarely used except in very crude devices, nevertheless the advantage of uniform velocity is an important one and by modifying the start and finish of the follower stroke this form of cam motion can be utilized. The modification is explained in Chapter 3.

Parabolic Motion (Fig. 2-5)

$$\left. \begin{array}{l} y = 2h\left(\dfrac{\theta}{\beta_1}\right)^2 \\[2ex] v = 4\,\dfrac{h}{T_1}\,\dfrac{\theta}{\beta_1} \\[2ex] a = 4\,\dfrac{h}{T_1^2} \end{array} \right\} \begin{array}{c} 0 \leqslant \theta \leqslant \dfrac{\beta_1}{2} \\[2ex] \text{(rise)} \end{array} \tag{2.2a}$$

$$\left. \begin{array}{l} y = h\left[1 - 2\left(\dfrac{\beta_1 - \theta}{\beta_1}\right)^2\right] \\[2ex] v = 4\,\dfrac{h}{T_1}\,\dfrac{\beta_1 - \theta}{\beta_1} \\[2ex] a = -4\,\dfrac{h}{T_1^2} \end{array} \right\} \begin{array}{c} \dfrac{\beta_1}{2} \leqslant \theta \leqslant \beta_1 \\[2ex] \text{(rise)} \end{array} \tag{2.2b}$$

$$\left. \begin{array}{l} y = h\left[1 - 2\left(\dfrac{\theta}{\beta_2}\right)^2\right] \\[2ex] v = -4\,\dfrac{h}{T_2}\,\dfrac{\theta}{\beta_2} \\[2ex] a = -4\,\dfrac{h}{T_2^2} \end{array} \right\} \begin{array}{c} 0 \leqslant \theta \leqslant \dfrac{\beta_2}{2} \\[2ex] \text{(return)} \end{array} \tag{2.2c}$$

$$\left. \begin{array}{l} y = 2h\left(\dfrac{\beta_2 - \theta}{\beta_2}\right)^2 \\[2ex] v = -4\,\dfrac{h}{T_2}\,\dfrac{\beta_2 - \theta}{\beta_2} \\[2ex] a = 4\,\dfrac{h}{T_2^2} \end{array} \right\} \begin{array}{c} \dfrac{\beta_2}{2} \leqslant \theta \leqslant \beta_2 \\[2ex] \text{(return)} \end{array} \tag{2.2d}$$

The most important advantage of this curve is that for a given angle of rotation and rise it produces the smallest possible acceleration. However, because of the sudden changes in acceleration at the beginning, middle, and end of the stroke, shocks are produced. If the follower system were perfectly rigid with no backlash or flexibility, this would be of little significance. But such systems are mechanically impossible to build and a certain amount of impact is caused at each of these change-over points.

Therefore this curve is not recommended for high speed.

Simple Harmonic Motion (Fig. 2-6)

$$
\left. \begin{array}{l}
y = \dfrac{h}{2}\left[1 - \cos\left(\pi\,\dfrac{\theta}{\beta_1}\right)\right] \\[2ex]
v = \dfrac{h}{2}\cdot\dfrac{\pi}{T_1}\sin\left(\pi\,\dfrac{\theta}{\beta_1}\right) \\[2ex]
a = \dfrac{h}{2}\cdot\dfrac{\pi^2}{T_1^2}\cos\left(\pi\,\dfrac{\theta}{\beta_1}\right)
\end{array} \right\}
\begin{array}{c}
0 \leqslant \theta \leqslant \beta_1 \\
(\text{rise})
\end{array}
\qquad (2.3a)
$$

$$
\left. \begin{array}{l}
y = \dfrac{h}{2}\left[1 + \cos\left(\pi\,\dfrac{\theta}{\beta_2}\right)\right] \\[2ex]
v = -\dfrac{h}{2}\dfrac{\pi}{T_2}\sin\left(\pi\,\dfrac{\theta}{\beta_2}\right) \\[2ex]
a = -\dfrac{h}{2}\dfrac{\pi^2}{T_2^2}\cos\left(\pi\,\dfrac{\theta}{\beta_2}\right)
\end{array} \right\}
\begin{array}{c}
0 \leqslant \theta \leqslant \beta_2 \\
(\text{return})
\end{array}
\qquad (2.3b)
$$

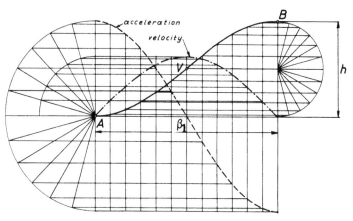

Fig. 2-6. Cam displacement, velocity, and acceleration curves for simple harmonic motion.

Smoothness in velocity and acceleration during the stroke is the advantage inherent in this curve. However, the instantaneous changes in acceleration at the beginning and end of the stroke tend to cause vibration, noise, and wear. As can be seen from Fig. 2-6, the maximum acceleration values occur at the ends of the stroke. Thus, if inertia loads are to be overcome by the follower, the resulting forces cause severe stresses in the members. These forces are in many cases much larger than the externally applied loads. This curve is not recommended for high speed.

Cycloidal Motion (Fig. 2-7)

$$
\left.
\begin{aligned}
y &= h\left[\frac{\theta}{\beta_1} - \frac{1}{2\pi}\sin\left(2\pi\,\frac{\theta}{\beta_1}\right)\right] \\[2mm]
v &= \frac{h}{T_1}\left[1 - \cos\left(2\pi\,\frac{\theta}{\beta_1}\right)\right] \\[2mm]
a &= \frac{2\pi h}{T_1^2}\sin\left(2\pi\,\frac{\theta}{\beta_1}\right)
\end{aligned}
\right\}
\quad
\begin{array}{c}
0 \leqslant \theta \leqslant \beta_1 \\
\text{(rise)}
\end{array}
\qquad (2.4a)
$$

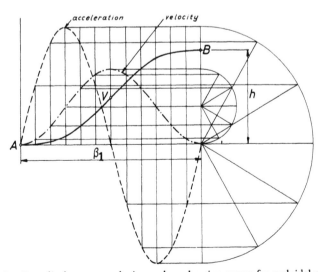

Fig. 2-7.   Cam displacement, velocity, and acceleration curves for cycloidal motion.

$$y = h\left[1 - \frac{\theta}{\beta_2} + \frac{1}{2\pi} \sin\left(2\pi \frac{\theta}{\beta_2} \right) \right]$$

$$v = \frac{h}{T_2}\left[\cos\left(2\pi \frac{\theta}{\beta_2} \right) - 1 \right] \qquad \left. \begin{array}{c} 0 \leqslant \theta \leqslant \beta_2 \\ \text{(return)} \end{array} \right. \qquad (2.4\,\text{b})$$

$$a = - \frac{2\pi h}{T_2^2} \sin\left(2\pi \frac{\theta}{\beta_2} \right)$$

This time-displacement curve has excellent acceleration characteristics; there are no abrupt changes in its associated acceleration curve. The maximum value of the acceleration of the follower for a given rise and time is somewhat higher than that of the simple harmonic motion curve. In spite of this the cycloidal curve is used often as a basis for designing cams for high-speed machinery because it results in low noise, vibration, and wear.

The cycloidal motion displacement curve is so called because it can be generated from a cycloid which is the locus of a point of a circle rolling on a straight line. Thus, in Fig. 2-8 P is a point on the circle which rolls on the straight line BC. The radius OP of the circle is made equal to $h/2\pi$ so that the circumference of the circle equals the distance BC. The circle starts its rolling when P is at C. When the center of the circle is in the position shown

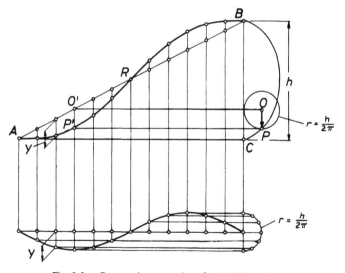

Fig. 2-8.   Geometric properties of cycloidal motion.

point P has moved to its present position. A horizontal line is drawn from O to O' which is on line AB and a vertical line is drawn from O' to intersect the horizontal line through P; the point of intersection is P' which is a point on the displacement curve.

Cycloidal motion can also be considered to be composed of a straight line AB on which is superposed a sine wave the amplitude of which is $r = h/2\pi$ and the amplitude being perpendicular to the base line AC.

In Chapter 3 more is said about this and a family of curves—the modified cycloids—is developed.

Double Harmonic Motion (Fig. 2-9)

$$
\left.
\begin{aligned}
y &= \frac{h}{2}\left[1 - \cos\left(\pi\,\frac{\theta}{\beta_1}\right) - \frac{1}{4}\left(1 - \cos\left(2\pi\,\frac{\theta}{\beta_1}\right)\right)\right] \\[2ex]
v &= \frac{h}{2}\frac{\pi}{T_1^2}\left[\sin\left(\pi\,\frac{\theta}{\beta_1}\right) - \frac{1}{2}\sin\left(2\pi\,\frac{\theta}{\beta_1}\right)\right] \\[2ex]
a &= \frac{h}{2}\frac{\pi^2}{T_1^2}\left[\cos\left(\pi\,\frac{\theta}{\beta_1}\right) - \cos\left(2\pi\,\frac{\theta}{\beta_1}\right)\right]
\end{aligned}
\right\}
\quad
\begin{aligned}
0 \leqslant \theta \leqslant \beta_1 \quad (2.5a)\\
\text{(rise)}
\end{aligned}
$$

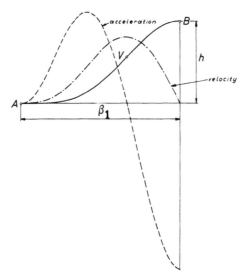

Fig. 2-9. Cam displacement, velocity, and acceleration curves for double harmonic motion.

$$y = \frac{h}{2}\left[1 + \cos\left(\pi\,\frac{\theta}{\beta_2}\right) + \frac{1}{4}\left(1 - \cos\left(2\pi\,\frac{\theta}{\beta_2}\right)\right)\right]$$

$$v = -\frac{h}{2}\,\frac{\pi}{T_2}\left[\sin\left(\pi\,\frac{\theta}{\beta_2}\right) - \frac{1}{2}\sin\left(2\pi\,\frac{\theta}{\beta_2}\right)\right] \qquad 0 \leqslant \theta \leqslant \beta_2 \quad (2.5b)$$
$$\text{(return)}$$

$$a = -\frac{h}{2}\,\frac{\pi^2}{T_2^2}\left[\cos\left(\pi\,\frac{\theta}{\beta_2}\right) - \cos\left(2\pi\,\frac{\theta}{\beta_2}\right)\right]$$

The negative acceleration is about double that of the positive and there is a sudden change in acceleration at the end of the rise. This curve is only good when used as part of a compound curve, that is, a curve made up of different or similar basic curves (parabolic, simple harmonic, etc.)

Cubic Curve No. 1 (Table 2-1)

$$y = 4h\left(\frac{\theta}{\beta_1}\right)^3$$

$$v = \frac{12h}{T_1}\left(\frac{\theta}{\beta_1}\right)^2 \qquad 0 \leqslant \theta \leqslant \frac{\beta_1}{2} \qquad (2.6a)$$
$$\text{(rise)}$$

$$a = \frac{24h}{T_1^2}\,\frac{\theta}{\beta_1}$$

$$y = h\left[1 - 4h\left(\frac{\beta_1 - \theta}{\beta_1}\right)^3\right]$$

$$v = \frac{12h}{T_1}(\beta_1 - \theta)^2 \qquad \frac{\beta_1}{2} \leqslant \theta \leqslant \beta_1 \qquad (2.6b)$$
$$\text{(rise)}$$

$$a = -\frac{24h}{T_1^2}(\beta_1 - \theta)$$

$$y = h\left[1 - 4\left(\frac{\theta}{\beta_2}\right)^3\right]$$

$$v = -\frac{12h}{T_2}\left(\frac{\theta}{\beta_2}\right)^2 \qquad 0 \leqslant \theta \leqslant \frac{\beta_2}{2} \qquad (2.6c)$$
$$\text{(return)}$$

$$a = -\frac{24h}{T_2^2}\,\frac{\theta}{\beta_2}$$

Table 2-1. Characteristics of Various Types of Cam Curves

TYPE OF CURVE	EQUATION NO.	ACCELERATION CURVE	VELOCITY FACTOR	ACCELERATION FACTOR	JERK FACTOR	CAM SPEED APPLICATION	PERFORMANCE AT HIGH SPEED	OTHER COMMENTS (See also chapter on cam manufacturing)
Constant Velocity Motion	2.1a-b		1.00	∞		Low speed and low masses	Poor	
Parabolic Motion	2.2a-d		2.00	4.00	3×∞	Medium speed	Good	
Simple Harmonic Motion	2.3a-b		1.57	4.93	2×∞	Medium speed	Good	
Cycloidal Motion	2.4a-b		2.00	6.28	61	High speed	Excellent	
Double Harmonic Motion	2.5a-b		2.00	5.5 9.9				
Cubic Curve No. 1	2.6a-d		3.00	12.00	1×∞	Low speed	Poor	
Cubic Curve No. 2	2.7a-b		1.50	6.00	2×∞	Low speed	Poor	
Cubic Curve No. 3	2.8a-f		2.00	8.00	32	Low speed	Poor	
3-4 Polynomial	2.9a-d		2.00	6.00	48	High speed	Excellent	
3-4-5 Polynomial	2.10a-b		1.88	5.77	60	High speed	Excellent	
4-5-6-7 Polynomial	2.11a-b		2.19	7.52	52.5	Medium to high speed	Good to excellent	
Trapezoidal Acceleration	2.12a-c		2.00	5.33	42.7	High speed	Good to excellent	
Modified Trapezoidal Acceleration	2.13a-l		2.00	4.89	61.4	High speed	Excellent	
Modified Sinusoidal Acceleration	2.14a-f		1.76	5.53	69.3	High speed	Best of all	Can be made on lathe with highest accuracy
Circular Cam Profile						High speed		
Circular Arc Profile						Medium speed	Good	Small cams can be made accurately
Circular Arc and Straight Line Profile	2.15.a 2.15.b			All values from 5.89 to ∞		Medium speed	Good	Small cams can be made accurately
Modified Cycloid						Medium to high speed	Good to excellent	

$$y = 4h\left(\frac{\beta_2 - \theta}{\beta_2}\right)^3$$

$$v = -\frac{12h}{T_2}(\beta_2 - \theta)^2$$

$$a = \frac{24h}{T_2^2}(\beta_2 - \theta)$$

$\left.\begin{array}{c} \\ \\ \\ \\ \end{array}\right\}$ $\dfrac{\beta_2}{2} \leqslant \theta \leqslant \beta_2$

(return)

(2.6d)

Because of the sudden change in acceleration at the middle of the rise and because of a rather high value of the maximum deceleration, this curve is only good when used as part of a compound curve.

Cubic Curve No. 2 (Table 2-1)

$$y = h\left[3\left(\frac{\theta}{\beta_1}\right)^2 - 2\left(\frac{\theta}{\beta_1}\right)^3\right]$$

$$v = 6\frac{h}{T_1}\left[\frac{\theta}{\beta_1} - \left(\frac{\theta}{\beta_1}\right)^2\right]$$

$$a = 6\frac{h}{T_1^2}\left(1 - 2\frac{\theta}{\beta_1}\right)$$

$\left.\begin{array}{c} \\ \\ \\ \\ \end{array}\right\}$ $0 \leqslant \theta \leqslant \beta_1$

(rise)

(2.7a)

$$y = h\left[1 - 3\left(\frac{\theta}{\beta_2}\right)^2 + 2\left(\frac{\theta}{\beta_2}\right)^3\right]$$

$$v = -6\frac{h}{T_2}\left[\frac{\theta}{\beta_2} - \left(\frac{\theta}{\beta_2}\right)^2\right]$$

$$a = -6\frac{h}{T_2^2}\left(1 - 2\frac{\theta}{\beta_2}\right)$$

$\left.\begin{array}{c} \\ \\ \\ \\ \end{array}\right\}$ $0 \leqslant \theta \leqslant \beta_2$

(return)

(2.7b)

This curve is usable for low speed only.

Cubic Curve No. 3 (Table 2-1)

$$y = \frac{16}{3}h\left(\frac{\theta}{\beta_1}\right)^3$$

$$v = 16\frac{h}{T_1}\left(\frac{\theta}{\beta_1}\right)^2$$

$$a = 32\frac{h}{T_1^2}\left(\frac{\theta}{\beta_1}\right)$$

$\left.\begin{array}{c} \\ \\ \\ \\ \end{array}\right\}$ $0 \leqslant \theta \leqslant \dfrac{\beta_1}{4}$

(rise)

(2.8a)

$$y = h\left[\frac{1}{6} - 2\frac{\theta}{\beta_1} + 8\left(\frac{\theta}{\beta_1}\right)^2 - \frac{16}{3}\left(\frac{\theta}{\beta_1}\right)^3\right]$$

$$v = \frac{h}{T_1}\left[-2 + 16\frac{\theta}{\beta_1} - 16\left(\frac{\theta}{\beta_1}\right)^2\right] \qquad \frac{\beta_1}{4} \leqslant \theta \leqslant \frac{3\beta_1}{4} \qquad (2.8\,b)$$

(rise)

$$a = \frac{h}{T_1^2}\left(16 - 32\frac{\theta}{\beta_1}\right)$$

$$y = h\left[-\frac{13}{3} + 16\frac{\theta}{\beta_1} - 16\left(\frac{\theta}{\beta_1}\right)^2 + \frac{16}{3}\left(\frac{\theta}{\beta_1}\right)^3\right]$$

$$v = \frac{h}{T_1}\left[16 - 32\frac{\theta}{\beta_1} + 16\left(\frac{\theta}{\beta_1}\right)^2\right] \qquad \frac{3\beta_1}{4} \leqslant \theta \leqslant \beta_1 \quad (2.8\,c)$$

(rise)

$$a = \frac{h}{T_1^2}\left(-32 + 32\frac{\theta}{\beta_1}\right)$$

$$y = h\left[\frac{16}{3} - 16\frac{\theta}{\beta_2} + 16\left(\frac{\theta}{\beta_2}\right)^2 - \frac{16}{3}\left(\frac{\theta}{\beta_2}\right)^3\right]$$

$$v = \frac{h}{T_2}\left[-16 + 32\frac{\theta}{\beta_2} - 16\left(\frac{\theta}{\beta_2}\right)^2\right] \qquad \frac{3\beta_2}{4} \leqslant \theta \leqslant \beta_2 \qquad (2.8\,d)$$

(return)

$$a = \frac{h}{T_2^2}\left(32 - 32\frac{\theta}{\beta_2}\right)$$

$$y = h\left[1 - \frac{16}{3}\left(\frac{\theta}{\beta_2}\right)^3\right]$$

$$v = -16\frac{h}{T_2}\left(\frac{\theta}{\beta_2}\right)^2 \qquad 0 \leqslant \theta \leqslant \frac{\beta_2}{4} \qquad (2.8\,e)$$

(return)

$$a = -32\frac{h}{T_2^2}\frac{\theta}{\beta_2}$$

$$y = h\left[\frac{5}{6} + 2\frac{\theta}{\beta_2} - 8\left(\frac{\theta}{\beta_2}\right)^2 + \frac{16}{3}\left(\frac{\theta}{\beta_2}\right)^3\right]$$

$$v = \frac{h}{T_2}\left[2 - 16\left(\frac{\theta}{\beta_2}\right) + 16\left(\frac{\theta}{\beta_2}\right)^2\right] \qquad \frac{\beta_2}{4} \leqslant \theta \leqslant \frac{3\beta_2}{4} \qquad (2.8\,f)$$

(return)

$$a = \frac{h}{T_2^2}\left(-16 + 32\frac{\theta}{\beta_2}\right)$$

This curve is usable only for low speed or in a compound curve.

3-4 Polynomial (Table 2-1)

$$y = h\left[8\left(\frac{\theta}{\beta_1}\right)^3 - 8\left(\frac{\theta}{\beta_1}\right)^4\right]$$

$$v = \frac{h}{T_1}\left[24\left(\frac{\theta}{\beta_1}\right)^2 - 32\left(\frac{\theta}{\beta_1}\right)^3\right] \qquad 0 < \theta < \frac{\beta_1}{2}$$

$$\text{(rise)}$$

$$a = \frac{h}{T_1^2}\left[48\frac{\theta}{\beta_1} - 96\left(\frac{\theta}{\beta_1}\right)^2\right]$$

$$(2.9a)$$

$$y = h\left[1 - 8\frac{\theta}{\beta_1} + 24\left(\frac{\theta}{\beta_1}\right)^2 - 24\left(\frac{\theta}{\beta_1}\right)^3 + 8\left(\frac{\theta}{\beta_1}\right)^4\right]$$

$$v = \frac{h}{T_1}\left[-8 + 48\frac{\theta}{\beta_1} - 72\left(\frac{\theta}{\beta_1}\right)^2 + 32\left(\frac{\theta}{\beta_1}\right)^3\right] \qquad \frac{\beta_1}{2} \leq \theta \leq \beta_1$$

$$\text{(rise)}$$

$$a = \frac{h}{T_1^2}\left[48 - 144\frac{\theta}{\beta_1} + 96\left(\frac{\theta}{\beta_1}\right)^2\right]$$

$$(2.9b)$$

$$y = h\left[1 - 8\left(\frac{\theta}{\beta_2}\right)^3 + 8\left(\frac{\theta}{\beta_2}\right)^4\right]$$

$$v = \frac{h}{T_2}\left[-24\left(\frac{\theta}{\beta_2}\right)^2 + 32\left(\frac{\theta}{\beta_2}\right)^3\right] \qquad 0 < \theta < \frac{\beta_2}{2}$$

$$\text{(return)}$$

$$a = \frac{h}{T_2^2}\left[-48\frac{\theta}{\beta_2} + 96\left(\frac{\theta}{\beta_2}\right)^2\right]$$

$$(2.9c)$$

$$y = h\left[8\frac{\theta}{\beta_2} - 24\left(\frac{\theta}{\beta_2}\right)^2 + 24\left(\frac{\theta}{\beta_2}\right)^3 - 8\left(\frac{\theta}{\beta_2}\right)^4\right]$$

$$v = \frac{h}{T_2}\left[-48\frac{\theta}{\beta_2} + 72\left(\frac{\theta}{\beta_2}\right)^2 - 32\left(\frac{\theta}{\beta_2}\right)^3\right] \qquad \beta_2 < \theta < \beta_2 \qquad (2.9d)$$

$$a = \frac{h}{T_2^2}\left[-48 + 144\frac{\theta}{\beta_2} - 96\left(\frac{\theta}{\beta_2}\right)^2\right]$$

This curve has characteristics very similar to that of cycloidal motion; it is a simple polynomial and in Chapter 10 more advanced polynomials are discussed.

3-4-5 Polynomial (Fig. 2-10)

$$y = h\left[10\left(\frac{\theta}{\beta_1}\right)^3 - 15\left(\frac{\theta}{\beta_1}\right)^4 + 6\left(\frac{\theta}{\beta_1}\right)^5\right]$$

$$v = \frac{h}{T_1}\left[30\left(\frac{\theta}{\beta_1}\right)^2 - 60\left(\frac{\theta}{\beta_1}\right)^3 + 30\left(\frac{\theta}{\beta_1}\right)^4\right] \left.\begin{array}{c} \\ \\ \\ \end{array}\right\} \quad 0 \leqslant \theta \leqslant \beta_1 \qquad (2.10a)$$

$$a = \frac{h}{T_1^2}\left[60\frac{\theta}{\beta_1} - 180\left(\frac{\theta}{\beta_1}\right)^2 + 120\left(\frac{\theta}{\beta_1}\right)^3\right]$$

(rise)

$$y = h\left[1 - 10\left(\frac{\theta}{\beta_2}\right)^3 + 15\left(\frac{\theta}{\beta_2}\right)^4 - 6\left(\frac{\theta}{\beta_2}\right)^5\right]$$

$$v = \frac{h}{T_2}\left[-30\left(\frac{\theta}{\beta_2}\right)^2 + 60\left(\frac{\theta}{\beta_2}\right)^3 - 30\left(\frac{\theta}{\beta_2}\right)^4\right] \left.\begin{array}{c} \\ \\ \\ \end{array}\right\} \quad 0 \leqslant \theta \leqslant \beta_2 \qquad (2.10b)$$

$$a = \frac{h}{T_2^2}\left[-60\frac{\theta}{\beta_2} + 180\left(\frac{\theta}{\beta_2}\right)^2 - 120\left(\frac{\theta}{\beta_2}\right)^3\right]$$

(return)

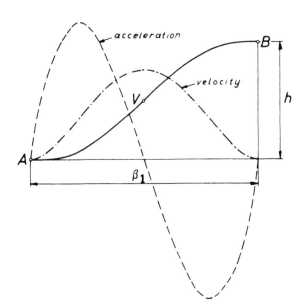

Fig. 2-10. Cam displacement, velocity, and acceleration curves for 3-4-5 polynomial motion.

This curve has good acceleration characteristics.

4-5-6-7 Polynomial (Table 2-1)

$$
\left.
\begin{aligned}
y &= h\left[35\left(\frac{\theta}{\beta_1}\right)^4 - 84\left(\frac{\theta}{\beta_1}\right)^5 + 70\left(\frac{\theta}{\beta_1}\right)^6 - 20\left(\frac{\theta}{\beta_1}\right)^7\right] \\
v &= \frac{h}{T_1}\left[140\left(\frac{\theta}{\beta_1}\right)^3 - 420\left(\frac{\theta}{\beta_1}\right)^4 + 420\left(\frac{\theta}{\beta_1}\right)^5 - 140\left(\frac{\theta}{\beta_1}\right)^6\right] \\
a &= \frac{h}{T_1^2}\left[420\left(\frac{\theta}{\beta_1}\right)^2 - 1680\left(\frac{\theta}{\beta_1}\right)^3 + 2100\left(\frac{\theta}{\beta_1}\right)^4 - 840\left(\frac{\theta}{\beta_1}\right)^5\right]
\end{aligned}
\right\}
\begin{array}{c}
0 < \theta < \beta_1 \\
\text{(rise)} \\[1em]
(2.11a)
\end{array}
$$

$$
\left.
\begin{aligned}
y &= h\left[1 - 35\left(\frac{\theta}{\beta_2}\right)^4 + 84\left(\frac{\theta}{\beta_2}\right)^5 - 70\left(\frac{\theta}{\beta_2}\right)^6 + 20\left(\frac{\theta}{\beta_2}\right)^7\right] \\
v &= \frac{h}{T_2}\left[-140\left(\frac{\theta}{\beta_2}\right)^3 + 420\left(\frac{\theta}{\beta_2}\right)^4 - 420\left(\frac{\theta}{\beta_2}\right)^5 + 140\left(\frac{\theta}{\beta_2}\right)^6\right] \\
a &= \frac{h}{T_2^2}\left[-420\left(\frac{\theta}{\beta_2}\right)^2 + 1680\left(\frac{\theta}{\beta_2}\right)^3 - 2100\left(\frac{\theta}{\beta_2}\right)^4 + 840\left(\frac{\theta}{\beta_2}\right)^5\right]
\end{aligned}
\right\}
\begin{array}{c}
0 < \theta < \beta_2 \\
\text{(return)} \\[1em]
(2.11b)
\end{array}
$$

This curve is sometimes used for high speed cams because of good acceleration characteristics.

Trapezoidal Acceleration; $C = 0.25\,\dfrac{\beta_1}{2}$ by rise and $0.25\,\dfrac{\beta_2}{2}$ by return (Table 2-1)

$$
\left.
\begin{aligned}
y &= \frac{64}{9}h\left(\frac{\theta}{\beta_1}\right)^3 \\
v &= \frac{64}{3}\frac{h}{T_1}\left(\frac{\theta}{\beta_1}\right)^2 \\
a &= \frac{128}{3}\frac{h}{T_1^2}\frac{\theta}{\beta_1}
\end{aligned}
\right\}
\quad 0 < \theta < \frac{\beta_1}{8}
\qquad (2.12a)
$$

$$y = h\left[\frac{1}{72} - \frac{1}{3}\frac{\theta}{\beta_1} + \frac{8}{3}\left(\frac{\theta}{\beta_1}\right)^2\right]$$

$$v = \frac{h}{T_1}\left(-\frac{1}{3} + \frac{16}{3}\frac{\theta}{\beta_1}\right) \qquad \left.\begin{array}{c}\\ \\ \\ \end{array}\right\} \qquad \frac{\beta_1}{8} \leqslant \theta \leqslant \frac{3\beta_1}{8} \qquad (2.12b)$$

$$a = \frac{16}{3}\frac{h}{T_1^2}$$

$$y = h\left[\frac{7}{18} - \frac{10}{3}\frac{\theta}{\beta_1} + \frac{32}{3}\left(\frac{\theta}{\beta_1}\right)^2 - \frac{64}{9}\left(\frac{\theta}{\beta_1}\right)^3\right]$$

$$v = \frac{h}{T_1}\left[-\frac{10}{3} + \frac{64}{3}\frac{\theta}{\beta_1} - \frac{64}{3}\left(\frac{\theta}{\beta_1}\right)^2\right] \qquad \left.\begin{array}{c}\\ \\ \end{array}\right\} \qquad \frac{3\beta_1}{8} \leqslant \theta \leqslant \frac{\beta_1}{2} \quad (2.12c)$$

$$a = \frac{h}{T_1^2}\left(\frac{64}{3} - \frac{128}{3}\frac{\theta}{\beta_1}\right)$$

This curve is good for high speed cams if it is machined accurately.

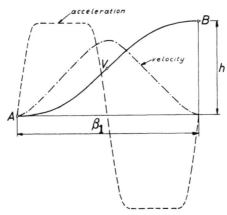

Fig. 2-11. Modified trapezoidal acceleration curve and associated cam displacement and velocity curves.

Modified Trapezoidal Acceleration (Fig. 2-11)

$$y = \frac{h}{\pi + 2}\left[2\frac{\theta}{\beta_1} - \frac{1}{2\pi}\sin\left(4\pi\frac{\theta}{\beta_1}\right)\right]$$

$$v = \frac{h}{T_1(\pi + 2)}\left[2 - 2\cos\left(4\pi\frac{\theta}{\beta_1}\right)\right] \qquad 0 < \theta < \frac{\beta_1}{8}$$

$$\text{(rise)}$$

$$a = \frac{8\pi h}{T_1^2(\pi + 2)}\sin\left(4\pi\frac{\theta}{\beta_1}\right)$$

$$\text{(2.13a)}$$

$$y = \frac{h}{\pi + 2}\left[4\pi\left(\frac{\theta}{\beta_1}\right)^2 + (2 - \pi)\frac{\theta}{\beta_1} + \frac{\pi^2 - 8}{16\pi}\right]$$

$$v = \frac{h}{T_1(\pi + 2)}\left(8\pi\frac{\theta}{\beta_1} + 2 - \pi\right) \qquad \frac{\beta_1}{8} < \theta < \frac{3\beta_1}{8}$$

$$\text{(rise)}$$

$$a = \frac{8\pi h}{T_1^2(\pi + 2)}$$

$$\text{(2.13b)}$$

$$y = \frac{h}{\pi + 2}\left[2(\pi + 1)\frac{\theta}{\beta_1} - \frac{1}{2\pi}\sin\left(4\pi\left(\frac{\theta}{\beta_1} - \frac{1}{4}\right)\right) - \frac{\pi}{2}\right]$$

$$v = \frac{h}{T_1(\pi + 2)}\left[2(\pi + 1) - 2\cos\left(4\pi\left(\frac{\theta}{\beta_1} - \frac{1}{4}\right)\right)\right] \qquad \frac{3\beta_1}{8} < \theta < \frac{\beta_1}{2}$$

$$\text{(rise)}$$

$$a = \frac{8\pi h}{T_1^2(\pi + 2)}\sin\left(4\pi\left(\frac{\theta}{\beta_1} - \frac{1}{4}\right)\right)$$

$$\text{(2.13c)}$$

Replacing θ by $\beta_1 - \theta$ and y by $h - y$,

$$y = \frac{h}{\pi + 2}\left[-\frac{\pi}{2} + 2(\pi + 1)\frac{\theta}{\beta_1} + \frac{1}{2\pi}\sin\left(4\pi\left(\frac{3}{4} - \frac{\theta}{\beta_1}\right)\right)\right]$$

$$v = \frac{h}{T_1(\pi + 2)}\left[2(\pi + 1) - 2\cos\left(4\pi\left(\frac{3}{4} - \frac{\theta}{\beta_1}\right)\right)\right] \qquad \frac{\beta_1}{2} < \theta < \frac{5\beta_1}{8}$$

$$\text{(rise)}$$

$$a = \frac{8\pi h}{T_1^2(\pi + 2)}\sin\left(4\pi\left(\frac{3}{4} - \frac{\theta}{\beta_1}\right)\right)$$

$$\text{(2.13d)}$$

$$y = \frac{h}{\pi + 2}\left[-2\,\frac{1}{16}\,\pi + \frac{1}{2\pi} + (2 + 7\pi)\,\frac{\theta}{\beta_1} - 4\pi\left(\frac{\theta}{\beta_1}\right)^2\right]$$

$$v = \frac{h}{T_1(\pi + 2)}\left(2 + 7\pi - 8\pi\,\frac{\theta}{\beta_1}\right)$$

$$a = -\frac{8\pi\,h}{T_1^2(\pi + 2)}$$

$$\left.\begin{array}{c} \frac{5\beta_1}{8} \leqslant \theta \leqslant \frac{7\beta_1}{8} \\ \text{(rise)} \end{array}\right\} \qquad (2.13\text{e})$$

$$y = \frac{h}{\pi + 2}\left[\pi + 2\,\frac{\theta}{\beta_1} + \frac{1}{2\pi}\sin\left(4\pi\left(1 - \frac{\theta}{\beta_1}\right)\right)\right]$$

$$v = \frac{h}{T_1(\pi + 2)}\left[2 - 2\cos\left(4\pi\left(1 - \frac{\theta}{\beta_1}\right)\right)\right]$$

$$a = -\frac{8\pi\,h}{T_1^2(\pi + 2)}\sin\left(4\pi\left(1 - \frac{\theta}{\beta_1}\right)\right)$$

$$\left.\begin{array}{c} \frac{7\beta_1}{8} \leqslant \theta \leqslant \beta_1 \\ \text{(rise)} \end{array}\right\} \qquad (2.13\text{f})$$

$$y = \frac{h}{\pi + 2}\left[\pi + 2 - 2\,\frac{\theta}{\beta_2} + \frac{1}{2\pi}\sin\left(4\pi\,\frac{\theta}{\beta_2}\right)\right]$$

$$v = \frac{h}{T_2(\pi + 2)}\left[-2 + 2\cos\left(4\pi\,\frac{\theta}{\beta_2}\right)\right]$$

$$a = -\frac{8\pi\,h}{T_2^2(\pi + 2)}\sin\left(4\pi\,\frac{\theta}{\beta_2}\right)$$

$$\left.\begin{array}{c} 0 \leqslant \theta \leqslant \frac{\beta_2}{8} \quad (2.13\text{g}) \\ \text{(return)} \end{array}\right\}$$

$$y = \frac{h}{\pi + 2}\left[\pi + 2 - 4\pi\left(\frac{\theta}{\beta_2}\right)^2 - (2 - \pi)\,\frac{\theta}{\beta_2} - \frac{\pi^2 - 8}{16\pi}\right]$$

$$v = \frac{h}{T_2(\pi + 2)}\left[-8\pi\,\frac{\theta}{\beta_2} - 2 + \pi\right]$$

$$a = -\frac{8\pi\,h}{T_2^2(\pi + 2)}$$

$$\left.\begin{array}{c} \frac{\beta_2}{8} \leqslant \theta \leqslant \frac{3\beta_2}{8} \\ \text{(return)} \\ (2.13\text{h}) \end{array}\right\}$$

$$y = \frac{h}{\pi + 2}\left[\frac{3\pi}{2} + 2 - 2(\pi + 1)\,\frac{\theta}{\beta_2} + \frac{1}{2\pi}\sin\left(4\pi\left(\frac{\theta}{\beta_2} - \frac{1}{4}\right)\right)\right]$$

$$v = \frac{h}{T_2(\pi + 2)}\left[-2(\pi + 1) + 2\cos\left(4\pi\left(\frac{\theta}{\beta_2} - \frac{1}{4}\right)\right)\right]$$

$$a = -\frac{8\pi\,h}{T_2^2(\pi + 2)}\sin\left(4\pi\left(\frac{\theta}{\beta_2} - \frac{1}{4}\right)\right)$$

$$\left.\begin{array}{c} \frac{3\beta_2}{8} \leqslant \theta \leqslant \frac{\beta_2}{2} \\ \text{(return)} \\ (2.13\text{i}) \end{array}\right\}$$

$$y = \frac{h}{\pi + 2}\left[\frac{3\pi}{2} + 2 - 2(\pi + 1)\frac{\theta}{\beta_2} - \frac{1}{2\pi}\sin\left(4\pi\left(\frac{3}{4} - \frac{\theta}{\beta_2}\right)\right)\right]$$

$$v = \frac{h}{T_2(\pi + 2)}\left[-2(\pi + 1) + 2\cos\left(4\pi\left(\frac{3}{4} - \frac{\theta}{\beta_2}\right)\right)\right]$$

$$a = \frac{8\pi h}{T_2^2(\pi + 2)}\sin\left(4\pi\left(\frac{3}{4} - \frac{\theta}{\beta_2}\right)\right)$$

$$\left.\begin{array}{c}\\ \\ \\ \end{array}\right\} \quad \frac{\beta_2}{2} < \theta < \frac{5\beta_2}{8}$$

(return)

(2.13j)

$$y = \frac{h}{\pi + 2}\left[3\frac{1}{16}\pi + 2 - \frac{1}{2\pi} - (2 + 7\pi)\frac{\theta}{\beta_2} + 4\pi\left(\frac{\theta}{\beta_2}\right)^2\right]$$

$$v = \frac{h}{T_2(\pi + 2)}\left(-2 - 7\pi + 8\pi\frac{\theta}{\beta_2}\right)$$

$$a = \frac{8\pi h}{T_2^2(\pi + 2)}$$

$$\left.\begin{array}{c}\\ \\ \\ \end{array}\right\} \quad \frac{5\beta_2}{8} < \theta < \frac{7\beta_2}{8}$$

(return)

(2.13k)

$$y = \frac{h}{\pi + 2}\left[2 - 2\frac{\theta}{\beta_2} - \frac{1}{2\pi}\sin\left(4\pi\left(1 - \frac{\theta}{\beta_2}\right)\right)\right]$$

$$v = \frac{h}{T_2(\pi + 2)}\left[-2 + 2\cos\left(4\pi\left(1 - \frac{\theta}{\beta_2}\right)\right)\right]$$

$$a = \frac{8\pi h}{T_2^2(\pi + 2)}\sin\left(4\pi\left(1 - \frac{\theta}{\beta_2}\right)\right)$$

$$\left.\begin{array}{c}\\ \\ \\ \end{array}\right\} \quad \frac{7\beta_2}{8} < \theta < \beta_2 \text{ (2.13l)}$$

(return)

This curve offers a slight improvement as compared with the trapezoidal acceleration curve.

Modified Sinusoidal Acceleration; $C = 0.25\frac{\beta_1}{2}$ by rise and $0.25\frac{\beta_2}{2}$ by return (Table 2-1)

$$y = \frac{h}{4 + \pi}\left[\pi\frac{\theta}{\beta_1} - \frac{1}{4}\sin\left(4\pi\frac{\theta}{\beta_1}\right)\right]$$

$$v = \frac{\pi h}{T_1(4 + \pi)}\left[1 - \cos\left(4\pi\frac{\theta}{\beta_1}\right)\right]$$

$$a = \frac{4\pi^2 h}{T_1^2(4 + \pi)}\sin\left(4\pi\frac{\theta}{\beta_1}\right)$$

$$\left.\begin{array}{c}\\ \\ \\ \end{array}\right\} \quad 0 < \theta < \frac{\beta_1}{8}$$

(rise)

(2.14a)

$$y = \frac{h}{4 + \pi}\left[2 + \pi\frac{\theta}{\beta_1} - \frac{9}{4}\sin\left(\frac{\pi}{3} + \frac{4\pi}{3}\frac{\theta}{\beta_1}\right)\right]$$

$$v = \frac{\pi h}{T_1(4 + \pi)}\left[1 - 3\cos\left(\frac{\pi}{3} + \frac{4\pi}{3}\frac{\theta}{\beta_1}\right)\right] \qquad \left.\begin{array}{c} \end{array}\right\} \frac{\beta_1}{8} \leqslant \theta \leqslant \frac{7\beta_1}{8} \quad (2.14\,b)$$

$$a = \frac{4\pi^2 h}{T_1^2(4 + \pi)}\sin\left(\frac{\pi}{3} + \frac{4\pi}{3}\frac{\theta}{\beta_1}\right) \qquad \text{(rise)}$$

$$y = \frac{h}{4 + \pi}\left[4 + \pi\frac{\theta}{\beta_1} - \frac{1}{4}\sin\left(4\pi\frac{\theta}{\beta_1}\right)\right]$$

$$v = \frac{\pi h}{T_1(4 + \pi)}\left[1 - \cos\left(4\pi\frac{\theta}{\beta_1}\right)\right] \qquad \left.\begin{array}{c} \end{array}\right\} \frac{7\beta_1}{8} \leqslant \theta \leqslant \beta_1 \quad (2.14\,c)$$

$$a = \frac{4\pi^2 h}{T_1^2(4 + \pi)}\sin\left(4\pi\frac{\theta}{\beta_1}\right) \qquad \text{(rise)}$$

$$y = \frac{h}{4 + \pi}\left[4 + \pi - \pi\frac{\theta}{\beta_2} + \frac{1}{4}\sin\left(4\pi\frac{\theta}{\beta_2}\right)\right]$$

$$v = \frac{\pi h}{T_2(4 + \pi)}\left[-1 + \cos\left(4\pi\frac{\theta}{\beta_2}\right)\right] \qquad \left.\begin{array}{c} \end{array}\right\} 0 \leqslant \theta \leqslant \frac{\beta_2}{8} \quad (2.14\,d)$$

$$a = -\frac{4\pi^2 h}{T_2^2(4 + \pi)}\sin\left(4\pi\frac{\theta}{\beta_2}\right) \qquad \text{(return)}$$

$$y = \frac{h}{4 + \pi}\left[2 + \pi - \pi\frac{\theta}{\beta_2} + \frac{9}{4}\sin\left(\frac{\pi}{3} + \frac{4\pi}{3}\frac{\theta}{\beta_2}\right)\right]$$

$$v = \frac{\pi h}{T_2(4 + \pi)}\left[-1 + 3\cos\left(\frac{\pi}{3} + \frac{4\pi}{3}\frac{\theta}{\beta_2}\right)\right] \qquad \left.\begin{array}{c} \end{array}\right\} \frac{\beta_2}{8} \leqslant \theta \leqslant \frac{7\beta_2}{8}$$

$$a = -\frac{4\pi^2 h}{T_2^2(4 + \pi)}\sin\left(\frac{\pi}{3} + \frac{4\pi}{3}\frac{\theta}{\beta_2}\right) \qquad \text{(return)}$$

(2.14\,e)

$$y = \frac{h}{4 + \pi}\left[\pi - \pi\frac{\theta}{\beta_2} + \frac{1}{4}\sin\left(4\pi\frac{\theta}{\beta_2}\right)\right]$$

$$v = \frac{\pi h}{T_2(4 + \pi)}\left[-1 + \cos\left(4\pi\frac{\theta}{\beta_2}\right)\right] \qquad \left.\begin{array}{c} \end{array}\right\} \frac{7\beta_2}{8} \leqslant \theta \leqslant \beta_2 \quad (2.14\,f)$$

$$a = -\frac{4\pi^2 h}{T_2^2(4 + \pi)}\sin\left(4\pi\frac{\theta}{\beta_2}\right) \qquad \text{(return)}$$

This curve has a low maximum velocity and good acceleration characteristics. High speed performance is good to excellent.

The Modified Cycloid

$$y = h\left[\frac{t_x}{T} - \frac{1}{2\pi}\sin\left(2\pi\,\frac{t_x}{T}\right)\right]$$

(2.15a)

$$\text{where } t = t_x + \frac{T}{2\pi}\,\frac{\tan\delta}{\tan v}\,\sin\left(2\pi\,\frac{t_x}{T}\right) \qquad 0 < t < T$$

(2.15b)

This curve is treated in Chapter 3, where it is shown how easily it can be fitted to fulfill the given requirements.

Comparison of Cam Displacement Curves

In Table 2-1 the velocity and acceleration characteristics of the aforementioned curves are given. The values in the velocity and acceleration factor columns are obtained by setting $h = 1$ and $T = 1$. With the help of these values one can immediately tell how much larger the velocity or acceleration is for one curve as compared with another for the same time and stroke. For example, it can be seen that for a given h and T, the maximum velocity for the cubic curve No. 1 is 50 percent larger than that for the parabolic motion curve (the values are given as 3 and 2, respectively.)

In this chart there is also a column of comparative factors for what is called *jerk* or *pulse;* i.e., the rate of change of acceleration or third time derivative of y:

$$\frac{da}{dt} = \frac{d^2v}{dt^2} = \frac{d^3y}{dt^3}$$

The magnitude of the jerk factor tells us whether the curve in question is suited for high speed or not. In general, jerk curves that are continuous and have low maximum values indicate good performance at high speeds.

Displacement Diagram Synthesis

In Chapter 2 a number of curves were described which can be used for the rise and return portions of both simple and complex cams. It was also pointed out in connection with Fig. 2-2 that the diagram derived represents in each case the time-displacement graph of a point *fixed* on the follower. Thus in Fig. 2-2, the knife edge of the follower will have exactly the motion indicated by the graph. Since the pressure between the cam and the follower is usually severe, a knife edge as illustrated is very impractical, and roller followers are universally used. Reflection will reveal that the only point on a rotating contact (roller) that has a motion exactly as required by the follower is the center of the roller itself. Later on the process necessary to develop the working surface for any cam to produce the required follower motion will be shown, but meanwhile the curves discussed in Chapter 2 must all be regarded as being, in effect, the time-displacement graphs of the centers of rollers.

Also, in Chapter 2, it was pointed out that the straight-line graph has the important advantage of uniform velocity. This is so desirable that many cams based on this graph are used. To avoid an instantaneous acceleration at the beginning and end of the stroke, a modification is introduced at these points. There are many different types of modifications possible, ranging from a simple circular arc to much more complicated curves. One of the curves used for this purpose is the parabolic curve given by Eq. (2.2). As seen from the derived time-graphs, this curve causes the follower to begin its stroke with zero velocity but having a finite and constant acceleration. We must accept the necessity of acceleration, but effort should be made to hold it to a minimum or to use the cycloidal curve.

Matching of Constant Velocity and
Parabolic Motion Curves

Our problem now is to match a sloping straight line to a parabola in such a way that the two are exactly tangent to each other. As illustrated in Fig. 3-1, it can be shown that for any parabola the vertex of which is

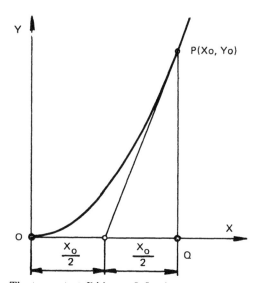

Fig. 3-1. The tangent at P bisects O-Q, when curve is a parabola.

at O, the tangent to the curve at the point P intersects the line OQ at its midpoint. This means that the tangent at P represents the velocity of the follower at time X_0 as shown in Fig. 3-1. Since the tangent also represents the velocity of the follower over the constant velocity portion of the stroke, the transition from rest to the maximum velocity is accomplished with smoothness. The application of this is illustrated in the following problem:

(a) A cam follower is to rise $\frac{1}{4}$ inch with constant acceleration; $1\frac{1}{4}$ inches with constant velocity, over an angle of 50 degrees; and then $\frac{1}{2}$ inch with constant deceleration.

In Fig. 3-2 the three rise distances are laid out, $y_1 = \frac{1}{4}$ in., $y_2 = 1\frac{1}{4}$ in., $y_3 = \frac{1}{2}$ in., and horizontals drawn. Next, an arbitrary horizontal distance ϕ_2 proportional to 50 degrees is measured off and points A and B are located. The line AB is prolonged to M_1 and M_2. By remembering that a tangent to a parabola, Fig. 3-1, will cut the abscissa axis at point $\left(\dfrac{X_0}{2}, 0\right)$, where X_0 is the abscissa of the point of tangency, the two values. $\phi_1 = 20°$ and $\phi_3 = 40°$ will be found. Analytically,

$$\frac{M_1E}{\phi_2} = \frac{y_1}{y_2}; \quad \frac{\frac{1}{2}\phi_1}{50°} = \frac{0.25}{1.25} \quad \therefore \phi_1 = 20°$$

$$\frac{FM_2}{\phi_2} = \frac{y_3}{y_2}; \quad \frac{\frac{1}{2}\phi_3}{50°} = \frac{0.50}{1.25} \quad \therefore \phi_3 = 40°$$

In Fig. 3-2, the portions of the parabola have been drawn in, and the details of this operation are as follows:

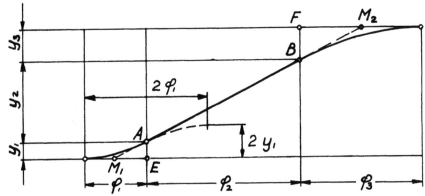

Fig. 3-2. Matching a parabola at each end of straight line displacement curve A-B to provide more acceptable acceleration and deceleration.

Assume that accuracy to the nearest thousandth of one inch is desired, and it is decided to plot values for every 5 degrees of cam rotation. Strictly computational methods are used in this development.

The formula for the parabolic curve has been given in Chapter 2 as:

$$y = 2h\left(\frac{\theta}{\beta_1}\right)^2 \qquad (2.2a)$$

To apply this formula to the above problem we must bear in mind that we are dealing with two different parabolas; one for accelerating the follower during a cam rotation of 20 degrees, the other for decelerating it in 40 degrees, these two being tangent to the same line AB.

The range of the curve described by the formula is from zero to maximum velocity. Since we are using the curve only to provide smoothness to and from the straight line representing constant velocity (A to B), the h value to be placed in the equation is $2y_1$ for angle ϕ_1, and $2y_3$ for angle ϕ_3.

In Formula 2.2, h is the full stroke when the acceleration and deceleration parabolas are joined together. In Fig. 3-2 the follower is accelerated over y_1, however, we must think of this parabola as being extended beyond A so that we have two tangent parábolas; Formula 2.2 then applies and $2y_1$ should be used for h. The same thing holds for the parabola at B.

Table (3-1) shows the computations and resulting values for the cam displacement diagram described. This table not only gives follower displacement values to the nearest thousandth of one inch for the curved portions of the rise diagram, but also for the straight line portion between A and B. Obviously, the intermediate points are not needed to draw the straight line but when the cam profile later is to be drawn or cut these values will be needed since they are to be measured on radial lines.

Table 3-1. Development of Modified Constant Velocity Cam
with Parabolic Matching

Divi-sion	ϕ Degrees	Computation	Follower displace-ment	Explanation
	0	0	0	$\beta = 40°, \quad h = 0.5$
	5	$0.000625 \times \ 5^2$	0.016	
ϕ_1	10	0.000625×10^2	0.063	$y = \dfrac{(2)\,(0.5)}{40^2}\,\theta^2$
$= 20°$	15	0.000625×15^2	0.141	
	20	0.000625×20^2	0.250	$= 0.000625\theta^2$
	25		0.375	
	30		0.500	
	35		0.625	
	40		0.750	
ϕ_2	45		0.875	1.250 in. divided into
	50		1.000	10 uniform divisions
$= 50°$	55		1.125	
	60		1.250	
	65		1.375	
	70		1.500	
	75	$2.000 - (0.0003125 \times 35^2)$	1.617	
	80	$2.000 - (0.0003125 \times 30^2)$	1.719	$\beta = 80°, \quad h = 1.000$
	85	$2.000 - (0.0003125 \times 25^2)$	1.805	
	90	$2.000 - (0.0003125 \times 20^2)$	1.875	$y = \dfrac{(2)\,(1.0)}{(80)^2}\,(110 - \theta)^2$
ϕ_3	95	$2.000 - (0.0003125 \times 15^2)$	1.930	
$= 40°$	100	$2.000 - (0.0003125 \times 10^2)$	1.969	$= 0.0003125\,(110 - \theta)^2$
	105	$2.000 - (0.0003125 \times \ 5^2)$	1.992	
	110	$2.000 - (0.0003125 \times \ 0)$	2.000	

(b) Often the problem arises that a displacement diagram similar to that of Fig. 3-2 is desired, but only the acceleration, constant velocity and deceleration portions of the stroke and the total angle are given. The same principles as in the foregoing examples apply.

Example:

$$y_1 = 0.15 \text{ in.,} \quad y_2 = 0.35 \text{ in.,} \quad y_3 = 0.20 \text{ in.,} \quad \text{and}$$
$$\phi_1 + \phi_2 + \phi_3 = 120°$$

Solution:

$$\frac{2M_1E}{2y_1} = \frac{\phi_2}{y_2} = \frac{2FM_2}{2y_3} = \frac{\phi_1 + \phi_2 + \phi_3}{2y_1 + y_2 + 2y_3} = \frac{120}{1.05}$$

$$\frac{M_1E}{0.15} = \frac{120}{1.05}; \quad 2M_1E = \phi_1 = \frac{(120)(0.15)(2)}{1.05} = 34.3°$$

$$\phi_2 = \frac{120}{1.05} \cdot 0.35 = 40.0°$$

$$\frac{FM_2}{y_3} = \frac{120}{1.05}; \quad 2FM_2 = \phi_3 = \frac{(120)(0.20)(2)}{1.05} = 45.7°$$

$$\overline{120.0°}$$

If A and B fall on the same point, the following simple equation results:

$$\frac{y_1}{y_3} = \frac{\phi_1}{\phi_3}$$

And this equation says that the matching point of the two parabolas must be on the line connecting the point where the rise starts with the point where the rise ends. This, of course, also holds for the return.

(c) *Example:*

$$y_1 = 0.35 \text{ in.,} \quad y_3 = 0.45 \text{ in.,} \quad \text{and} \quad \phi_1 + \phi_3 = 90°$$

Solution:

$$\frac{\phi_1}{y_1} = \frac{\phi_3}{y_3} = \frac{\phi_1 + \phi_3}{y_1 + y_3} = \frac{90}{0.80};$$

$$\phi_1 = 0.35 \cdot \frac{90}{0.80} = 39.4°$$

$$\phi_3 = 0.45 \cdot \frac{90}{0.80} = 50.6°$$

$$\overline{90.0°}$$

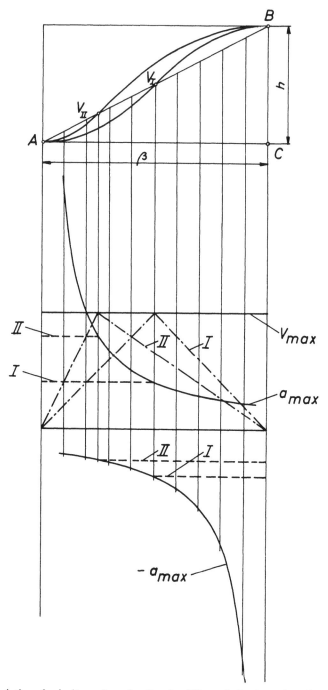

Fig. 3-3. Variation of velocity and acceleration for different inflection points V_I and V_{II}.

Significance of the Position of
the Matching Point

It is of interest to know what influence the changing positions of the matching point of the two parabolas has on the maximum velocity and on the maximum acceleration. Figure 3-3 shows these variations. Here the variations of V_{max} and a_{max} for a given stroke h and time β_1 are represented graphically. For instance, the maximum velocity when the curve passes through V_I is found by going vertically down to the horizontal line which is designated V_{max}. The maximum acceleration a_{max} is found by going down to the curve a_{max} and the maximum deceleration is given by the $-a_{max}$-curve. Thus it is seen that the maximum velocity remains unchanged, no matter whether V_I or V_{II}, is used as a matching point. It can also be seen that if the time for acceleration is cut in half by using V_{II} rather than V_I as a matching point; then the acceleration is doubled, but the deceleration is decreased two-thirds. The two hyperbolas shown indicate this variation, and it is seen that if the matching point is placed close to V_I, the middle of the stroke, then the increase and decrease in maximum accelerations are rather small.

(d) It is often desired that the machine part, when moving from A to B, Fig. 3-4, at a certain time should be at point P. This means that in the time-displacement diagram h, y_1, ϕ_1, and β_1 are given. If motion is to start at A and end at B, the location of P cannot be just anywhere, but must be within the area enclosed by the parabola with the apex at A and going through B, and the parabola with apex at B and going through A. Practically, of course, this area is further diminished by the fact that the acceleration cannot exceed a certain value and also that the radius of curvature cannot be less than a certain value. If P lies below straight line AB, P must lie on a parabola with apex at A, and if it lies above AB, P must lie on a parabola with apex at B.

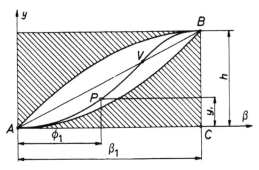

Fig. 3-4. Here the end points A and B and an intermediate point P are given.

Example:

$$h = 0.50 \text{ in.}, \quad \beta_1 = 120°, \quad y_1 = 0.15 \text{ in.}, \quad \text{and } \phi_1 = 54.5°$$

It follows from the given values that P lies below AB and the equation for the parabola having apex at A is used:

$$y = k\phi^2_1$$

Point P lies on this parabola so

$$y_1 = k\beta^2_1; \quad k = \frac{0.15}{54.5^2}$$

$y = \dfrac{0.15}{54.5^2} \cdot \phi_1$ is the equation for the parabola passing through P.

The equation for straight line AB:

$$y = \frac{h}{\beta_1} \cdot \beta = \frac{0.50}{120} \cdot \beta$$

The intersection of the parabola and AB gives the matching point V:

$$\frac{0.15}{54.5^2} \cdot \phi^2_1 = \frac{0.50}{120} \cdot \phi_1; \quad \phi_1 = \frac{(0.50)(54.5)^2}{(120)(0.15)} = 82.4°$$

The corresponding displacement is:

$$y = \frac{0.50}{120} \cdot 82.4 = 0.343 \text{ in.}$$

Because cycloidal motion, 3–4 polynomial, trapezoidal acceleration, and modified trapezoidal acceleration have the same velocity of the matching point as parabolic motion, the methods used above are in principal also valid for these curves.

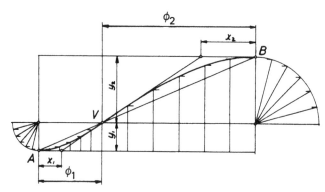

Fig. 3-5. Curve matching for different points of inflection, simple harmonic motion.

It is possible to extend the above described matching methods to simple harmonic motion, and it is done as follows:

In Fig. 3-5 the motion starts at A and ends at B. The matching point V must lie on AB.

The slope of the tangent at V is found by differentiating the equation for the simple harmonic motion curve connecting A with V:

$$y = y_1 \left[1 - \cos\left(\pi \frac{\phi}{\phi_1} \right) \right]$$

$$\frac{dy}{d\beta} = y_1 \cdot \frac{\pi}{\beta_1} \cdot \sin\left(\pi \frac{\phi}{\phi_1} \right)$$

$$\left(\frac{dy}{d\phi} \right)_{\max} = \frac{\pi}{2} \cdot \frac{y_1}{\phi_1}$$

The equation for the tangent at V is:

$$y - y_1 = \frac{\pi}{2} \cdot \frac{y_1}{\phi_1} (\phi - \phi_1)$$

And the intersection with the X-axis:

$$-y_1 = \frac{\pi}{2} \cdot \frac{y_1}{\phi_1} \phi - \frac{\pi}{2} y_1$$

$$\beta = \frac{(\phi_1)(\pi/2 - 1)(2)(\phi_1)}{\pi y_1} = \frac{(\pi/2 - 1)}{\pi/2} \phi_1 \approx 0.364 \beta_1$$

$$\therefore \qquad x_1 = 0.364 \beta_1$$

$$x_2 = 0.364 \beta_2$$

The exact value is $x_1 = \dfrac{\pi/2 - 1}{\pi/2} \phi_1 = \left(1 - \dfrac{2}{\pi} \right) \phi_1$

An unusually versatile cam curve is the modified cycloid, as will be shown in the following.

Graphical Construction of the Modified Cycloid for Given Requirements

---

In the displacement diagram is given point A (Fig. 3-6a), beginning of motion, point P, together with the slope of the displacement curve at that point, and point B, the end of motion. Construct the displacement curve using a modified cycloid, thereby obtaining a motion which has a continuous acceleration. Figure 3-6b shows the construction. DPE is the

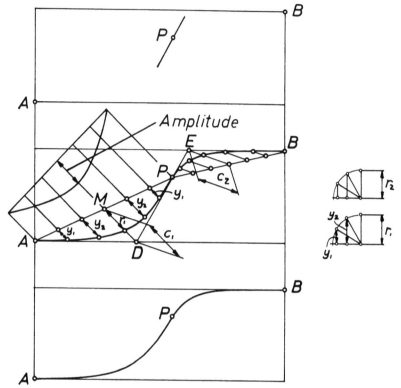

Fig. 3-6. (a) Hére the end points A and B and the velocity at an intermediate point P are given. (b) Graphical construction of modified cycloid. (c) Final displacement diagram.

line of slope at P. First, AP is divided into a number of equal parts, say six, and from the midpoint M of AP a line is drawn to D. This gives a distance c_1 and the relationship between c_1 and r_1 is given by $r_1 = \dfrac{2c_1}{\pi}$.

Now, a quarter circle is drawn with r_1 as radius, Fig. 3-6b (small diagram) and divided into three equal parts, whereby y_1 and y_2 are determined. Through the dividing points of AP lines parallel to MD are drawn and from these dividing points the distances y_1, y_2, and r_1 are laid out as shown. The points so determined are points on the modified cycloid. Any other number of dividing points can be chosen, and the method for finding points of the modified cycloid remains essentially the same.

The other part of the displacement curve from P to B is determined in exactly the same way with the aid of the other small diagram in which r_2 is the radius.

The acceleration curve derived from this displacement curve (by the method shown later) is continuous. The method is also easily used in

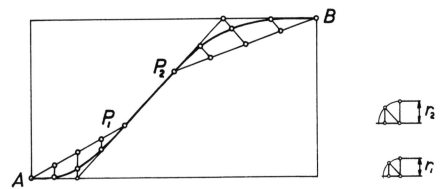

Fig. 3-7. Here end points A and B and two intermediate points P_1 and P_2 with constant velocity between them are given.

connection with a motion where a constant velocity is required over a certain distance. This is shown in Fig. 3-7.

Constant velocity motion is required from P_1 to P_2. Using the same method as described previously AP_1 and P_2B are connected with a modified cycloid, and again an acceleration curve is obtained which is continuous.

The flexibility of the modified cycloid is perhaps best demonstrated by the following:

Example: Given the first part A to B of the displacement diagram (Fig. 3-8), which is based on the technological requirements of the machine in which the cam is to be used, construct the rest of the curve so that motion will end at D. The displacement curve so constructed should have a continuous acceleration.

Solution: Connect B with D and draw the tangent to the curve at B. Choose a conveniently located point C and choose the line of maximum slope FCE which determines the maximum velocity during the return of the follower. The rest of the construction is carried out exactly as before.

It is, of course, important to know the values of the acceleration for these curves and an analytical treatment of these curves is now shown.

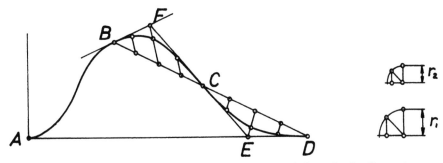

Fig. 3-8. Here the displacement curve from A to B and the end point D are given.

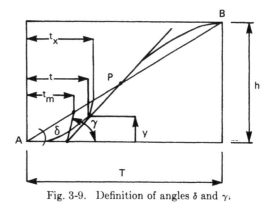

Fig. 3-9. Definition of angles δ and γ.

Determination of Velocity and Acceleration for the Modified Cycloid

In Fig. 3-9 the angle of slope for the straight line connecting A and B is δ and the direction of the amplitude for the superimposed sine wave is given by the angle γ. The coordinate t_m is to the midpoint of line AP. The coordinate t_x is to any point on line AB and is related to the curve coordinate t by equation 3.1b. The equation for the modified cycloid:

$$y = h\left[\frac{t_x}{T} - \frac{1}{2\pi}\sin\left(2\pi\frac{t_x}{T}\right)\right] \qquad 3.1a$$

where
$$t_m = t + \frac{T}{2\pi}\cdot\frac{\tan\delta}{\tan\gamma}\sin\left(2\pi\frac{t}{T}\right) \qquad 3.1b$$

The velocity v is found from

$$v = \frac{1 - \cos\left(2\pi\frac{t_x}{T}\right)}{1 - \dfrac{\tan\delta}{\tan\gamma}\cos\left(2\pi\frac{t_x}{T}\right)}\cdot\frac{h}{T} \qquad 3.2$$

$$a = \frac{2\pi\left(1 - \dfrac{\tan\delta}{\tan\gamma}\right)\sin\left(2\pi\frac{t_x}{T}\right)}{\left[1 - \dfrac{\tan\delta}{\tan\gamma}\cos\left(2\pi\frac{t_x}{T}\right)\right]^3}\cdot\frac{h}{T^2} \qquad 3.3a$$

or
$$a = \frac{2\pi(1 - x)\sin\left(2\pi\frac{t}{T}\right)}{\left[1 - x\cos\left(2\pi\frac{t}{T}\right)\right]^3}\cdot\frac{h}{T^2} \qquad 3.3b$$

where
$$x = \frac{\tan\delta}{\tan\gamma}$$

The acceleration a is obviously a function of x and t_x, and for a certain stroke h and time T for rise, a certain maximum value of a is obtained which is a function of x. For all these there is one value of $x = x$ (optimum) which will give the lowest possible maximum value of the acceleration, and it can be shown that

$$x \text{ optimum} = 1 - \tfrac{1}{2}\sqrt{3} = 0.134$$

For this value of x, the maximum acceleration becomes:

$$a_{max}(\text{optimum}) = 5.89 \frac{h}{T^2}$$

For parabolic motion

$$a_{max} = 4.00 \frac{h}{T^2}$$

For simple harmonic motion

$$a_{max} = 4.93 \frac{h}{T^2}$$

For cycloidal motion

$$a_{max} = 6.28 \frac{h}{T^2}$$

Example: A cam rotates with $N = 200$ rpm, the stroke of the follower is $h = 2.0$ in., and the corresponding cam shaft rotation is $\beta_1 = 60°$. Draw displacement, velocity and acceleration curves for the modified cycloid having the lowest maximum acceleration.

Solution: In Fig. 3-10 an arbitrary scale is chosen for the abscissa, namely $\beta_1 = 60°$ has a length of 4 in. The stroke is laid out to scale (laying the stroke out to a different scale would change nothing in the following procedure). Points A and B represent the start and end of the lift, respectively. Referring to Fig. 3-9 the angle δ is found from:

$$\tan \delta = \frac{h}{T} = \frac{2.00}{4.00} = 0.5$$

Because the lowest maximum acceleration is wanted, a value of $x = x$ (optimum) $= 0.134$ has to be used and

$$\tan \gamma = \frac{\tan \delta}{X(\text{optimum})} = \frac{0.5}{0.134} = 3.73; \quad \gamma \approx 75°$$

Referring again to Fig. 3-10, P is the middle point of AB and AP is divided into, say, six equal parts and the dividing points are numbered 0–6. Point D is situated so that line $3{-}D$ makes an angle of $\gamma = 75°$ with the horizontal. The line $3{-}D$ indicates the direction of the amplitude of the sine wave which is superimposed on AP as already explained in connection with Fig. 3-6b and the displacement curve can now be drawn (the curve from *6* to *B* is congruent with the curve from *6* to *A*).

The velocity at any point can be found from equation 3.2 but can also be found graphically the following way:

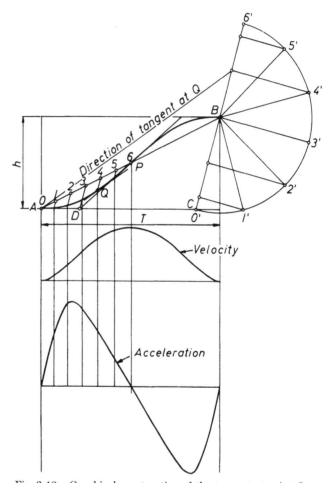

Fig. 3-10. Graphical construction of the tangent at point *Q*.

Through *B* draw a line parallel to *3–D*; the point of intersection with *AD* is *C*. With *B* as center and *BC* as radius, a half circle is drawn and divided into six equal parts, the points being designated $0', 1', 2', \cdots, 6'$. Now to find the velocity at point *Q* draw a perpendicular to *BC* from $4'$ and connect the point of intersection with *A*. This line is parallel to the tangent at *Q*. The procedure is repeated for the remaining points and the velocity curve is obtained.

In order to calculate the velocity at point Q, use equation 3.2. The value of $t_x = T/3$. (Referring to Fig. 3-10, t_x is determined by taking the ratio of the measured horizontal distance of point Q from the vertical passing through point A to the measured horizontal distance corresponding to T.)

$$v = \frac{1 - \cos\left(2\pi \dfrac{T}{3T}\right)}{1 - \dfrac{\tan \delta}{\tan \gamma} \cos\left(2\pi \dfrac{T}{3T}\right)} \cdot \frac{h}{T} \qquad\qquad 3.2$$

$$= \frac{1 - \cos 120°}{1 - 0.134 \cos 120°} \cdot \frac{h}{T}$$

where

$$T = \frac{60}{N} \cdot \frac{\beta}{360} = \frac{60}{200} \cdot \frac{60}{360} = 0.05 \text{ sec.}$$

or

$$v = \frac{1 - (-0.500)}{1 - (0.134)(-0.500)} \cdot \frac{2}{0.05} = 53.1 \text{ in./sec.}$$

The acceleration is found with the help of equation 3.3a:

$$a = \frac{2\pi \left(1 - \dfrac{\tan \delta}{\tan \gamma}\right) \sin\left(2\pi \dfrac{t_y}{T}\right)}{\left[1 - \dfrac{\tan \delta}{\tan \gamma} \cos\left(2\pi \dfrac{t_y}{T}\right)\right]^3} \cdot \frac{h}{T^2}$$

For point Q:

$$a = \frac{2\pi(1 - 0.134) \sin 120°}{(1 - 0.134 \cos 120°)^3} \cdot \frac{2}{0.05^2} = 3,102 \text{ in./sec}^2$$

In order to find the maximum acceleration:

$$(a_{max}) \text{ optimum} = 5.89 \frac{h}{T^2} = 5.89 \frac{2}{0.05^2} = 4,710 \text{ in./sec}^2$$

It is also of interest to notice that the maximum pressure angle of the modified cycloid is lower than that of the cycloid; in Fig. 3-10 the angles are approximately 41 and 45 degrees respectively. Therefore, use of the modified cycloid is recommended.

JENSEN, PREBEN W.: "Konstruktion, Berechnung und Herstellung von Kurven-scheiben," *Technica*, Nr. 23, Nov. 8, 1957, pp. 1319–1322.
———"Konstruktion, Berechnung und Herstellung von Kurvenscheiben," *Technica*, Nr. 24, Nov. 22, 1957, pp. 1391–1393.

Cam Profile Determination

In the material treated in this chapter and in the ones to follow it is assumed that the sequence of events and the magnitudes of the follower movements have been prescribed. Thus we have as a specification for a particular cam its own unique time-displacement diagram. Let it be understood, however, that the designer of the machine itself and the cam designer are not necessarily two individuals each working in his own compartment. The cam specialist must be in consultation with all the related designers and must make it clear when unreasonable requirements are assigned to the mechanisms. It is not contributory to good machine design to present the cam specialist with a sequence of events which fail to take into account the many side effects of mass and motion, and then to expect him by some magic to evolve a satisfactory cam. His presence in the early stages of the over-all design is a necessity.

In *all* of the cam constructions that follow an artificial device called an *inversion* is used. This represents a mental concept which is very helpful in performing the graphical work. The construction of a cam profile requires the drawing of many positions of the cam with the follower in each case in its related location. However, instead of revolving the cam, it is assumed that the follower rotates around the *fixed* cam. It requires the drawing of many follower positions, but since this is done more or less diagrammatically it is relatively simple.

As part of the inversion process, the direction of rotation is very important. In order to preserve the correct sequence of events, the artificial rotation of the follower must be the reverse of the cam's prescribed rotation. Thus, in Fig. 4-1b, the prescribed cam rotation is counterclockwise whereas the artificial rotation of the follower is clockwise.

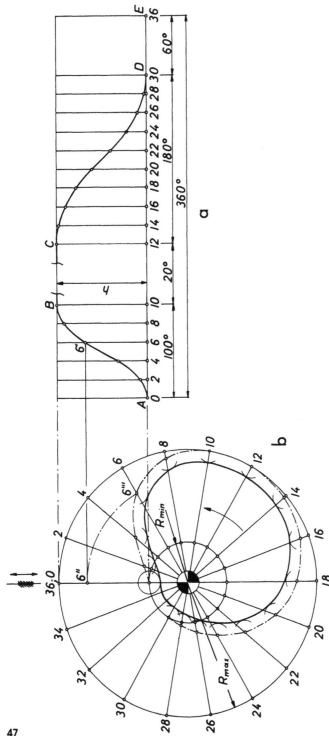

Fig. 4-1. (a) Time displacement diagram for cam to be laid out. (b) Construction of contour of cam with radial translating roller follower.

Radial Translating Roller Follower

The time displacement diagram for a cam with a radial translating roller follower is shown in Fig. 4-1a. This diagram is read from left to right as follows: For 100 degrees of cam shaft rotation the follower rises h inches $(A-B)$, dwells in its upper position for 20 degrees $(B-C)$, returns over 180 degrees $(C-D)$, and finally dwells in its lowest position for 60 degrees $(D-E)$. Then the entire cycle is repeated.

Figure 4-1b shows the cam construction layout with the cam pitch curve as a dot and dash line. To locate a point on this curve take a point on the displacement curve, as $6'$ at the 60-degree position, and project this horizontally to point $6''$ on the 0-degree position of the cam construction diagram. Using the center of cam rotation, an arc is struck from point $6''$ to intercept the 60-degree position radial line which gives point $6'''$ on the cam pitch curve. It will be seen that the smaller circle in the cam construction layout has a radius $R_{\min}$ equal to the smallest distance from the center of cam rotation to the pitch curve and, similarly, the larger circle has a radius $R_{\max}$ equal to the largest distance to the pitch curve. Thus, the difference in radii of these two circles is equal to the maximum rise h of the follower.

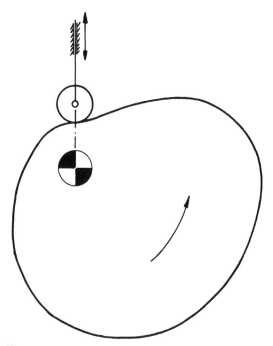

Fig. 4-2. Actual shape of cam constructed in Fig. 4-1.

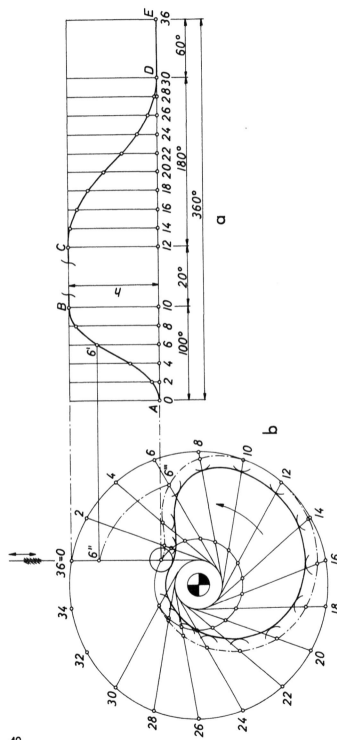

Fig. 4-3. (a) Time displacement diagram for cam to be laid out. (b) Construction of contour of cam with offset translating roller follower.

The cam pitch curve is also the actual profile or working surface when a knife-edged follower is used. To get the profile or working surface for a cam with a roller follower, a series of arcs with centers on the pitch curve and radii equal to the radius of the roller are drawn and the inner envelope drawn tangent to these arcs is the cam working surface or profile shown as a solid line in Fig. 4-1b. The profile of this cam is also shown, by itself, in Fig. 4-2.

Offset Translating Roller Follower

Given the time-displacement diagram Fig. 4-3a and an offset follower. The construction of the cam in this case is very similar to the foregoing case and is shown in Fig. 4-3b. In this construction it will be noted that the angular position lines are not drawn radially from the cam shaft center but tangent to a circle having a radius equal to the amount of offset of the center line of the cam follower from the center of the cam shaft. For counterclockwise rotation of the cam, points $6'$, $6''$, and $6'''$ are located in succession as indicated. Figure 4-4 shows the actual profile.

Swinging Roller Follower

Given the time-displacement diagram Fig. 4-5a and the length of the swinging follower arm L_f it is required that the displacement of the follower

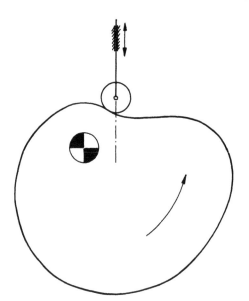

Fig. 4-4.   Actual shape of cam constructed in Fig. 4-3.

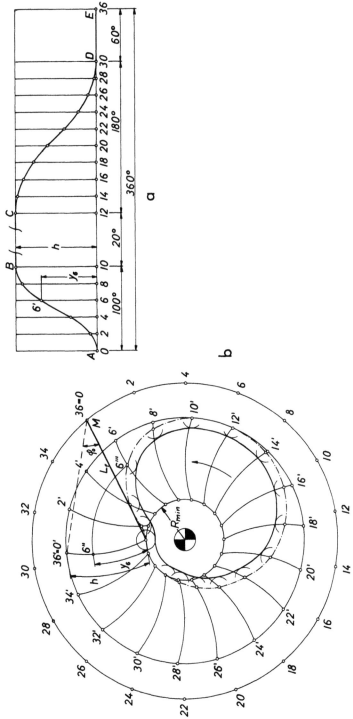

Fig. 4-5. (a) Time displacement diagram for cam to be laid out. (b) Construction of contour of cam with swinging roller follower.

center along the circular are that it describes be equal to the corresponding displacements in the time-displacement diagram. Knowing L_f and ϕ_0, the displacement h of Fig. 4-5a can be found from the formula $h = 2\pi L_f \cdot \dfrac{\phi_0}{360°}$.

Point M is the actual position of the pivot center of the swinging follower with respect to the cam shaft center. It is again required that the rotation of the cam be counterclockwise and therefore M is considered to have been rotated clockwise around the cam shaft center, whereby the points *2, 4, 6*, etc. are obtained as shown in Fig. 4-5b. Around each of the pivot points, *2, 4, 6*, etc. circular arcs whose radii equal L_f are drawn between the R_{min} and R_{max} circles giving the points *2′, 4′, 6′*, etc. The R_{min} circle with center at the cam shaft center is drawn through the lowest position of the center of the roller follower and the R_{max} circle through the highest position as in Fig. 4-1b. The different points on the pitch curve are now located. Point *6′′′* for instance, is found by: (1) taking the y_6 ordinate of point *6′* from the displacement diagram; (2) laying out this length along arc *O′* from the intersection of the R_{min} circle with this arc, whereby point *6′′* is obtained; and (3) drawing an arc through *6′′* with center at the cam shaft, the intersection of this arc with arc *6′* locating *6′′′*. Figure 4-6 shows the final cam; *O* is the cam shaft center and M is the fixed pivot for the swinging follower arm L.

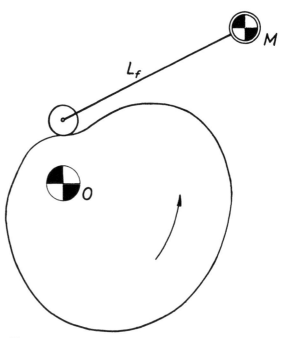

Fig. 4-6. Actual shape of cam constructed in Fig. 4-5.

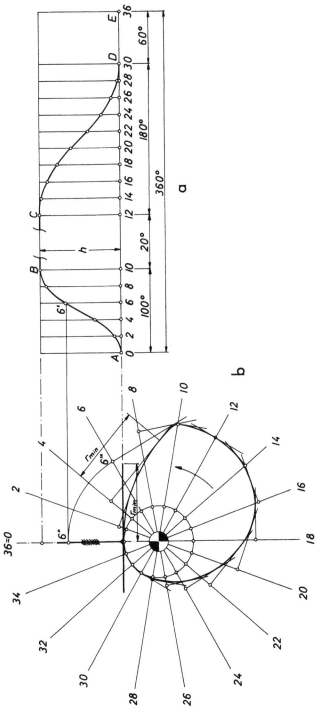

Fig. 4-7. (a) Time displacement diagram of cam to be laid out. (b) Construction of contour of cam with flat-faced translating follower.

Flat-faced Translating Follower

Given the displacement diagram in Fig. 4-7a, for a cam with a flat-faced translating follower. The points on the rays *2, 4, 6,* etc. are obtained in exactly the same way as for the radial translating follower (See point *6'''*) but through each point so obtained is drawn a perpendicular to the ray. The final cam profile is obtained by drawing a curve such that these perpendiculars are tangent to the cam profile, Fig. 4-7b. It should be noticed that the longest distance from a ray to the point of tangency determines the size of the flat face. If $r_{\min}$ is the longest perpendicular distance to the right from a ray to the point of tangency, then the flat-faced follower has to have at least the same distance to the right plus an increment sufficiently large to prevent sharp edge effects.

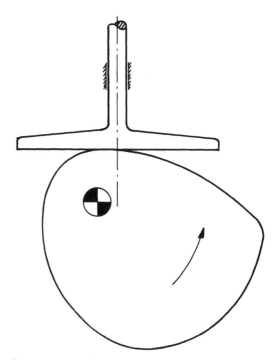

Fig. 4-8. Actual shape of cam constructed in Fig. 4-7.

Without actually altering anything the system can be made offset; the cam shape will remain exactly the same but the offset can be adjusted so that the flat-faced follower will have the same size on both sides of the vertical shaft and this is shown in Fig. 4-8.

It is to be noticed that the cam profile shown in Figs. 4-7b, and 4-8 has a rather abrupt change in curvature at position 10. Had the minimum

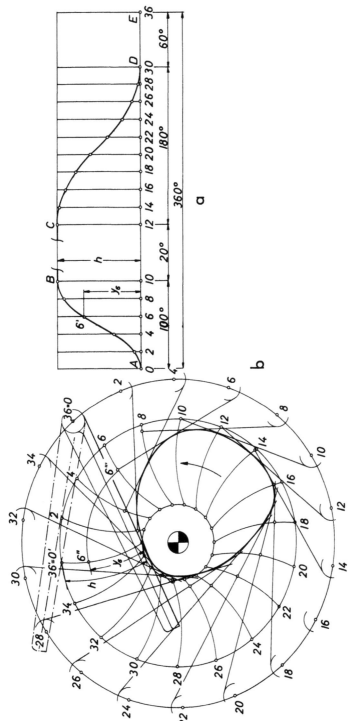

Fig. 4-9. (a) Time displacement diagram of cam to be laid out. (b) Construction of contour of cam with flat-faced swinging follower.

circle (R_{min}) been made smaller, this abruptness would have been more pronounced even to the degree that a cusp would be produced. This condition is detrimental to the action of the follower and causes it to deviate from the motion prescribed by the cam displacement diagram. This abortive condition is called "undercutting" and must be avoided. The remedy is to make R_{min} larger.

Oscillating Flat-faced Follower

The time-displacement diagram for a cam with an oscillating flat-faced follower is shown in Fig. 4-9a. Actually, the displacement is in degrees but it is more convenient when drawing the profile to work with distances. The initial construction is similar to that for a cam with swinging roller follower in that the distances y_1, y_2, etc. are laid off *along* the arcs (see point $6''$).

In Fig. 4-9b the flat-faced follower is first drawn in its two extreme positions and a point is found on the face of the follower such that when the follower moves from one extreme position to the other then this point will travel the distance h. (See the arc on which point $6''$ is located.) The displacement is found in a similar manner as explained in connection with

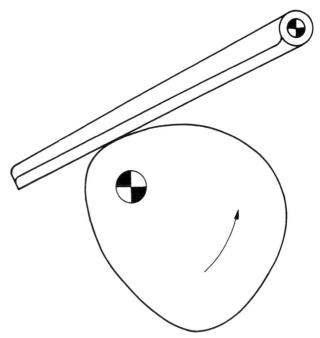

Fig. 4-10. Actual shape of cam constructed in Fig. 4-9.

Fig. 4-5a. Then points on the cam profile are constructed exactly as if this point on the face of the follower was the center of a roller as in Fig. 4-5b; the only difference is that when the points on the rays are found, the corresponding positions of the follower face are drawn as tangents from each of these points to the corresponding small circle at the center of rotation of the follower (as from point 6''' to the circle at point 6). The radii of these small circles is equal to the perpendicular distance from the face of the follower to its center of rotation. Finally the cam profile itself is constructed tangent to the successive positions of the face of the follower.

The actual cam is shown in Fig. 4-10 and also in this case the cam was almost undercut as is evident from the relatively small radius of curvature.

Calculation of Cam Profile

It has been shown how to determine the cam profile graphically. This method is in many cases sufficiently accurate but often it is desired to determine the cam profile by calculation—for instance when the cam has to be cut by incremental cutting (see Chapter 9 on cam manufacture).

The following equations give points on the pitch curve of the cam or points on the cam profile itself in either polar or rectangular coordinates for the following cases:

6. Radial translating roller follower
7. Offset translating roller follower
8. Swinging roller follower
9. Translating flat-faced follower
10. Swinging centric flat-faced follower
11. Swinging eccentric flat-faced follower

Radial Translating Roller Follower (Fig. 4-11)

Polar coordinates:

$$\left. \begin{aligned} r &= R_{\min} + y \\ y &= y\ (\theta) \end{aligned} \right\} \qquad 4.1$$

Rectangular coordinates:

$$\left. \begin{aligned} X &= r \sin \theta = (R_{\min} + y) \sin \theta \\ Y &= r \cos \theta = (R_{\min} + y) \cos \theta \end{aligned} \right\} \qquad 4.2$$

Example: $R_{\min} = 2$ in., $h = 1$ in., $\beta = 60°$ (rise), radial translating roller follower. Calculate the cam profile for $\theta = 0°,\ 10°,\ 20°, \ldots, 60°$, if the motion is simple harmonic, using both polar and rectangular coordinates.

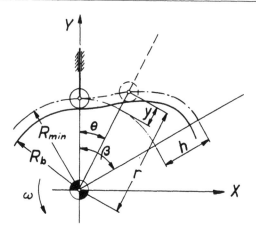

Fig. 4-11. Diagram for cam profile calculation—radial translating roller follower.

Solution: The equation for simple harmonic motion is:

$$y = \frac{h}{2}\left[1 - \cos\left(\pi\frac{\theta}{\beta_1}\right)\right] \qquad 0 \leqq \theta \leqq \beta_1 \qquad (2.3)$$

and

$$y = \frac{h}{2}\left[1 - \cos\left(\pi\frac{\theta}{\beta_1}\right)\right] \qquad 0 \leqq \theta \leqq \beta_1$$

Polar coordinates:

$$y(0°) = \tfrac{1}{2}\left[1 - \cos\left(\pi\frac{0}{60}\right)\right] = 0.5 - 0.500 = 0.000$$

$$y(10°) = \tfrac{1}{2}\left[1 - \cos\left(\pi\frac{10}{60}\right)\right] = 0.5 - 0.433 = 0.067$$

$$y(20°) = \tfrac{1}{2}\left[1 - \cos\left(\pi\frac{20}{60}\right)\right] = 0.5 - 0.250 = 0.250$$

$$y(30°) = \tfrac{1}{2}\left[1 - \cos\left(\pi\frac{30}{60}\right)\right] = 0.5 - 0.000 = 0.500$$

$$y(40°) = \tfrac{1}{2}\left[1 - \cos\left(\pi\frac{40}{60}\right)\right] = 0.5 + 0.250 = 0.750$$

$$y(50°) = \tfrac{1}{2}\left[1 - \cos\left(\pi\frac{50}{60}\right)\right] = 0.5 + 0.433 = 0.933$$

$$y(60°) = \tfrac{1}{2}\left[1 - \cos\left(\pi\frac{60}{60}\right)\right] = 0.5 + 0.500 = 1.000$$

Table 4-1. Polar Coordinates for Simple Harmonic Motion Cam

Case I. Angle of Rise $\beta = 100°$

θ, Deg.	y, Inch	θ, Deg.	y, Inch	θ, Deg.	y, Inch	θ, Deg.	y, Inch
.0	-.00000	30.0	.20610	60.0	.65450	90.0	.97552
1.0	.00024	31.0	.21895	61.0	.66936	91.0	.98014
2.0	.00098	32.0	.23208	62.0	.68406	92.0	.98429
3.0	.00221	33.0	.24547	63.0	.69857	93.0	.98795
4.0	.00394	34.0	.25912	64.0	.71288	94.0	.99114
5.0	.00615	35.0	.27300	65.0	.72699	95.0	.99384
6.0	.00885	36.0	.28711	66.0	.74087	96.0	.99605
7.0	.01204	37.0	.30142	67.0	.75452	97.0	.99778
8.0	.01570	38.0	.31593	68.0	.76791	98.0	.99901
9.0	.01985	39.0	.33063	69.0	.78104	99.0	.99975
10.0	.02447	40.0	.34549	70.0	.79389	100.0	1.00000
11.0	.02955	41.0	.36050	71.0	.80645		
12.0	.03511	42.0	.37565	72.0	.81871		
13.0	.04112	43.0	.39092	73.0	.83065		
14.0	.04758	44.0	.40630	74.0	.84227		
15.0	.05449	45.0	.42178	75.0	.85355		
16.0	.06184	46.0	.43733	76.0	.86448		
17.0	.06962	47.0	.45294	77.0	.87505		
18.0	.07783	48.0	.46860	78.0	.88525		
19.0	.08645	49.0	.48429	79.0	.89507		
20.0	.09549	50.0	.50000	80.0	.90450		
21.0	.10492	51.0	.51570	81.0	.91354		
22.0	.11474	52.0	.53139	82.0	.92216		
23.0	.12494	53.0	.54705	83.0	.93037		
24.0	.13551	54.0	.56266	84.0	.93815		
25.0	.14644	55.0	.57821	85.0	.94550		
26.0	.15772	56.0	.59369	86.0	.95241		
27.0	.16934	57.0	.60907	87.0	.95887		
28.0	.18128	58.0	.62434	88.0	.96488		
29.0	.19354	59.0	.63949	89.0	.97044		

Table 4-1 (Cont.). Polar Coordinates for Simple Harmonic Motion Cam

θ, Deg.	y, Inch	θ, Deg.	y, Inch	θ, Deg.	y, Inch	θ, Deg.	y, Inch
.0	-.00000	30.0	.14644	60.0	.50000	90.0	.85355
1.0	.00017	31.0	.15582	61.0	.51308	91.0	.86268
2.0	.00068	32.0	.16543	62.0	.52616	92.0	.87157
3.0	.00154	33.0	.17527	63.0	.53922	93.0	.88020
4.0	.00273	34.0	.18533	64.0	.55226	94.0	.88857
5.0	.00427	35.0	.19561	65.0	.56526	95.0	.89667
6.0	.00615	36.0	.20610	66.0	.57821	96.0	.90450
7.0	.00837	37.0	.21679	67.0	.59111	97.0	.91206
8.0	.01092	38.0	.22768	68.0	.60395	98.0	.91933
9.0	.01381	39.0	.23875	69.0	.61672	99.0	.92632
10.0	.01703	40.0	.25000	70.0	.62940	100.0	.93301
11.0	.02059	41.0	.26142	71.0	.64200	101.0	.93940
12.0	.02447	42.0	.27300	72.0	.65450	102.0	.94550
13.0	.02867	43.0	.28474	73.0	.66690	103.0	.95129
14.0	.03320	44.0	.29663	74.0	.67918	104.0	.95677
15.0	.03806	45.0	.30865	75.0	.69134	105.0	.96193
16.0	.04322	46.0	.32081	76.0	.70336	106.0	.96679
17.0	.04870	47.0	.33309	77.0	.71525	107.0	.97132
18.0	.05449	48.0	.34549	78.0	.72699	108.0	.97552
19.0	.06059	49.0	.35799	79.0	.73857	109.0	.97940
20.0	.06698	50.0	.37059	80.0	.75000	110.0	.98296
21.0	.07367	51.0	.38327	81.0	.76124	111.0	.98618
22.0	.08066	52.0	.39604	82.0	.77231	112.0	.98907
23.0	.08793	53.0	.40888	83.0	.78320	113.0	.99162
24.0	.09549	54.0	.42178	84.0	.79389	114.0	.99384
25.0	.10332	55.0	.43473	85.0	.80438	115.0	.99572
26.0	.11142	56.0	.44773	86.0	.81466	116.0	.99726
27.0	.11979	57.0	.46077	87.0	.82472	117.0	.99845
28.0	.12842	58.0	.47383	88.0	.83456	118.0	.99931
29.0	.13731	59.0	.48691	89.0	.84417	119.0	.99982
						120.0	1.00000

Table 4-2.   Polar Coordinates for Cycloidal Motion Cam

Case I. Angle of Rise $\beta = 100^\circ$							
θ, Deg.	y, Inch	θ, Deg.	y, Inch	θ, Deg.	y, Inch	θ, Deg.	y, Inch
.0	−.00000	30.0	.14863	60.0	.69354	90.0	.99354
1.0	.00000	31.0	.16202	61.0	.71144	91.0	.99527
2.0	.00005	32.0	.17599	62.0	.72894	92.0	.99667
3.0	.00017	33.0	.19053	63.0	.74601	93.0	.99776
4.0	.00041	34.0	.20562	64.0	.76263	94.0	.99858
5.0	.00081	35.0	.22124	65.0	.77875	95.0	.99918
6.0	.00141	36.0	.23736	66.0	.79437	96.0	.99958
7.0	.00223	37.0	.25398	67.0	.80946	97.0	.99982
8.0	.00332	38.0	.27105	68.0	.82400	98.0	.99994
9.0	.00472	39.0	.28855	69.0	.83797	99.0	.99999
10.0	.00645	40.0	.30645	70.0	.85136	100.0	.99999
11.0	.00855	41.0	.32472	71.0	.86415		
12.0	.01105	42.0	.34332	72.0	.87633		
13.0	.01398	43.0	.36223	73.0	.88789		
14.0	.01736	44.0	.38141	74.0	.89884		
15.0	.02124	45.0	.40081	75.0	.90915		
16.0	.02562	46.0	.42041	76.0	.91884		
17.0	.03053	47.0	.44017	77.0	.92789		
18.0	.03599	48.0	.46005	78.0	.93633		
19.0	.04202	49.0	.48000	79.0	.94415		
20.0	.04863	50.0	.50000	80.0	.95136		
21.0	.05584	51.0	.51999	81.0	.95797		
22.0	.06366	52.0	.53994	82.0	.96400		
23.0	.07210	53.0	.55982	83.0	.96946		
24.0	.08115	54.0	.57958	84.0	.97437		
25.0	.09084	55.0	.59918	85.0	.97875		
26.0	.10115	56.0	.61858	86.0	.98263		
27.0	.11210	57.0	.63776	87.0	.98601		
28.0	.12366	58.0	.65667	88.0	.98894		
29.0	.13584	59.0	.67527	89.0	.99144		

Table 4-2 (Cont.). Polar Coordinates for Cycloidal Motion Cam

| \multicolumn{8}{c}{Case II. Angle of Rise $\beta = 120°$} |||||||| |

θ, Deg.	y, Inch	θ, Deg.	y, Inch	θ, Deg.	y, Inch	θ, Deg.	y, Inch
.0	-.00000	30.0	.09084	60.0	.50000	90.0	.90915
1.0	.00000	31.0	.09939	61.0	.51666	91.0	.91727
2.0	.00003	32.0	.10838	62.0	.53330	92.0	.92494
3.0	.00010	33.0	.11780	63.0	.54989	93.0	.93219
4.0	.00024	34.0	.12765	64.0	.56642	94.0	.93901
5.0	.00047	35.0	.13793	65.0	.58285	95.0	.94539
6.0	.00081	36.0	.14863	66.0	.59918	96.0	.95136
7.0	.00129	37.0	.15974	67.0	.61536	97.0	.95691
8.0	.00193	38.0	.17127	68.0	.63140	98.0	.96206
9.0	.00274	39.0	.18319	69.0	.64725	99.0	.96680
10.0	.00375	40.0	.19550	70.0	.66291	100.0	.97116
11.0	.00498	41.0	.20818	71.0	.67834	101.0	.97514
12.0	.00645	42.0	.22124	72.0	.69354	102.0	.97875
13.0	.00817	43.0	.23464	73.0	.70849	103.0	.98201
14.0	.01017	44.0	.24839	74.0	.72316	104.0	.98494
15.0	.01246	45.0	.26246	75.0	.73753	105.0	.98753
16.0	.01505	46.0	.27683	76.0	.75160	106.0	.98982
17.0	.01798	47.0	.29150	77.0	.76535	107.0	.99182
18.0	.02124	48.0	.30645	78.0	.77875	108.0	.99354
19.0	.02485	49.0	.32165	79.0	.79181	109.0	.99501
20.0	.02883	50.0	.33708	80.0	.80449	110.0	.99624
21.0	.03319	51.0	.35274	81.0	.81680	111.0	.99725
22.0	.03793	52.0	.36859	82.0	.82872	112.0	.99806
23.0	.04308	53.0	.38463	83.0	.84025	113.0	.99870
24.0	.04863	54.0	.40081	84.0	.85136	114.0	.99918
25.0	.05460	55.0	.41714	85.0	.86206	115.0	.99952
26.0	.06098	56.0	.43357	86.0	.87234	116.0	.99975
27.0	.06780	57.0	.45010	87.0	.88219	117.0	.99989
28.0	.07505	58.0	.46669	88.0	.89161	118.0	.99996
29.0	.08272	59.0	.48333	89.0	.90060	119.0	.99999
						120.0	.99999

Table 4-3.   Polar Coordinates for 3-4-5 Polynomial Motion Cam

Case I. Angle of Rise $\beta = 100°$							
θ, Deg.	y, Inch	θ, Deg.	y, Inch	θ, Deg.	y, Inch	θ, Deg.	y, Inch
.0	−.00000	30.0	.16308	60.0	.68256	90.0	.99144
1.0	.00000	31.0	.17655	61.0	.69969	91.0	.99365
2.0	.00007	32.0	.19052	62.0	.71650	92.0	.99547
3.0	.00025	33.0	.20496	63.0	.73298	93.0	.99692
4.0	.00060	34.0	.21985	64.0	.74910	94.0	.99802
5.0	.00115	35.0	.23516	65.0	.76483	95.0	.99884
6.0	.00196	36.0	.25089	66.0	.78014	96.0	.99939
7.0	.00307	37.0	.26701	67.0	.79503	97.0	.99974
8.0	.00452	38.0	.28349	68.0	.80947	98.0	.99992
9.0	.00634	39.0	.30030	69.0	.82344	99.0	.99999
10.0	.00856	40.0	.31744	70.0	.83692	100.0	1.00000
11.0	.01121	41.0	.33485	71.0	.84989		
12.0	.01431	42.0	.35254	72.0	.86235		
13.0	.01790	43.0	.37045	73.0	.87427		
14.0	.02200	44.0	.38857	74.0	.88565		
15.0	.02661	45.0	.40687	75.0	.89648		
16.0	.03175	46.0	.42531	76.0	.90674		
17.0	.03745	47.0	.44388	77.0	.91644		
18.0	.04370	48.0	.46253	78.0	.92556		
19.0	.05052	49.0	.48125	79.0	.93411		
20.0	.05792	50.0	.50000	80.0	.94208		
21.0	.06588	51.0	.51874	81.0	.94947		
22.0	.07443	52.0	.53746	82.0	.95629		
23.0	.08355	53.0	.55611	83.0	.96254		
24.0	.09325	54.0	.57468	84.0	.96824		
25.0	.10351	55.0	.59312	85.0	.97338		
26.0	.11434	56.0	.61142	86.0	.97799		
27.0	.12572	57.0	.62954	87.0	.98209		
28.0	.13764	58.0	.64745	88.0	.98568		
29.0	.15010	59.0	.66514	89.0	.98878		

Table 4-3 (Cont.).   Polar Coordinates for 3-4-5 Polynomial Motion Cam

\$\theta\$, Deg.	\$y\$, Inch	\$\theta\$, Deg.	\$y\$, Inch	\$\theta\$, Deg.	\$y\$, Inch	\$\theta\$, Deg.	\$y\$, Inch
			Case II. Angle of Rise \$\beta = 120°\$				
.0	−.00000	30.0	.10351	60.0	.50000	90.0	.89648
1.0	.00000	31.0	.11249	61.0	.51562	91.0	.90507
2.0	.00004	32.0	.12186	62.0	.53122	92.0	.91327
3.0	.00015	33.0	.13161	63.0	.54679	93.0	.92107
4.0	.00035	34.0	.14174	64.0	.56231	94.0	.92847
5.0	.00068	35.0	.15223	65.0	.57776	95.0	.93547
6.0	.00115	36.0	.16308	66.0	.59312	96.0	.94208
7.0	.00181	37.0	.17427	67.0	.60838	97.0	.94828
8.0	.00267	38.0	.18581	68.0	.62352	98.0	.95408
9.0	.00375	39.0	.19768	69.0	.63852	99.0	.95948
10.0	.00508	40.0	.20987	70.0	.65338	100.0	.96450
11.0	.00668	41.0	.22237	71.0	.66806	101.0	.96913
12.0	.00856	42.0	.23516	72.0	.68256	102.0	.97338
13.0	.01073	43.0	.24824	73.0	.69685	103.0	.97726
14.0	.01323	44.0	.26159	74.0	.71093	104.0	.98078
15.0	.01605	45.0	.27520	75.0	.72479	105.0	.98394
16.0	.01921	46.0	.28906	76.0	.73840	106.0	.98676
17.0	.02273	47.0	.30314	77.0	.75175	107.0	.98926
18.0	.02661	48.0	.31744	78.0	.76483	108.0	.99144
19.0	.03086	49.0	.33193	79.0	.77762	109.0	.99331
20.0	.03549	50.0	.34661	80.0	.79012	110.0	.99491
21.0	.04051	51.0	.36147	81.0	.80231	111.0	.99624
22.0	.04591	52.0	.37647	82.0	.81418	112.0	.99732
23.0	.05171	53.0	.39161	83.0	.82572	113.0	.99818
24.0	.05792	54.0	.40687	84.0	.83692	114.0	.99884
25.0	.06452	55.0	.42223	85.0	.84776	115.0	.99932
26.0	.07152	56.0	.43768	86.0	.85825	116.0	.99964
27.0	.07892	57.0	.45320	87.0	.86838	117.0	.99984
28.0	.08672	58.0	.46877	88.0	.87813	118.0	.99995
29.0	.09492	59.0	.48437	89.0	.88750	119.0	.99999
						120.0	1.00000

Next, r is calculated from $r = R_{min} + y\ (\theta) = 2.0 + y\ (\theta)$ and the following results are obtained:

θ	0°	10°	20°	30°	40°	50°	60°
r	2.000	2.067	2.250	2.500	2.750	2.933	3.000

In order to facilitate the calculation of y, tables have been compiled for $\theta = 0° - 100°$ and $0° - 120°$ with intervals of 1° for simple harmonic, cycloidal and 3-4-5 polynomial motion, Tables 4-1, 4-2, and 4-3.

When $\beta = 50°$, for instance, then the values of y are taken for intervals of $100/50 = 2°$ in Table 4-1. Actually, the plot will be for every 1 degree of cam rotation. When h is some value other than 1, multiply the y value by h.

Rectangular coordinates:

The rectangular coordinates are obtained with the help of equation 4.2,

for instance, for $\theta = 20°$:
$$X = r \sin \theta = 2.250 \cdot \sin 20° = 0.769 \text{ in.}$$
$$Y = r \cos \theta = 2.250 \cdot \cos 20° = 2.12 \text{ in.}$$

The following results are obtained:

θ	0°	10°	20°	30°	40°	50°	60°
X	0.000	0.358	0.769	1.250	1.77	2.24	2.60
Y	2.00	2.04	2.12	2.16	2.11	1.88	1.50

Offset Translating Roller Follower

Polar coordinates: (Fig. 4-12)

$$y_0 = \sqrt{R_{min}^2 - e^2} \qquad 4.3$$
$$r = \sqrt{e^2 + (y + y_0)^2} \qquad 4.4$$
$$\theta' = \theta - \delta \qquad 4.5$$

$$\tan (\theta_0 + \delta) = \frac{y_0 + y}{e} = \frac{\tan \theta_0 + \tan \delta}{1 - \tan \theta_0 \tan \delta}$$

$$\tan \theta_0 = \frac{y_0}{e}$$

$$\frac{y_0 + y}{e} = \frac{\dfrac{y_0}{e} + \tan \delta}{1 - \dfrac{y_0}{e} \tan \delta}$$

Rectangular coordinates: (Fig. 4-13)

$$\sin \theta_0 = \frac{e}{R_{min}} \qquad\qquad 4.7$$

$$X = R_{min} \sin (\theta + \theta_0) + y \sin \theta \qquad\qquad 4.8$$

$$Y = R_{min} \cos (\theta + \theta_0) + y \cos \theta \qquad\qquad 4.9$$

Example: $R_{min} = 2$ in., $h = 1$ in., $\beta = 60°$ (rise), offset translating roller follower with $e = 1$ in., parabolic motion, and cam rotates counterclockwise. Calculate polar and rectangular coordinates for $\theta = 0°$ [10°] 60°, i.e. from 0° to 60° at intervals of 10°. The calculations below are for $\theta = 20°$.

Solution: The equation for parabolic motion is:

$$y = 2h \left(\frac{\theta}{\beta_1}\right)^2 \qquad 0 \leq \theta \leq \frac{\beta_1}{2}$$

and replacing t by θ and T by β:

$$y = 2h \left(\frac{\theta}{\beta_1}\right)^2 \qquad 0 \leq \theta \leq \frac{\beta_1}{2}$$

Polar coordinates:

$$y(20°) = (2)(1) \left(\frac{20}{60}\right)^2 = \frac{2}{9} \text{ in.} = 0.222 \text{ in.}$$

$$y_0 = \sqrt{4 - 1} = \sqrt{3} = 1.732 \text{ in.} \qquad\qquad (4.3)$$

$$r = \sqrt{1 + (0.222 + 1.732)^2}$$
$$= \sqrt{1 + 3.82} = \underline{\underline{2.2 \text{ in.}}} \qquad\qquad (4.4)$$

$$\tan \delta = \frac{(0.222)(1)}{1 + 1.732^2 + (0.222)(1.732)} \qquad\qquad (4.6)$$

$$= \frac{0.222}{4.384} = 0.0506$$

$$\delta = 2.88°$$

$$\theta' = 20 - 2.88 = \underline{\underline{17.12°}} \qquad\qquad (4.5)$$

Rectangular coordinates:

$$\sin \theta_0 = \frac{1.0}{2.0} \qquad\qquad (4.7)$$

$$\theta_0 = 30°$$

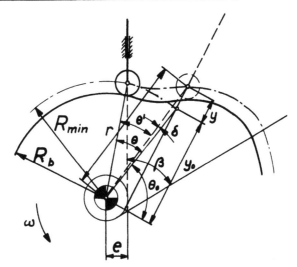

Fig. 4-12.   Diagram for cam profile calculation—offset translating roller follower
(polar coordinates).

$$y_0 + e \tan \delta = y_0 - \frac{y_0^2}{e} \tan \delta + y - \frac{y \cdot y_0}{e} \tan \delta$$

$$\tan \delta \left(e + \frac{y_0^2 + y \cdot y_0}{e} \right) = y$$

$$\tan \delta = \frac{y \cdot e}{e^2 + y_0^2 + y \cdot y_0} \qquad\qquad 4.6$$

which together with eq. 4.5 gives the value of θ'.

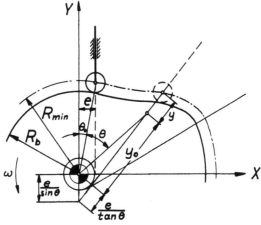

Fig. 4-13.   Diagram for cam profile calculation—offset translating roller follower
(rectangular coordinates).

$$X = 2.0 \sin (20° + 30°) + 0.222 \cdot \sin 20°$$
$$= (2.0)(0.766) + (0.222)(0.342)$$
$$= 1.532 + 0.075 = 1.607 \text{ in.}$$

$$Y = (2.0)(\cos 50°) + (0.222)(\cos 20°)$$
$$= (2.0)(0.6428) + (0.222)(0.9397)$$
$$= 1.286 + 0.208 = 1.494 \text{ in.}$$

Swinging Roller Follower

Polar coordinates: (Fig. 4-14)

In the equations developed so far the roller follower has been translating so that the equation for the time-displacement curve could be used in the form already given.

The equations must now be written in a form usable for swinging followers:

Let $y = f(t)$ be the equation for the time-displacement curve: then

$$\frac{y}{\phi} = \frac{h}{\phi_0} \qquad \text{or} \qquad \phi = \phi_0 \frac{y}{h} \qquad\qquad 4.10$$

For instance, for simple harmonic motion:

$$y = \frac{h}{2}\left[1 - \cos\left(\pi \frac{t}{T}\right)\right] \text{ and with } \phi_0 = 35° \text{ and } h = 2 \text{ in.}$$

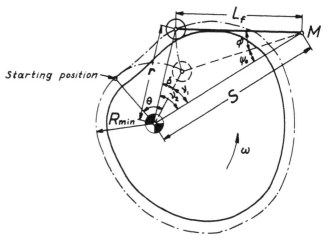

Fig. 4-14.   Diagram for cam profile calculation—swinging roller follower (polar coordinates).

$$\phi = \phi_0 \frac{\frac{h}{2}\left[1 - \cos\left(\pi\frac{t}{T}\right)\right]}{h} = 35\,\frac{1 - \cos\left(\pi\frac{t}{T}\right)}{2} \qquad (4.10)$$

$$\cos\psi_0 = \frac{L_f^2 + S^2 - R_{min}^2}{2\cdot L_f\cdot S} \qquad 4.11$$

$$r = \sqrt{L_f^2 + S^2 - 2\cdot L_f\cdot S\cdot \cos\left(\psi_0 + \phi\right)} \qquad 4.12$$

$$\delta = \gamma_2 - \gamma_1 \qquad 4.13$$

$$\cos\gamma_2 = \frac{r^2 + S^2 - L_f^2}{2\cdot r\cdot S} \qquad 4.14$$

$$\cos\gamma_1 = \frac{R_{min}^2 + S^2 - L_f^2}{2\cdot R_{min}\cdot S} \qquad 4.15$$

$$\theta' = \theta - \delta \quad \text{(counterclockwise rotation of cam)} \qquad 4.16a$$

$$\theta' = \theta + \delta \quad \text{(clockwise rotation of cam)} \qquad 4.16b$$

Rectangular coordinates: (Fig. 4-15)

If the X–Y axis are placed as shown:

$$X = X_1 - X_2$$
$$Y = Y_1 + Y_2$$

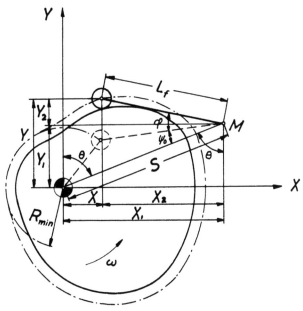

Fig. 4-15. Diagram for cam profile calculation—swinging roller follower (rectangular coordinates).

$$X_1 = S \sin \theta$$

$$X_2 = L_f \sin (\psi_0 + \theta + \phi)$$

$$Y_1 = S \cos \theta$$

$$Y_2 = L_f \cos (\psi_0 + \theta + \phi)$$

for counterclockwise rotation of cam

$$X = S \sin \theta - L_f \sin (\psi_0 + \theta + \phi) \qquad 4.17$$

$$Y = S \cos \theta - L_f \cos (\psi_0 + \theta + \phi) \qquad 4.18$$

for clockwise rotation of cam

$$X = S \sin (-\theta) - L_f \sin (\psi_0 - \theta + \phi) \qquad 4.19$$

$$Y = S \cos (-\theta) - L_f \cos (\psi_0 - \theta + \phi) \qquad 4.20$$

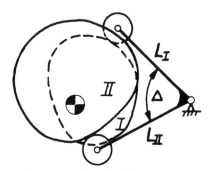

Fig. 4-16. Diagram for cam profile calculation—complementary cam profile (rectangular coordinates).

Equations 4.17 through 4.20 are especially suited for determining the complementary cam profile when a double lever is employed. Thus, in Chapter 1 a number of positive displacement cams were illustrated. The basic feature of these cams is the use of a pair of constraining surfaces to rigidly control the movement of the follower thus preventing inertia effects from interfering with the prescribed follower motions. Figures 1-2a, b, c, illustrate three types of grooved cams based on this idea. It is clear that in each of these cams, a second cam surface has been provided against which the roller has contact. In Fig. 1-2b, two rollers are employed. Since these rollers have their roller axes coincident it follows that the angular distance between the rollers is zero degrees. In Fig. 4-16 is illustrated a follower with two rollers separated by the angle Δ. While the two cams in this illustration appear quite different they are both derived from the same displacement diagram.

The following equations apply: (Fig. 4-16)

$$X = S \sin \theta - L_I \sin (\psi_0 + \theta + \phi)$$
$$Y = S \cos \theta - L_I \cos (\psi_0 + \theta + \phi)$$
for cam I

$$X = S \sin \theta - L_{II} \sin (\psi_0 - \Delta + \theta + \phi)$$
$$Y = S \cos \theta - L_{II} \cos (\psi_0 - \Delta + \theta + \phi)$$
for cam II

It can be seen that the equations for cam I are identical to equations 4.17 and 4.18. The equations for cam II are derived by subtracting the angle Δ and using the length L_{II} of the second follower arm.

Translating Flat-faced Follower

Polar coordinates of profile points:

$$r = \sqrt{(R_b + y)^2 + \left(\frac{dy}{d\theta}\right)^2} \qquad 4.21$$

Rectangular coordinates of cam profile: (Fig. 4-17)

$$X = (R_b + y) \cos \theta - \frac{dy}{d\theta} \sin \theta \qquad 4.22$$

$$Y = (R_b + y) \sin \theta + \frac{dy}{d\theta} \cos \theta \qquad 4.23$$

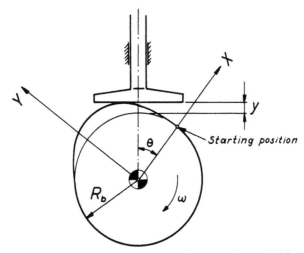

Fig. 4-17. Diagram for cam profile calculation—translating flat-faced follower (polar and rectangular coordinates).

Rectangular coordinates to produce cam profile with cutter or grinder of radius R_g:

$$X = (R_b + y) \cos \theta - \frac{dy}{d\theta} \sin \theta + R_g \cdot \cos \theta \qquad 4.24$$

$$Y = (R_b + y) \sin \theta + \frac{dy}{d\theta} \cos \theta + R_g \cdot \sin \theta \qquad 4.25$$

Swinging Centric Flat-faced Follower

Rectangular coordinates of cam profile: (Fig. 4-18)

$$X = S \left[\cos \theta + \frac{\cos (\psi_0 + \phi) \cos (\psi_0 - \theta + \phi)}{\dfrac{d\phi}{d\theta} - 1} \right] \qquad 4.26$$

$$Y = S \left[\sin \theta + \frac{\cos (\psi_0 + \phi) \cos (\psi_0 - \theta + \phi)}{\dfrac{d\phi}{d\theta} - 1} \right] \qquad 4.27$$

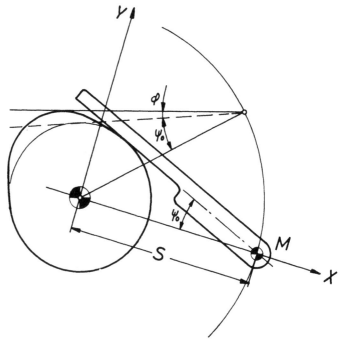

Fig. 4-18. Diagram for cam profile calculation—swinging centric flat-faced follower (polar and rectangular coordinates).

The rectangular cutter coordinates are:

$$X_c = S \left[\cos \theta + \frac{\cos (\psi_0 + \phi) \cos (\psi_0 - \theta + \phi)}{\dfrac{d\phi}{d\theta} - 1} \right] + R_g \sin (\psi_0 - \theta + \phi)$$

4.28

$$Y_c = S \left[\sin \theta + \frac{\cos (\psi_0 + \phi) \cos (\psi_0 - \theta + \phi)}{\dfrac{d\phi}{d\theta} - 1} \right] + R_g \cos (\psi_0 - \theta + \phi)$$

4.29

The polar cutting coordinates are found from equations 4.34 and 4.35.

Swinging Eccentric Flat-faced Follower (Fig. 4-19)

The equations 4.26 and 4.27 are modified slightly in the case of an eccentric flat-faced follower:

$$X = S \left[\cos \theta + \frac{\cos (\psi_0 + \phi) \cos (\psi_0 - \theta + \phi)}{\dfrac{d\phi}{d\theta} - 1} \right] \pm f \sin (\psi_0 - \theta + \phi)$$

4.30

$$Y = S \left[\sin \theta + \frac{\cos (\psi_0 + \phi) \cos (\psi_0 - \theta + \phi)}{\dfrac{d\phi}{d\theta} - 1} \right] \pm f \cos (\psi_0 - \theta + \phi)$$

4.31

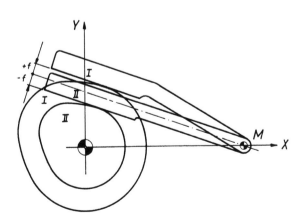

Fig. 4-19. Diagram for cam profile calculation—swinging eccentric flat-faced follower (polar and rectangular coordinates).

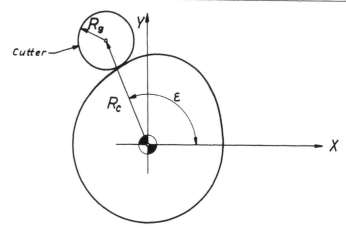

Fig. 4-20. Diagram for calculation of cutting coordinates for swinging eccentric flat-faced follower.

The + sign in front of f is used for follower I, the − sign for follower II. The rectangular cutting coordinates are: (Fig. 4-20)

$$X_c = S\left[\cos\theta + \frac{\cos(\psi_0 + \phi)\cos(\psi_0 - \theta + \phi)}{\dfrac{d\phi}{d\theta} - 1}\right]$$

$$+ (\pm f + R_g)\sin(\psi_0 - \theta + \phi) \quad 4.32$$

$$Y_c = S\left[\cos\theta + \frac{\cos(\psi_0 + \phi)\cos(\psi_0 - \theta + \phi)}{\dfrac{d\phi}{d\theta} - 1}\right]$$

$$+ (\pm f + R_g)\cos(\psi_0 - \theta + \phi) \quad 4.33$$

and the polar cutting coordinates:

$$R_c = \sqrt{X_c^2 + Y_c^2} \qquad 4.34$$

$$\tan\epsilon = \frac{Y_c}{X_c} \qquad 4.35$$

Pressure Angle and Radius of Curvature

The pressure angle at any point on the profile of a cam may be defined as the angle between the direction where the follower wants to go at that point and where the cam wants to push it. It is the angle between the tangent to the path of follower motion and the line perpendicular to the tangent of the cam profile at the point of cam-roller contact. (See Fig. 5-4.)

In this chapter we present material which permits us to analyze true pressure angle conditions to a greater depth than was indicated in the earlier chapters. In particular, we introduce work done by Karl Alexander Flocke,* a German kinematician. His method, expanded upon by the author, permits us to find with mathematical exactness the pressure angle in any disk cam with roller follower provided we know the equation upon which the follower motion is based.

The size of the pressure angle is important because:

(1) Increasing the pressure angle increases the side thrust and this increases the forces exerted on cam and follower.
(2) Reducing the pressure angle increases the cam size and often this is not desirable because:
   (a) The size of the cam determines, to a certain extent, the size of the machine.
   (b) Large cams require more precise cutting points in manufacturing and, therefore, an increase in cost.
   (c) Large cams mean that the circumferential speed of the cam is high and small deviations from the theoretical path of the follower cause additional acceleration, the size of which increases with the square of the cam size.

*See Ref. 336 in the Bibliography.

75

(d) Larger cams mean more revolving weight and in high-speed machines this leads to increased vibrations in the machine.

(e) The inertia of a large cam may interfere with quick starting and stopping.

The maximum pressure angle α_m should, in general, be kept to or below 30 degrees for translating type followers and to or below 45 degrees for swinging type followers. These values are on the conservative side and in many cases can be increased considerably, but beyond these limits troubles could develop and an analysis will be necessary.

In the following graphical methods are described by which a cam mechanism can be designed with translating or swinging roller follower having given maximum pressure angles for rise *and* return. These methods are applicable to any kind of time-displacement diagram and has been used here to make nomographs by means of which the pressure angle can be determined for *any* instant during the stroke. These nomographs also indicate what changes should be made to alter the maximum pressure angle. See also Chapter 13.

Determination of Cam Size for a Radial or an Offset Translating Follower

Figure 5-1 shows the time-displacement diagram. The maximum displacement is preferably made to scale, but the length of the abscissa can be chosen arbitrarily. The distance L from 0 to 360 degrees is measured and is set equal to $2\pi k$ from which

$$k = \frac{L}{2\pi}$$

is calculated and laid out as length E to M in Fig. 5-2.

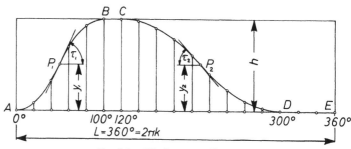

Fig. 5-1. Displacement diagram.

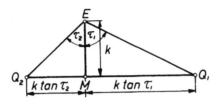

Fig. 5-2. Construction to find $k \tan \tau_1$ and $k \tan \tau_2$.

In the time-displacement diagram of Fig. 5-1 the two points P_1 and P_2 having the maximum angles of slope, τ_1 and τ_2, are located. The corresponding ordinates are designated y_1 and y_2 respectively. In this example they are of equal length.

Angles τ_1 and τ_2 are laid out as shown in Fig. 5-2 and the points of intersection with a perpendicular to EM at M determine Q_1 and Q_2. The distances

$$MQ_1 = k \tan \tau_1$$

$$MQ_2 = k \tan \tau_2$$

can now be found and are laid out in Fig. 5-3 the following way: If the rotation of the cam is counterclockwise, then lay out $k \tan \tau_1$, to the left, $k \tan \tau_2$ to the right from points R_{y1} and R_{y2}, respectively, where $R_u R_{y_1} = y_1$ and $R_u R_{y_2} = y_2$, R_{y_1} and R_{y_2} being the same point in this case. $R_u R_0$ is made equal to the stroke h with R_u indicating the lowest position and R_0 the highest position of the center of the translating roller follower.

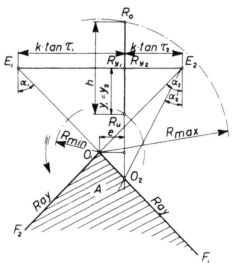

Fig. 5-3. Finding proportions of cam; offset translating follower.

The prescribed pressure angle α_1 ia laid out at E_1 as shown and a ray $E_1 F_1$ is determined. Any point on this ray chosen as the cam shaft center will proportion the cam so that the pressure angle at a point on the cam profile corresponding to point P_1 of the displacement diagram will be exactly α_1.

The angle α_2 is laid out at E_2 as shown and another ray $E_2 F_2$ is determined. Similarly, any point on this ray chosen as the cam shaft center will proportion the cam so that the pressure angle at a point on the cam profile corresponding to point P_2 of the displacement diagram will be exactly α_2.

Any point chosen within the cross-hatched area A as the cam center will yield a cam whose pressure angles at points corresponding to P_1 and P_2 will not exceed the specified values α_1 and α_2, respectively. If O_1 is chosen as the cam shaft center, the pressure angles on the cam profile corresponding to points P_1 and P_2 are exactly α_1 and α_2 respectively. Selection of point O_1 also yields the smallest possible cam for the given requirements and requires an offset follower in which e is the offset distance.

If O_2 is chosen as the cam shaft center, a radial translating follower is obtained (zero offset). In that case, the pressure angle α_1 for the rise is unchanged, whereas the pressure angle for the return is changed from α_2 to α_2'. That is, the pressure angle on the return stroke is reduced at the point P_2.

Figure 5-4 shows the shape of the cam when O_1 from Fig. 5-3 is chosen as the cam shaft center and it is seen that the pressure angle at a point on the cam profile corresponding to point P_1 is α_1 and at a point corresponding to point P_2 is α_2.

In the foregoing, a cam mechanism has been so proportioned that the pressure angles α_1 and α_2 at points on the cam corresponding to points P_1 and P_2 were obtained. By choosing P_1 and P_2 at the points of greatest slope, the maximum pressure angles which will occur at other points are, in general, only slightly greater than α_1 and α_2, respectively.

However, if the pressure angles α_1 and α_2 are not to be exceeded at any point—i.e., they are to be maximum pressure angles—the procedure described above must be repeated by letting P_1 take different positions on the curve for rise $(A-B)$ and P_2 different positions on the curve for return $(C-D)$.

This has been done for a number of different curves and the nomograms in Figs. 5-5 to 5-10 were the result. These nomograms are to be used the following way: R_u and R_0 indicate the two extreme positions of the translating follower and the distance between them is unity. For counter-clockwise rotation of the cam, go to the left in the nomogram to that curve which is indicated by the same number of degrees as that of the cam angle rotation for rise. (If the cam rotates clockwise, go to the right for rise and to the left for return.) For constant velocity motion, this curve is a vertical

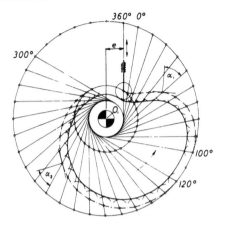

Fig. 5-4.   Construction of cam contour; offset translating follower.

straight line as shown in Fig. 5-5 and if α_1 is to be the maximum pressure angle for the rise, a line is drawn from the lowest point of this straight line and having an angle of α_1 with the vertical. (In all the other diagrams the line is drawn as a tangent to a curve.) Next, go to the right to that straight line (or curve) which is indicated by the same number of degrees as that of the cam angle rotation for return. In the same way as for rise, a line is drawn from the lowest point in the constant velocity nomogram (or as a tangent to the curve in all the other nomograms) having an angle of α_2 with the vertical. The point where these two lines intersect is the desired cam shaft center.

In all the nomograms the following example is solved:

$$h = 1 \text{ in.,} \ \beta_1 = 25°, \ \beta_2 = 80°, \ \alpha_{m1} = \alpha_{m2} = 30°$$

and rotation of cam is counterclockwise.

The results are given in Table 5-1.

Table 5-1

	$R_{max}{}^*$	e
Constant velocity motion	3.65	0.8
Parabolic motion	5.90	1.58
Simple harmonic motion	4.80	1.24
Cycloidal motion	5.90	1.58
3-4-5 Polynomial	5.65	1.48
Modified Trapezoidal Acceleration	5.90	1.57

$^*R_{max} = O_1 R_0$

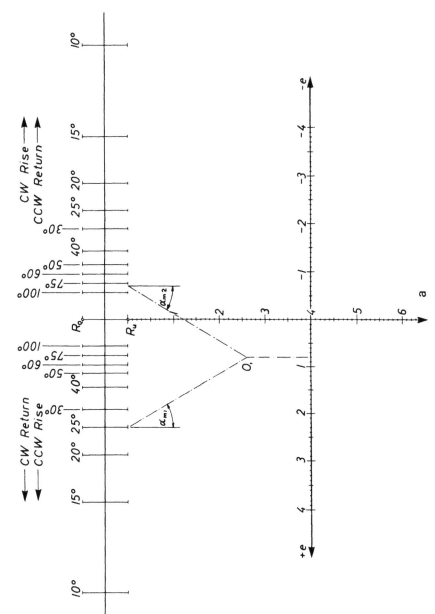

Fig. 5-5. Pressure angle nomogram; constant velocity motion.

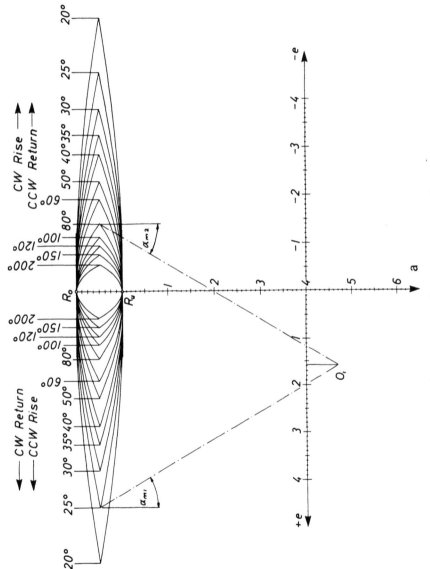

Fig. 5-6. Pressure angle nomogram; parabolic motion.

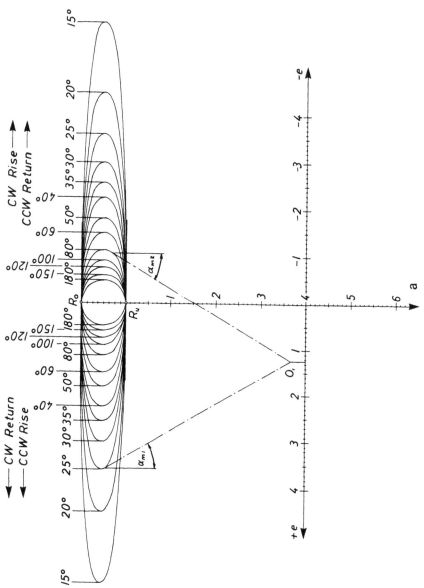

Fig. 5-7. Pressure angle nomogram; simple harmonic motion.

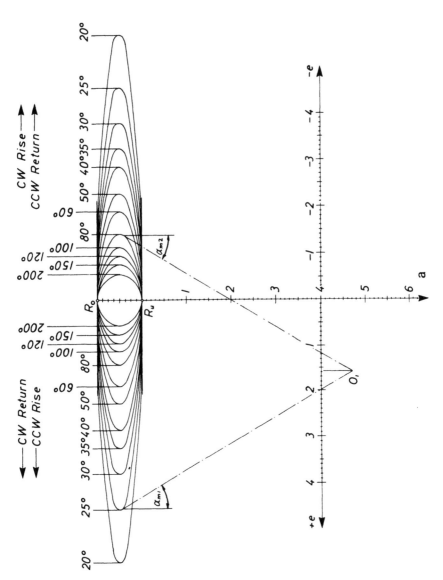

Fig. 5-8. Pressure angle nomogram; cycloidal motion.

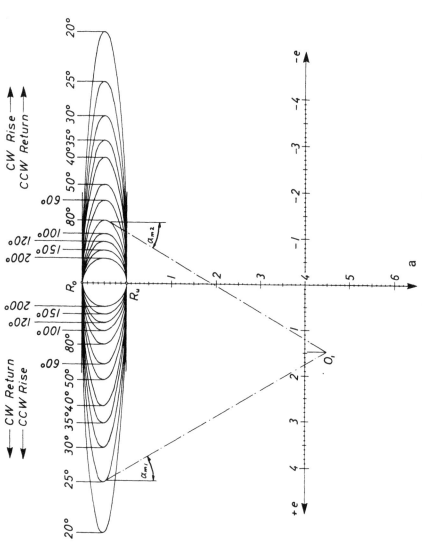

Fig. 5-9.   Pressure angle nomogram; 3-4-5 polynomial.

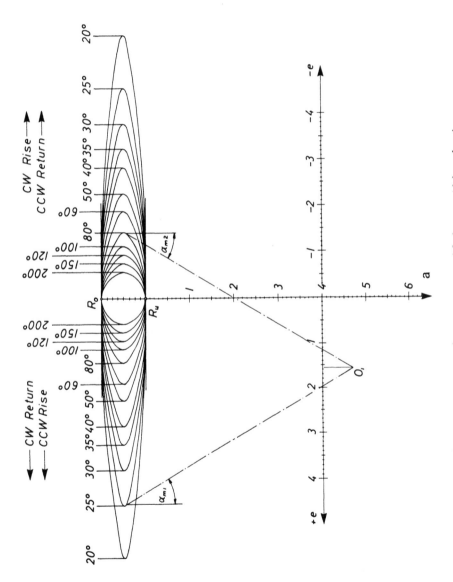

Fig. 5-10. Pressure angle nomogram; modified trapezoidal acceleration.

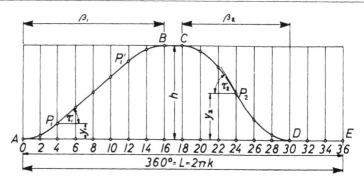

Fig. 5-11.   Displacement diagram.

In the case that h is different from one inch, say 1.45 in., all values in Table 5-1 are multiplied by 1.45.

Determination of Cam Size for Swinging Roller Follower

For any given instant a swinging roller follower can be replaced by a translating roller follower, and if the velocity of the roller center remains the same in magnitude and direction, then exactly the same pressure angle is obtained at the point in question.

The proportioning of a cam with swinging roller follower having specified pressure angles at selected points follows the same procedure as that for a translating follower as shown by the following:

Example: Given the diagram for the roller displacement along its circular arc, Fig. 5-11 with $h = 1.95$ in., the periods of rise and fall, respectively, $\beta_1 = 160°$ and $\beta_2 = 120°$, the length of the swinging follower arm $L_f = 3.52$ in., rotation of the cam away from pivot point M, and pressure angles $\alpha_1 = \alpha_2 = 45°$ (corresponding to the points P_1 and P_2 in the displacement diagram). Find the cam proportions.

Solution: Figs. 5-12 and 5-13 show the construction. Distances $k \tan \tau_1$ and $k \tan \tau_2$ are determined in Fig. 5-12. (See previous example.) R_{y1} is determined by making the distance $R_u R_{y_1} = y_1$ along the arc $R_u R_{y_1}$ and R_{y_2} by

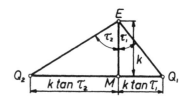

Fig. 5-12.   Construction to find $k \tan \tau_1$ and $k \tan \tau_2$.

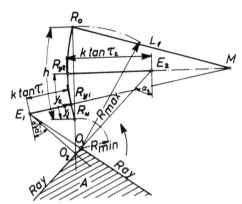

Fig. 5-13. Finding proportions of cam; swinging roller follower (CCW rotation).

making $R_u R_{y_2} = y_2$. The arc $R_u R_0 = h$ and R_u indicates the lowest position of the center of the swinging roller follower and R_0 the highest position.

Because the cam (i.e., the surface of the cam as it passes under the follower roller) rotates away from pivot point M, $k \tan \tau_1$ is laid out away from M, that is from R_{y_1} to E_1 and $k \tan \tau_2$ is laid out toward M from R_{y_2} to E_2. Angle α_1 at E_1 determines one ray and α_2 at E_2 another ray, which together subtend an area A having the property that if the cam shaft center is chosen inside this area the pressure angles at the points of the cam corresponding to P_1 and P_2 in the displacement diagram will not exceed the given values α_1 and α_2 respectively. If the cam shaft center is chosen on the ray drawn from E_1 at an angle of $\alpha_1 = 45°$, the pressure angle α_1 on the cam profile corresponding to point P_1 will be exactly 45 degrees, and if chosen on the ray from E_2 the pressure angle α_2 corresponding to P_2 will be exactly 45 degrees. If another point, O_2 for example, is chosen as the cam shaft center, the pressure angle corresponding to P_1 will be α_1' and that corresponding to P_2 will be α_2.

Figure 5-14 shows the construction for rotation toward pivot point M (clockwise rotation of the cam in this case). The layout of the cam curve is made in a manner similar to that shown previously in Fig. 4-5.

Figure 5-15 shows the construction of the cam profile for counterclockwise rotation if O_2 from Fig. 5-13 is chosen as cam shaft center.

In the example, the cam mechanism was so proportioned that the pressure angles at certain points (corresponding to P_1 and P_2) do not exceed certain prescribed values (namely α_1 and α_2).

To make sure that the pressure angle at *no point* along the cam profile exceeds the specified value, the previous procedure should be repeated for a series of points along the profile. This is done in the following example:

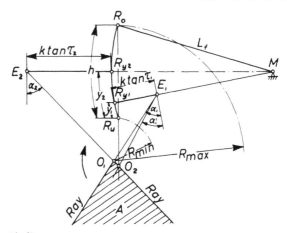

Fig. 5-14.   Finding proportions of cam; swinging roller follower (CW rotation).

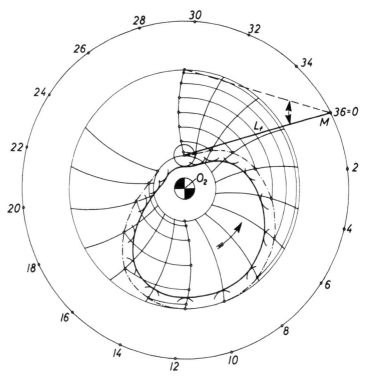

Fig. 5-15.   Construction of cam contour; swinging roller follower.

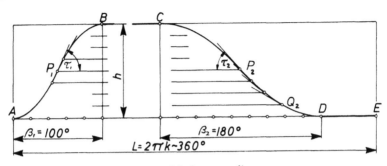

Fig. 5-16.   Displacement diagram.

Example: Given $\beta_1 = 100°$, $\beta_2 = 180°$, $h = 1.95$ in., parabolic motion for both rise and return, $L_f = 3.5$ in., rotation of cam away from pivot point M, and the maximum pressure angles $\alpha_1 = \alpha_2 = 45°$. Find the smallest possible cam proportions.

Solution: First, the displacement diagram, Fig. 5-16, is drawn with $h = 1.95$ in. and $360° = 7$ in. (arbitrarily).

From this $\qquad k = \dfrac{7}{2\pi} = 1.12$ in. $= EM$, in Fig. 5-17.

The total displacement of Fig. 5-16 is now divided into 8 equal parts and at each intersection of these horizontal dividing lines with the curve for rise and return the slope is found. The construction is shown for points P_1 and P_2 with angles τ_1 and τ_2, and by using Fig. 5-17 to find $k \tan \tau_1$ and $k \tan \tau_2$ the points E_1 and E_2 can be determined (Fig. 5-18).

The procedure is shown repeated for several points and in this way the curves along which the successive $k \tan \tau_1$ and $k \tan \tau_2$ lengths are determined. At each of the points indicated the respective angles α_1 and α_2 are laid out and an area A is determined so that if the cam shaft is chosen within this are the maximum pressure angles α_1 for the rise and α_2 for the return are not exceeded.

Point O_1 is the location of the cam shaft center for the smallest possible cam for the given conditions. R_{max} is measured and

$$R_{max} = 3.05 \text{ in.}$$

To get the largest cam radius, the roller radius r_f must, of course, be subtracted for open track cams and added for closed track cams (plus, of course, the thickness of material outside the track). By examining Fig. 5-18, it is also possible to tell at what points on the cam profile the maximum pressure angle occurs. The maximum pressure angle for rise occurs when the follower has moved half the stroke, whereas the maximum pressure

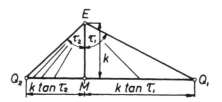

Fig. 5-17.   Construction to find $k \tan \tau_1$ and $k \tan \tau_2$.

angle for return occurs when the follower has moved back to a position $h/8$ from the bottom, corresponding to point Q_2' in Fig. 5-18 and point Q_2 in Fig. 5-16.

In many cases a quick check on pressure angles is desired and although the nomogram in Fig. 5-19 is exact only for radial translating roller followers and then only at the middle of the stroke, it can be used for swinging roller followers with a good approximation provided that the chord $R_u R_o$ passes through the cam shaft center.

Example: Given $R_{min} = 3.38$ in., $h = 2.8$ in., $\beta = 123°$, and a radial translating follower. Find the maximum pressure angle.

Solution: Drawing the lines Nos. 1, 2 and 3 in Fig. 5-19 gives a maximum pressure angle of 23 degrees for simple harmonic motion, and a maximum pressure angle of 28 degrees for parabolic and cycloidal motion.

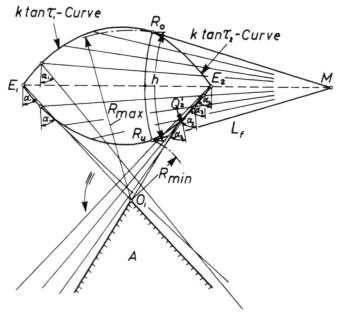

Fig. 5-18.   Finding proportions of cams; swinging roller follower (CCW rotation).

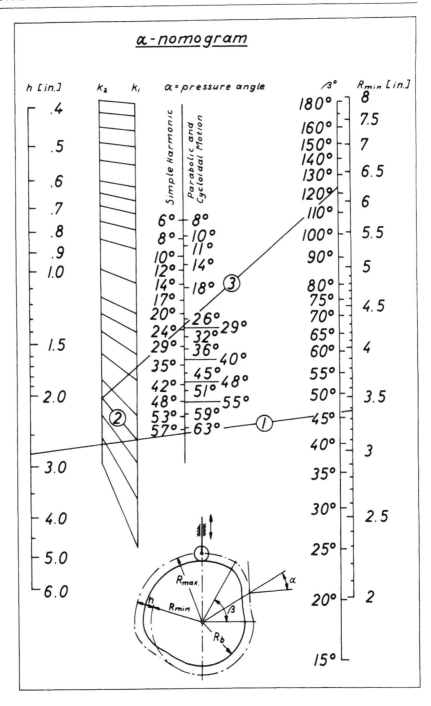

Fig. 5-19. Pressure angle nomogram; radial translating follower.

It has now been shown how to proportion a cam with respect to pressure angles for a given displacement diagram. What are the possibilities of diminishing the size of the cam further? An obvious solution is either to extend the time for rise and/or return; to choose another displacement diagram having a lower maximum velocity; or to decrease h by increasing the lever-ratio. This sort of investigation is often necessary and requires that one know the functioning of the machine in question very well. It is likely that improvements in this direction are possible. Some suggestions for mechanisms which will lower the maximum pressure angle are given in the chapter on cam mechanisms.

Theory of the Flocke Method

For those interested in the theory underlying the Flocke method, which the author has perfected so that an optimum cam size with respect to pressure angles can be obtained, the following proof is provided.

A translating follower with knife edge is shown in Figure 5-20. A_0 is the center of rotation of the cam; B indicates the knife edge on the follower; A is a point on the cam coincident with point B on the follower; and α is the pressure angle. In the position shown the velocity of the follower is V_B and the point A on the cam has the velocity $V_A \perp A_0 A$.

For a given time-displacement diagram the displacement of the follower relative to the cam is independent of the rotational speed of the cam which is set equal to $\omega = 1$ rad/sec. The scale of the velocity vector for A is chosen so that the length of $V_A = A_0 A$. The direction of the tangent to the cam at the point under consideration is the same as the velocity of point B relative to A namely $V_{B/A}$. The vector equation

$$V_B = V_A \longmapsto V_{B/A}$$

can now be solved (triangle ACD shown in Figure 5-20).

Next, the velocity vectors V_A, V_B, and $V_{B/A}$ are rotated $90°$ CCW to positions $V_A^\daleth$, $V_B^\daleth$, and $V_{B/A}^\daleth$, respectively. The head of $V_A^\daleth$ must coincide with A_0 because the scale factor was chosen so that $V_A = A_0 A$. It is obvious that the two triangles $V_A V_{b/A} V_B$ and $V_A^\daleth V_{B/A}^\daleth V_B^\daleth$ are congruent.

A new center of rotation is chosen for the cam somewhere on the extension of velocity vector $V_{B/A}^\daleth$, say O_2. It is seen from the new set of congruent triangles that $V_{B/A}^\daleth$ does not change its direction and V_B remains unchanged since B is intended to move with the same velocity as before to maintain the same time-displacement diagram as before. But the direction of $V_{B/A}^\daleth$ determines the pressure angle α at the point under consideration and for this position of the cam it can be said that choosing the center of

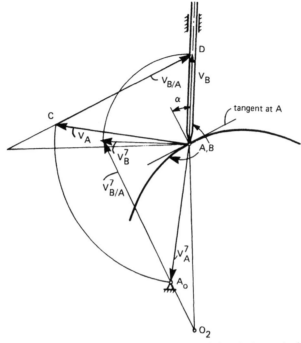

Fig. 5-20. Relationship between velocities and the Flocke method.

rotation of the cam anywhere on $V_{B/A}^{\rightharpoonup}$ results in a cam having the same specified pressure angle α at the point A.

It now remains to find the length of V_B. In the time-displacement diagram, Figure 5-21, the length of the abscissa, L in inches, is equal to the product of 1 revolution of the cam, $\beta = 2\pi$ rad, and a proportionality constant:

$$L = \beta k = 2\pi k \text{ inches}$$

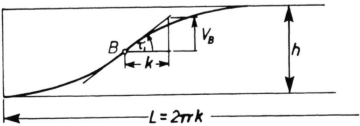

Fig. 5-21. Relationship between the velocity V_B of the roller follows and the displacement diagram.

But, L also represents the time for 1 revolution, which is $\beta = 2\pi$rad Therefore, the time T for one revolution is:

$$T = \frac{\beta}{\omega} = \frac{2\pi}{\omega} = 2\pi \text{ sec} \quad \text{since } \omega = 1 \text{ rad/sec}$$

The horizontal velocity at point B in Figure 5-21 is therefore:

$$V_H = \frac{L}{T} = \frac{2\pi\kappa}{2\pi} = k \text{ in./sec}$$

and $\tan \tau_1, = \dfrac{V_B}{V_H} = \dfrac{V_B}{k}$ so that $V_B = k \tan \tau_1$, which is the factor used in the

Flocke method, and it is also seen that the angle between $\overrightarrow{V_{B/A}}$ and $V_B = \alpha.$

Radius of Curvature

The minimum radius of curvature of a cam should be kept as large as possible (1) to prevent undercutting of the convex portion of the cam and (2) to prevent too high surface stresses. Figures 5-22a, b, and c illustrate how undercutting occurs.

In Fig. 5-22a the radius of curvature of the relative path of follower is ρ_{min} and the cam will at that point have a radius of curvature $R_c = \rho_{min} - r_f$.

In Fig. 5-22b $\rho_{min} = r_f$ and $R_c = 0$. Therefore, the actual cam will have a sharp corner which in most cases will result in too high surface stresses.

In Fig. 5-22c is shown the case where $\rho_{min} < r_f$. This case is not possible because undercutting will occur and the actual motion of the roller follower will deviate from the desired one as shown.

Undercutting cannot occur at the concave portion of the cam profile (working surface), but caution should be exerted in not making the radius

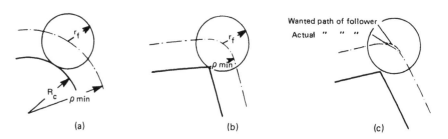

Fig. 5-22.   (a) No undercutting. (b) Sharp corner on cam. (c) Undercutting.

of curvature equal to the radius of the roller follower. This condition would occur if there is a cusp on the displacement diagram which, of course, should always be avoided.

A standard formula for radius of curvature found in calculus books is:

$$\rho = \frac{\left[r^2 + \left(\dfrac{dr}{d\theta}\right)^2\right]^{3/2}}{r^2 + 2\left(\dfrac{dr}{d\theta}\right)^2 - r\,\dfrac{d^2r}{d\theta^2}}$$

Letting

$$r = R_{min} + y = f(\theta)$$

we get

$$\rho = \frac{[(f(\theta))^2 + (f'(\theta))^2]^{3/2}}{(f(\theta))^2 + 2(f'(\theta))^2 - f(\theta)f''(\theta)}$$

ρ_{min} can be found from this formula either by differentiating once and finding the value for which $d\rho/d\theta = 0$ or by plotting the curve ρ as a function of θ and finding where this curve has its minimum.

For parabolic motion:

$$r = f(\theta) = R_{min} + h - 2h\left(\frac{\beta_1 - \theta}{\beta}\right)^2$$

$$\frac{dr}{d\theta} = f'(\theta) = -4\,\frac{h}{\beta_1^2}\,(\beta_1 - \theta)$$

$$\frac{d^2r}{d\theta^2} = f''(\theta) = -4\,\frac{h}{\beta_1^2}$$

$$0 \leqslant \theta \leqslant \frac{\beta_1}{2}$$

For simple harmonic motion:

$$r = f(\theta) = R_{min} + \frac{h}{2}\left(1 - \cos\left(\pi\,\frac{\theta}{\beta_1}\right)\right)$$

$$\frac{dr}{d\theta} = f'(\theta) = \frac{\pi h}{2\beta}\sin\left(\pi\,\frac{\theta}{\beta_1}\right)$$

$$\frac{d^2r}{d\theta^2} = f''(\theta) = \frac{\pi^2 h}{2\beta_1^2}\cos\left(\pi\,\frac{\theta}{\beta_1}\right)$$

$$0 \leqslant \theta \leqslant \beta_1$$

For cycloidal motion:

$$r = f(\theta) = R_{min} + h\left(\frac{\theta}{\beta_1} - \frac{1}{2\pi}\sin\left(2\pi\frac{\theta}{\beta_1}\right)\right)$$

$$\frac{dr}{d\theta} = f'(\theta) = \frac{h}{\beta_1}\left(1 - \cos\left(2\pi\frac{\theta}{\beta_1}\right)\right) \qquad\qquad 0 \leqslant \theta \leqslant \beta_1$$

$$\frac{d^2r}{d\theta^2} = f''(\theta) = \frac{2\pi h}{\beta_1^2}\sin\left(2\pi\frac{\theta}{\beta_1}\right)$$

For double harmonic motion:

$$r = f(\theta) = R_{min} + \frac{h}{2}\left[1 - \cos\left(\pi\frac{\theta}{\beta_1}\right) - \frac{1}{4}\left(1 - \cos\left(2\pi\frac{\theta}{\beta_1}\right)\right)\right]$$

$$\frac{dr}{d\theta} = f'(\theta) = \frac{h\pi}{2\beta_1}\left[\sin\left(\pi\frac{\theta}{\beta_1}\right) - \frac{1}{2}\sin\left(2\pi\frac{\theta}{\beta_1}\right)\right] \qquad\qquad 0 \leqslant \theta \leqslant \beta_1$$

$$\frac{d^2r}{d\theta^2} = f''(\theta) = \frac{h}{2}\left(\frac{\pi}{\beta_1}\right)^2\left[\cos\left(\pi\frac{\theta}{\beta_1}\right) - \cos\left(2\pi\frac{\theta}{\beta_1}\right)\right]$$

In the preceding formulas θ must be in radians. Usually it is the upper part of the lift curve that must be investigated in order to prevent undercutting.

The ρ_{min} has been calculated from the above formulas and the four nomograms, Figs. 5-23, 5-24, 5-25 and 5-26, were obtained.

Example: Given $h = 1$ in., $R_{min} = 2.9$ in., and $\beta = 60°$. Find ρ_{min} for parabolic motion, simple harmonic motion, cycloidal motion and double harmonic motion.

Solution: The nomograms show that:

$$\rho_{min} = 2.03 \text{ in. for parabolic motion}$$
$$\rho_{min} = 1.8 \text{ in. for simple harmonic motion}$$
$$\rho_{min} = 1.6 \text{ in. for cycloidal motion}$$
$$\rho_{min} = 1.2 \text{ in. for double harmonic motion}$$

In case h is, say, 1.7 in., the above values are multiplied by 1.7.

Although the nomograms are valid only for a radial translating follower, they can be used with good approximation for offset and swinging roller followers also.

If an accurate value of ρ_{min} is desired for an offset or swinging roller follower, use the computer programs in Chapter 13.

$\mathcal{G}_{min}$-NOMOGRAM
PARABOLIC MOTION, $h=1$ in.

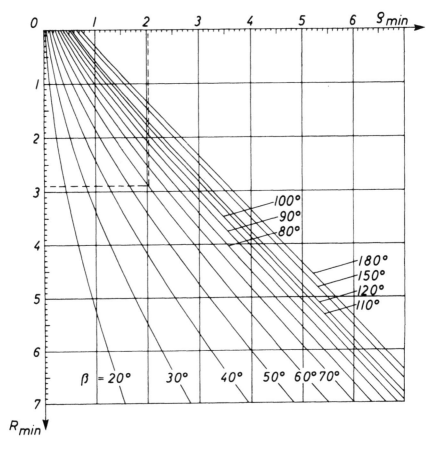

Fig. 5-23.   Nomogram for radius of curvature; parabolic motion.

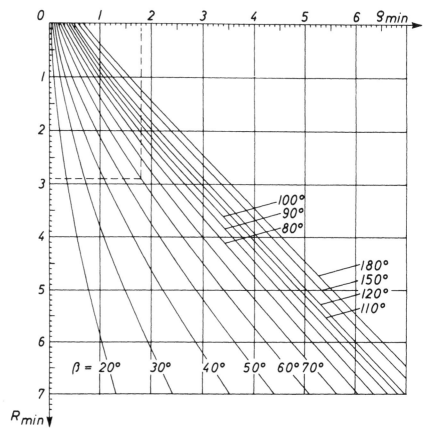

Fig. 5-24.   Nomogram for radius of curvature; simple harmonic motion.

Finding Radius of Curvature of Plate Cams Using Complex Vectors

Finding the radius of curvature of cams at various points is important to avoid undercutting and to check Hertz' pressure.

The use of complex vectors, based on the graphical method, is an effective tool to establish the necessary equations.

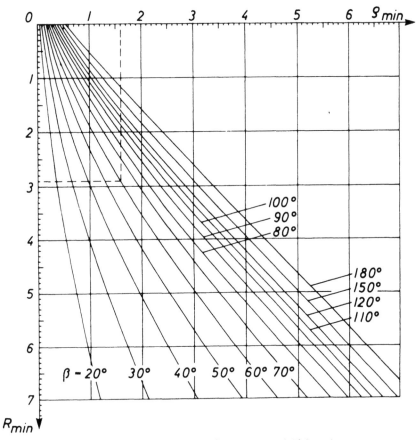

Fig. 5-25.   Nomogram for radius of curvature; cycloidal motion.

Radius of Curvature, Swinging Roller Follower

The procedure will be shown by first solving the problem using graphical methods and then, based on this approach analytical expressions are developed.

Example: A swinging roller follower, Fig. 5-27, has a simple harmonic motion, R_{min} = 1.5 in., L_f = 2.5 in., ϕ_0 = 30° and β = 60°. The cam rotates

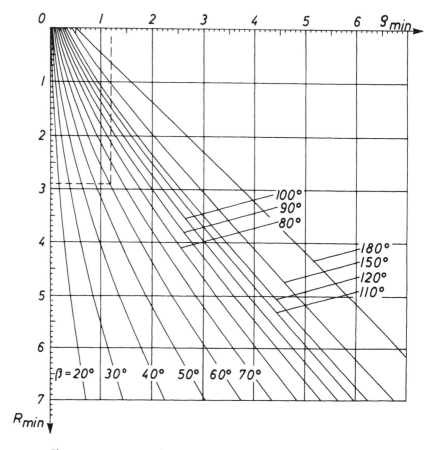

$\mathcal{S}_{min}$-NOMOGRAM
DOUBLE-HARMONIC MOTION, h=1 in.

Fig. 5-26.   Nomogram for radius of curvature; double-harmonic motion.

CCW and the radius of curvature R_c of the cam is to be determined in the position, where the cam has rotated 50° from the starting position where the roller center is at R_u the distance $MO = 3$ in. Roller radius $r_f = .25$ in.

Graphical Solution: The arc through which the center of the roller moves is

$$R_u R_0 = h = L_f \times \phi_0 = 2.5 \times 30 \times \frac{\pi}{180} = 1.31 \text{ in.}$$

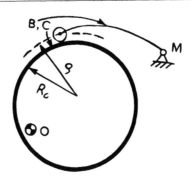

Fig. 5-27. Swinging roller follower.

$$y = \frac{h}{2}\left[1 - \cos\left(\pi \times \frac{\theta}{\beta}\right)\right] = 1.22 \text{ in.}$$

$$\phi = \frac{y}{h} \times \phi_0 = 27.99°$$

$$\psi_0 = \cos^{-1} \frac{MO^2 + L_f^2 - R_{min}^2}{2 \times MO \times L_f} = 29.93°$$

$$\theta_1 = 180 - \psi_0 - \phi = 122.1°$$

$$OC = \sqrt{MO^2 + L_f^2 - 2 \times MO \times L_f \cos\left(\psi_0 + \phi\right)} = 2.7 \text{ in.}$$

$$\theta_2 = \cos^{-1} \frac{OC^2 + MO^2 - L_f^2}{2 \times OC \times MO} = 51.7°$$

$$T = \frac{\pi}{3} \sec\ (\omega = 1/\sec \text{ is assumed})$$

$$V_B = \frac{h}{2} \times \frac{\pi}{T} \times \sin\left(\pi \times \frac{\theta}{\beta}\right) = .98 \text{ in.}$$

$$V_C = OC \times \omega = 2.7 \text{ in.}$$

The velocity vector equation

$$V_B = V_C \longmapsto V_{B/C}$$

is drawn in Fig. 5-28 and measuring on the drawing:

$$V_{B/C} = 3.15 \text{ in.}$$

The acceleration vector equation is

$$a_B^n \longmapsto a_B^t = a_C^n \longmapsto a_C^t \longmapsto a_{B/C}^n \longmapsto a_{B/C}^t \longmapsto 2 \times V_{B/C} \times \omega$$

where

$$a_B^n = \frac{V_B^2}{L_f} = .39 \text{ in.}$$

$$a_B^t = \frac{h}{2} \times \frac{\pi^2}{T^2} \times \cos\left(\pi \times \frac{\theta}{\beta}\right) = -5.1 \text{ in.}$$

$$a_C^n = OC \times \omega^2 = 2.7 \text{ in./sec}^2$$

$$a_C^t = 0 \ (\omega = \text{const.})$$

$$2 \times V_{B/C} \times \omega = 6.3 \text{ in.}$$

$$a_B^n = \frac{V_B^2}{L_f} = .39 \text{ in./sec}^2$$

$$a_B^t = \frac{h}{2} \times \frac{\pi^2}{T^2} \times \cos\left(\pi \times \frac{\theta}{\beta}\right) = -5.1 \text{ in./sec}^2$$

$$a_C^n = OC \times \omega^2 = 2.7 \text{ in./sec}^2$$

$$a_C^t = 0 \ (\omega = \text{const.})$$
$$2 \times V_{B/C} \times \omega = 6.3 \text{ in/sec}^2$$

From the acceleration vector polygon, Fig. 5-28, and measuring on the drawing:-

$$a_{B/C}^n = 8.1 \text{ in./sec}^2$$

$$\rho = \frac{V_{B/C}^2}{a_{B/C}^n} = 1.22 \text{ in.}$$

$$R_c = \rho - r_f = .97 \text{ in. (closed-track cam)}$$

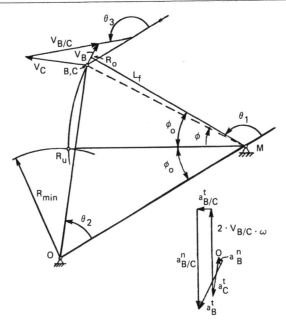

Fig. 5-28. Determination of radius of curvature; velocity and acceleration vector polygons.

Analytical Solution Using Complex Vectors: The angles θ_1, θ_2 and θ_3 are indicated in fig. 5-28. First the velocity vector equation is written as usual:

$$V_B = V_C \longmapsto V_{B/C}$$

and in complex form:

$$-i\, V_B(\cos\theta_1 + i\sin\theta_2) = i\, V_C(\cos\theta_2 + i\sin\theta_2) + V_{B/C}(\cos\theta_3 + i\sin\theta_3)$$

$$V_B \sin\theta_1 = -V_C \sin\theta_2 + V_{B/C}\cos\theta_3$$

$$-V_B \cos\theta_1 = V_C \cos\theta_2 + V_{B/C}\sin\theta_3$$

$$V_{B/C} = \frac{V_B \sin\theta_1 + V_C \sin\theta_2}{\cos\theta_3} \qquad\qquad 5.1$$

$$\theta_3 = \tan^{-1}\left(\frac{V_B \cos \theta_1 + V_C \cos \theta_2}{V_B \sin \theta_1 + V_C \sin \theta_2}\right) \qquad 5.2$$

To find the angles:

$$\psi_0 = \cos^{-1}\frac{MO^2 + L_f^2 - R_{min}^2}{2 \times MO \times L_f} \qquad 5.3$$

$$\phi = \frac{y}{h}\phi_0 \qquad 5.4$$

$$\theta_1 = \pi - \psi - \phi \qquad 5.5$$

$$OC = \sqrt{MO^2 + L_f^2 - 2 \times MO \times L_f \cos(\psi_0 - \phi)} \qquad 5.6$$

$$\theta_2 = \cos^{-1}\frac{OC^2 + MO^2 - L_f^2}{2 \times OC \times MO} \qquad 5.7$$

and with the values given:

$$\phi = 27.99° \qquad (5.4)$$
$$\psi_0 = 29.93° \qquad (5.3)$$
$$\theta_1 = 122.1° \qquad (5.5)$$
$$OC = 2.7 \text{ in.} \qquad (5.6)$$
$$\theta_2 = 51.7° \qquad (5.7)$$
$$\theta_3 = 158.69° \qquad (5.2)$$
$$V_{B/C} = -3.16 \text{ in.} \qquad (5.1)$$

To find the accelerations:

$$a_B^n \longmapsto a_B^t = a_C^n \longmapsto a_C^t \longmapsto a_{B/C}^n \longmapsto a_{B/C}^t \longmapsto 2 \times V_{B/C} \times$$

but $\qquad a_C^t = 0 \; (\omega = \text{const.})$

and in complex form the vector equation may be written

$$i^2 a_B^n e^{i\theta_1} - i a_B^t e^{i\theta_1} = i^2 a_C^n e^{i\theta_2} + i a_{B/C}^n + a_{B/C}^t e^{i\theta_3} + i \, 2 \, V_{B/C}\omega e^{i\theta_3}$$

Separating real and complex terms results in two equations and solving for $a_{B/C}^n$:

$$a^n_{B/C} = a^n_B \sin(\theta_3 - \theta_1) - a^t_B \cos(\theta_3 - \theta_1) + a^n_C \sin(\theta_2 - \theta_3) - 2V_{B/C}\omega$$
$$(5.8)$$

With the values given or found the following is obtained from equation 5.8:

$$a^n_{B/C} = 8.08 \text{ in./sec}^2$$

$$\gamma = \frac{V^2_{B/C}}{a^n_{B/C}} = 1.24 \text{ in.}$$

$$R_c = \gamma - r_f = .99 \text{ in.}$$

The equations developed in the foregoing have been used for developing some of the computer programs in Chapter 13.

Circular Cams

A cam having the shape of a circle with a center displaced from the cam shaft center, Fig. 6-1, is excellent for high speed use because it produces a smoothly continuous change in acceleration; it can be manufactured accurately (see Chapter 9 on cam manufacture) with respect to kinematic proportions; and where a groove is used it can be cut to close tolerances and with an excellent surface finish. The reason why this type of cam is employed so little is that many designers are not aware of these advantages. Those that are aware of them have found it difficult to determine optimum proportions.

A graphical method for proportioning such a mechanism is presented here, and charts are given from which the optimum proportions can be immediately determined.

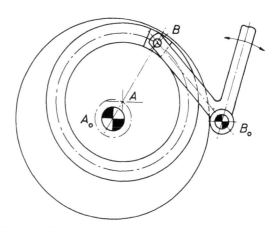

Fig. 6-1. High-speed, high-load circular cam mechanism.

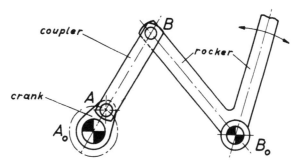

Fig. 6-2. Kinematic equivalent mechanism.

Although the time-displacement relationships of a circular cam have certain limitations, as for instance there can be no dwells or limited times for backward and forward motion; nevertheless, the circular cam and follower system can be proportioned so as to partially overcome these limitations. (See later section: Swinging Roller Follower.)

The kinematically equivalent mechanism to the high-speed, high-load circular cam shown in Fig. 6-1 is the four-bar linkage, shown in Fig. 6-2, with the crank A_0A, the coupler AB, and the rocker BB_0. This four-bar linkage can only be used when the crank A_0A is at the end of a shaft. If it were placed in the middle of a shaft the coupler AB would have to cut through the shaft and that is, of course, not possible.

It is possible to place a four-bar linkage at the middle of a shaft, if the sheave arrangement of Fig. 6-3 is used. The mechanisms of Figs. 6-1, 6-2, and 6-3 are all kinematically equivalent as long as the distances A_0A, AB, BB_0, and A_0B_0 are the same in all three figures. The design in Fig. 6-3 is used often in connection with both light and heavy machinery. As mentioned, the cam shown in Fig. 6-1 is capable of satisfactory operation at high speed and is easy to manufacture. A roller or sliding piece at B can be used.

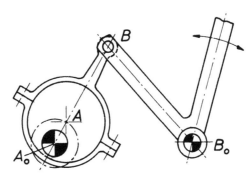

Fig. 6-3. Kinematic equivalent mechanism.

Proportioning of Circular Cams with Swinging Roller Follower

Because the circular cam of Fig. 6-1 and the four-bar linkage of Fig. 6-2 are kinematically equivalent mechanisms; i.e., for any given rpm of the input shaft, the output motion of the two mechanisms is exactly the same as long as the distances A_0A, AB, and A_0B_0 are kept equal, the problem of proportioning a circular cam with swinging roller follower is equivalent to proportioning a four-bar linkage.

Figure 6-4 shows a four-bar linkage A_0ABB_0 in its two dead-center positions; i.e., when the rocker B_0B is in its extreme left and extreme right positions. The total angle through which the rocker moves is ψ_0 and the angle through which the crank moves by clockwise rotation when the rocker

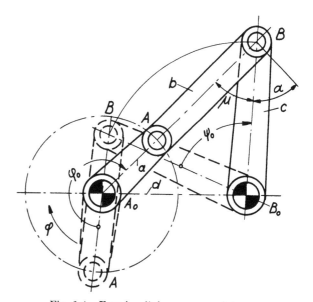

Fig. 6-4. Four-bar linkage, nomenclature.

goes from its extreme left to its extreme right position is ϕ_0. The angles ϕ_0 and ψ_0 are practical requirements imposed on the four-bar linkage; a large angle ϕ_0 proportions a four-bar linkage which will give a relatively slow motion of the rocker when it moves from left to right, but a rapid motion when moving from right to left.

When cams are being proportioned, it is desirable to aim for a low maximum pressure angle α_m. For linkages it is desirable to have a minimum transmission angle $\mu_{\min}$ which is as close to 90° as possible. The transmission angle μ is defined as the smallest angle between the centerlines of

circles are drawn through R. The circles are designated K_a and K_b. Next, draw a perpendicular from A_0 on B_0R to intersect K_b at L. LB_0 intersects K_b at E. Point B of the four-bar linkage A_0ABB_0 can be chosen anywhere on the arc LE. In order to get the four-bar linkage with the best force transmission characteristics, the angle β is determined from the chart in Fig. 6-9. For the given values of $\phi_0 = 160°$ and $\psi_0 = 40°$, the angle β is found to be 50.5 degrees. This angle is laid out at A_0 in a counterclockwise direction from A_0B_0. The intersection with circles K_a and K_b determines points A and B, respectively.

The four-bar linkage A_0ABB_0 will fulfill the given requirements, namely, that $\phi_0 = 160°$ and $\psi_0 = 40°$ and the linkage has the best μ_{min}, namely $\approx 30°$. This is, of course, a rather low value of μ_{min}; it corresponds to a pressure angle of 60 degrees so that a further analysis of this mechanism would be necessary if it has to run at a high speed.

The chart in Fig. 6-10 gives the ratio of crank-length to frame-length for a four-bar linkage.. The chart, Fig. 6-11, gives the ratio of coupler length to frame-length, and the chart, Fig. 6-12, gives the ratio of rocker-length to frame-length. These charts are helpful for a quick estimate of the overall linkage proportions.

In Fig. 6-5 the frame-length d was chosen as 4 inches.

The chart, Fig. 6-10 gives $\dfrac{a}{d} = 0.25$; hence $a = 1.00$ in.

The chart, Fig. 6-11 gives $\dfrac{b}{d} = 0.52$; hence $b = 2.08$ in.

The chart, Fig. 6-12 gives $\dfrac{c}{d} = 0.78$; hence $c = 3.12$ in.

The values found from the charts agree closely with the dimensions of Fig. 6-5.

Example: $\phi_0 = 208°$ and $\psi_0 = 64°$. Determine optimum proportions.

Solution: The procedure is the same as before (Fig. 6-5), but the difference from the previous example is in the possible locations of point B. Whereas the point of intersection of LB_0 with the circle K_b was E and the circular arc LE was the locus for point B, in this example (Fig. 6-6) point E is determined as the intersection of A_0B_0 with K_b and the circular arc LE is the locus for point B. Angle β is found to be 43 degrees. Angle μ_{min} in this case reaches a value of 33 degrees.

Example: $\phi_0 = 220°$ and $\psi_0 = 40°$. Determine optimum proportions.

Solution: Because $\phi_0 = 180° + \psi_0$, the angle $A_0RB_0 = 90°$ and the circle K_b degenerates to a straight line (Fig. 6-7). Therefore, A coincides with R and the locus of point B is on the line A_0R beyond L ($A_0R = RL$). Because β is the same for the whole family of linkages, a different approach is used to find the optimum proportions.

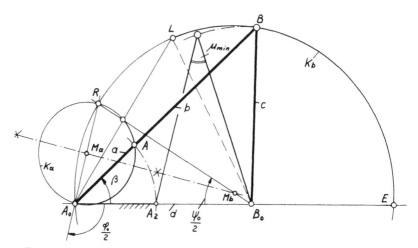

Fig. 6-6. Graphical construction to determine proportions of four-bar linkage.

Draw a circle with center at A_0 and radius A_0A; this circle intersects A_0B_0 at A_1 and A_2. Draw a perpendicular to A_0B_0 at A_0. With A_1B_0 as diameter, draw the half circle t. The intersection of the perpendicular to A_0B_0 at A_0 with t gives $B_{1,2}$. The optimum four-bar linkage is $A_0A_1B_{1,2}B_0$. The best μ_{min} occurs when the four-bar linkage is in position $A_0A_2B_{1,2}B_0$.

All linkages proportioned in this special way lie in the chart, Fig. 6-9, on the heavy dashed diagonal line. Angle μ_{min} is found to be 30 degrees.

Particular attention should be given to the centric four-bar linkage; i.e.; $\phi_0 = 180°$, because in this case for a given value of ψ_0 there is no other four-bar linkage that has better transmission angles. The chart, Fig. 6-9, indi-

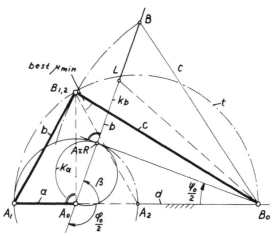

Fig. 6-7. Graphical construction to determine proportions of four-bar linkage.

cates that $\beta = 0°$ for $\phi_0 = 180°$. Thus, the most favorable centric four-bar linkage is without any practical significance and the chart shows only the theoretical limiting value of best μ_{min}. Therefore, for this special case the chart in Fig. 6-13, must be used.

Example: $\phi_0 = 180°$ and $\psi_0 = 90°$. Find optimum proportions.

Solution: The chart, Fig. 6-9, gives a limiting value of $\mu_{min} = 45°$. That, however, would result in the crank length being equal to zero.

By using the chart, Fig. 6-13, we find that if a value of $\mu_{min} = 40°$ can be tolerated, then $\beta = 23°$. With these values, the four-bar linkage is proportioned with the help of Fig. 6-8.

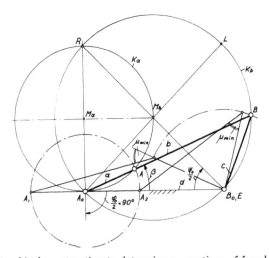

Fig. 6-8. Graphical construction to determine proportions of four-bar linkage.

Summary: The charts, Figs. 6-9 to 6-13, enable us to determine the proportions of a four-bar linkage with given angles ϕ_0 and ψ_0 without having to determine the lengths of the links by construction. The geometrical construction of such a four-bar linkage is easily done by drawing a few lines. Among the family of linkages obtained from the geometrical construction, the best one with respect to optimum transmission angles can be read from the charts. The charts show the best possible value of minimum transmission angle for given ϕ_0 and ψ_0.

For instance, a rocker angle $\psi_0 = 100°$ can be obtained only with poor transmission properties as can be seen immediately from the chart, Fig. 6-9.* Thus, the advantage of these charts is evident.

* For instance if $\mu_{min} = 30°$ then $176° < \phi_0 < 188°$, and if $\mu_{min} > 40°$ there is no solution.

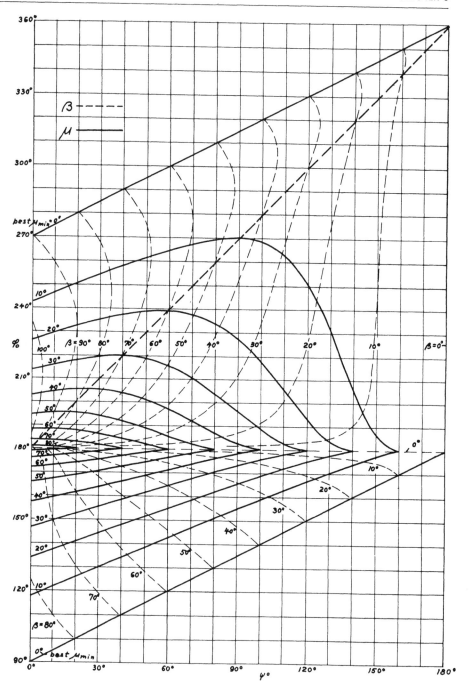

Fig. 6-9. Chart to find best μ and β.

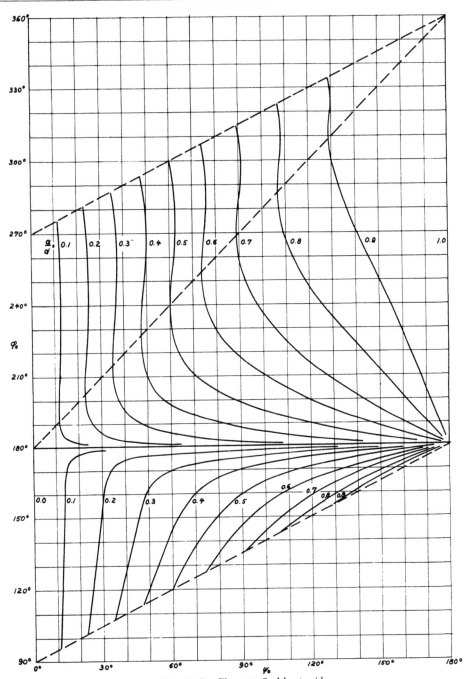

Fig. 6-10. Chart to find best a/d.

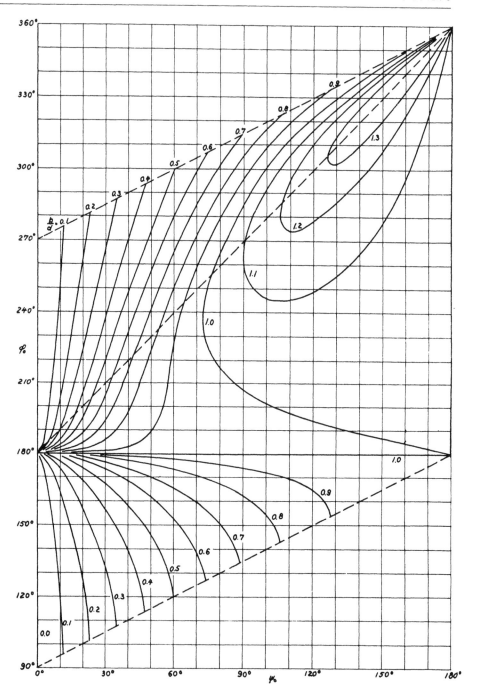

Fig. 6-11. Chart to find best b/d.

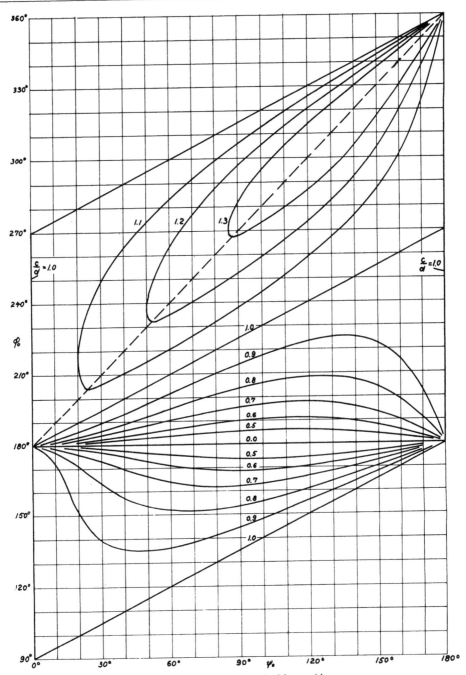

Fig. 6-12. Chart to find best c/d.

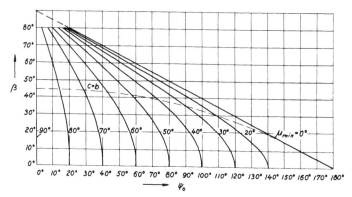

Fig. 6-13. Chart to find optimum proportions.

Proportioning of Circular Cams with Swinging Roller Follower—Additional Requirements

In the procedure so far described, the only criterions imposed on the design were angles ϕ_0, ψ_0, and the best possible transmission angle μ_{min}.

In certain cases, however, the value of μ_{min} is so favorable that we actually can choose other proportions of the four-bar linkage and still get good transmission angles. For example, with $45° \leqq \mu_{min}$ a change in proportions could be used to fulfill further requirements.

In automatic machinery, it is often required that the machine member to be moved be at a certain place at a certain time. The following procedure allows for change of the linkage proportions to make a certain adjustment of the time-displacement diagram possible.

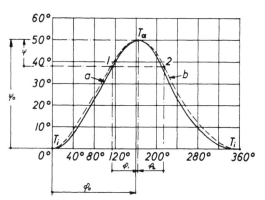

Fig. 6-14. Displacement diagram.

the coupler and the rocker; See Fig. 6-4. The relationship between α and μ is that $\alpha + \mu = 90°$. Practical limitations on μ are:

$$45° \leqq \mu$$

As in the case of limitations on the pressure angle, the limits imposed on μ can probably be exceeded but then an analysis of the mechanism must be made.

It is possible to obtain a number of four-bar linkages (a family of linkages) which have the same values of ϕ_0 and ψ_0. During a full rotation of the crank the transmission angle μ takes on different values. The value of μ when its deviation from 90° is a minimum is designated μ_{min} and if this value of μ_{min} is the largest possible for the given family of linkages, the best μ_{min} is obtained. If one had to find the one linkage with the best force transmission characteristic out of a whole family of linkages by a trial-and-error procedure, considerable work would have to be done. Charts have been worked out for this purpose by J. Volmer in Germany, however, and these charts save considerable work. (See Figs. 6-9, 6-10, 6-11, 6-12, and 6-13.)

Example: Given $\phi_0 = 160°$ and $\psi_0 = 40°$. Determine the proportions of a four-bar linkage for optimum force transmission.

Solution: Figure 6-5. Choose the length $A_0B_0 = 4.00$ in. Lay out $\dfrac{\psi_0}{2} = 20°$ in a clockwise direction at B_0 and lay out $\dfrac{\phi_0}{2} = 80°$ in a clockwise direction at A_0. The intersection of these two lines determines point R. Bisect A_0R; this determines M_a and M_b. With M_a and M_b as centers,

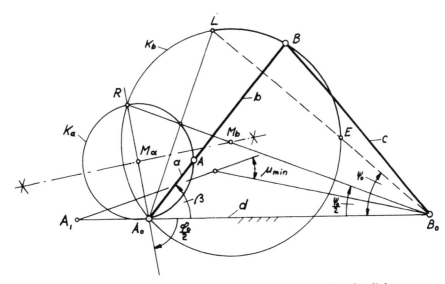

Fig. 6-5. Graphical construction to determine proportions of four-bar linkage.

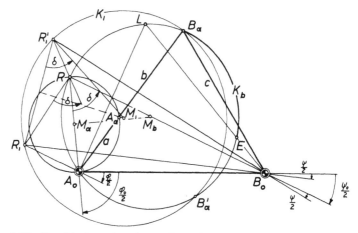

Fig. 6-15. Graphical construction to determine proportions of four-bar linkage.

In Fig. 6-14 the points T_i and T_a are given. T_i corresponds to the extreme position of the rocker when it is closest to A_0 and T_a corresponds to the other extreme position. Because ϕ_0 and ψ_0 are given, the usual construction is made in Figs. 6-15 and 6-16. In Fig. 6-15 a construction is shown which will let the rocker occupy a position during the forward stroke corresponding to point 1 of the a-curve (shown in full). In Fig. 6-16 a construction is shown which will let the rocker occupy a position during the backward stroke corresponding to point 2 of the b-curve (shown dotted). These additional constructions are carried out the following way:

In Fig. 6-15 $\phi_1/2$ and $\psi/2$ are laid out at A_0 and B_0 respectively, and the point of intersection is R_1. Angle $A_0RB_0 = \delta$ is laid out from RR_1 as

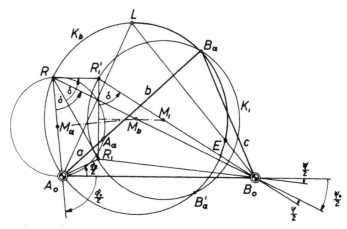

Fig. 6-16. Graphical construction to determine proportions of four-bar linkage.

shown and intersects the line drawn from B_0 at an angle of $(\psi_0 + \psi)/2$ at point R_1'. δ is laid out as shown from RR_1' and the intersection with the bisector of R_1R_1' yields M_1, the center of the circle K_1 with radius M_1R_1. K_1 intersects the circle K_b in two points; namely, B_a' and B_a. Of these two points, only B_a can be used because it lies on the arc LE. The four-bar linkage $A_0A_aB_aB_0$ is the desired four-bar linkage.

Figure 6-16 is similar to Fig. 6-15, the only difference being in the direction the angles are laid out.

The constructions were carried through for:

$$\phi_0 = 168°, \quad \psi_0 = 50°, \quad \phi_1 = 50°, \quad \text{and} \quad \psi = 12°.$$

Proportioning of Circular Cams with Translating Roller Follower (charts)

Figure 6-17 shows a circular cam with an offset translating roller follower. The kinematically equivalent mechanism is shown in the form of a slider-crank in Fig. 6-18. The proportioning of the cam mechanism, Fig. 6-17, is now changed to the problem of proportioning of the slider-crank.

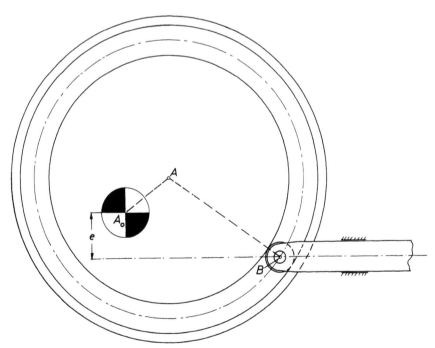

Fig. 6-17.   Offset translating roller follower; circular cam.

The Slider-Crank

The slider-crank is used for transforming a rotary motion to a back-and-forth motion along a straight line, or as in automobile engines, to transform the back and forward motion of the piston into a rotary output. This mechanism is also used in drawing presses, certain types of printing presses, and high-speed industrial sewing machines.

In most cases a centrical slider-crank is used; however, specific requirements can be met by using an offset slider-crank; i.e., a slider-crank where the path of point B, Fig. 6-18, does not pass through the crank-shaft center A_0 but is offset the distance e. In Fig. 6-18 is also shown the positions when the slider is in its inner dead position $(A_0A_iB_i)$, when it is in its outer dead position $(A_0A_aB_a)$ and finally it is shown in the position where the transmission angles reaches its minimum value, $\mu_{\min}$.

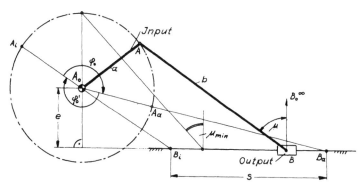

Fig. 6-18. Nomenclature.

When the crank rotates clockwise from A_0A_i to A_0A_a the slider moves from the extreme left position to the extreme right position and by motion of the crank during the remaining part of the revolution the slider moves back. Therefore, since the times for forward and backward stroke of the slider are different; the smaller angle $A_iA_0A_a$ is designated as ϕ_0 and the larger angle $A_1A_0A_a$ as ϕ_0'.

In order to proportion a slider-crank where the angle for the forward stroke ϕ_0 and the stroke s is given, we use the following construction, Fig. 6-19. Lay out $B_iB_a = s$ and at B_a lay out the angle $\phi_0 - 90°$ clockwise from B_iB_a. The intersection of the side of this angle with the bisector to B_iB_a gives point M. With M as a center draw the circle K_0 through B_i and B_a. The intersection of the circle K_0 with B_aM gives point L and with the bisector to B_iB_a, point N. The middle point of NB_a determines the center of the circle K_a, that passes through B_a. Point A_0 has been chosen arbitrarily anywhere on the circular arc B_iL. The intersection of A_0B_a

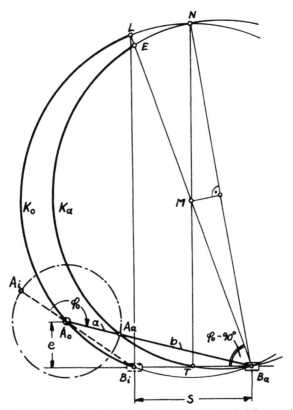

Fig. 6-19. Graphical construction to determine proportions of slider-crank mechanism.

with circle K_a determines point A_a and $A_0A_a = a$, the length of the crank-shaft. Depending upon where on arc B_iL point A_0 is chosen, different proportions of the slider-crank are obtained, and among these many solutions there is one solution having the best minimum transmission angle. These best solutions (with respect to μ_{min}) can be obtained from the charts in Figs. 6-20 to 6-24.

Example: Given: $\phi_0 = 160°$ $\dfrac{e}{s} = 0.17$. Determine the proportions of a slider-crank.

Solution: Fig. 6-20 gives $\mu_{min} = 40°$

$$\text{Fig. 6-21} \qquad \frac{a}{s} = 0.484$$

$$\text{Fig. 6-22} \qquad \frac{b}{s} = 0.85$$

$$\text{Fig. 6-23} \qquad \lambda = \frac{a}{b} = 0.57$$

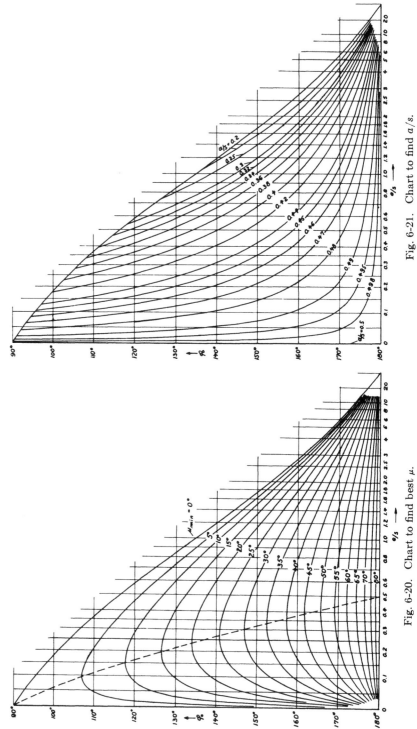

Fig. 6-21. Chart to find a/s.

Fig. 6-20. Chart to find best μ.

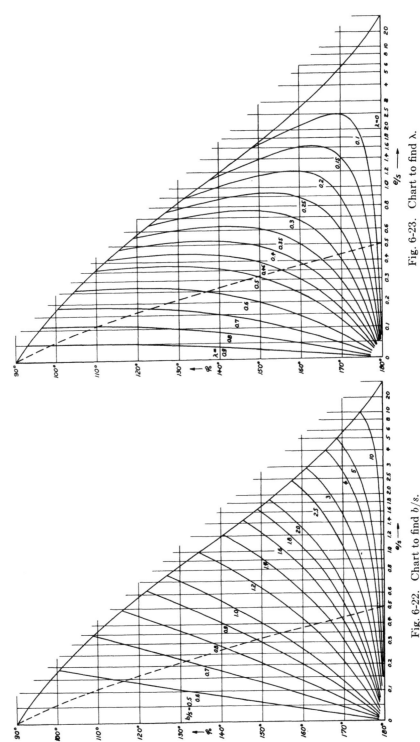

Fig. 6-23. Chart to find λ.

Fig. 6-22. Chart to find b/s.

If a larger eccentricity can be allowed, the transmission angle can be improved.

Example: Given $\phi_0 = 160°$. Determine the proportions of a slider-crank with the best transmission angle.

Solution: In Fig. 6-20 find the intersection of $\phi_0 = 160°$ and the dotted curve indicating the best transmission angle.

Fig. 6-20 $\mu_{min} = 43°$ $\dfrac{e}{s} = 0.375$

Fig. 6-21 $\dfrac{a}{s} = 0.467$

Fig. 6-22 $\dfrac{b}{s} = 1.15$

Fig. 6-23 $\lambda = \dfrac{a}{b} = 0.41$

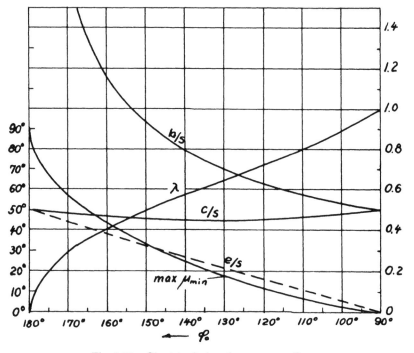

Fig. 6-24.   Chart to find optimum proportions.

Figure 6-24 gives the proportions of a slider-crank for given ϕ_0 and with the best μ_{min}.

If very accurate dimensions are desired, the following equations give the exact sizes:

$$2a = \sqrt{s\left(s - 2e \cos \frac{\phi_0}{2}\right)} \qquad 6.1$$

$$2b = \sqrt{s\left(s + 2e \tan \frac{\phi_0}{2}\right)} \qquad 6.2$$

$$\lambda = \frac{a}{b} = \sqrt{\frac{s - 2e \cos \dfrac{\phi_0}{2}}{s + 2e \tan \dfrac{\phi_0}{2}}} \qquad 6.3$$

$$\cos \mu_{min} = \frac{a + e}{b} = \frac{2e + \sqrt{s\left(s - 2e \cos \dfrac{\phi_0}{2}\right)}}{\sqrt{s\left(s + 2e \tan \dfrac{\phi_0}{2}\right)}} \qquad 6.4$$

And finally it is worthwhile to point out that by replacing the slider with a rack in mesh with a pinion, the slider-crank can be used to transform rotary input to oscillating output over more than 360 degrees and furthermore that it is possible to vary the time for the forward motion of the pinion relative to the backward stroke using the methods and charts described.

Circular-arc and Straight-line Cams

A cam profile composed of circular arcs and straight lines is a very good profile for many applications. The advantages are that the profile is determined by the coordinates of the centers of the arcs together with their respective radii and the straight lines are tangents to the circular arcs. Therefore, if the master cam is handmade, it is very easy to draw the contour on the plate and then cut and file it correspondingly, or to make special fixtures. For small cams a special technique where combined machining and filing is possible is described in the chapter about cam manufacturing.

The disadvantages are that although it is desirable in high-speed cams to have an acceleration that varies continuously, this cannot be obtained with circular arc straight line cams because at certain points along the profile the radius of curvature changes suddenly from one value to another.

Layout of a Circular-arc Straight-line Cam Profile

In order to get a good profile; i.e., a profile where the maximum acceleration is kept small, the cam contour is constructed first from the desired acceleration characteristic and the profile so obtained is then approximated with straight lines and circular arcs. This method has proven successful for designing cams used in mechanical calculators as will be explained later in this chapter.

Determination of Velocities and Accelerations in Circular-arc Straight-line Cams

The method used to find velocities and accelerations resulting from these profiles consists essentially of replacing the cam and follower system with a kinematically equivalent mechanism. As previously explained, two mechanisms are said to be kinematically equivalent if their output motions over a given (finite) range are exactly the same.

Figure 7-1 shows the most common types of cam and follower systems. In each diagram cam sections having a circular profile or a straight-line profile are shown together with their kinematically equivalent mechanisms.

Thus, in Fig. 7-1a at the left is shown a cam having the center of its circular arc at A. If this cam mechanism is replaced by a four-bar linkage A_0ABB_0 as shown to the right, then the output motion of B_0B will be exactly the same for the two mechanisms provided that A_0A rotates with the same angular velocity in both mechanisms.

As can be seen from Figs. 7-1b the equivalent mechanism of an offset translating roller follower in contact with a circular arc profile is an offset slider crank, because A, the center of the circular arc, is always at the same distance from B over the range of the arc. The same kind of reasoning is applied in the other cases shown in Fig. 7-1. It should be noted that a flat-faced follower cannot be used with a cam having a straight line as part of its contour because of the impact that would result when the contour changes from a curve to a straight line.

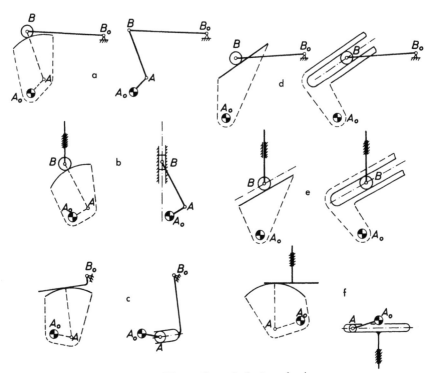

Fig. 7-1. Kinematic equivalent mechanisms.

In Chapter 12 it is explained how to find velocities and accelerations in linkages similar to those shown in Fig. 7-1 and in the following examples the accelerations will be determined by this method for the entire cycle.

Example: The essential dimensions of a circular-arc straight-line profile cam are given in Fig. 7-2; the roller follower is of the translating type, and the cam speed N is 1000 rpm counterclockwise. It is required to draw the acceleration curve for one full revolution of the cam and on the basis of this curve to suggest ways of improving the characteristics of this profile.

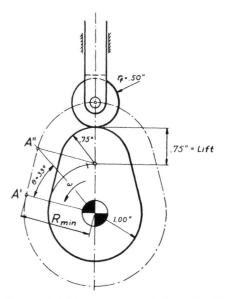

Fig. 7-2. Circular-arc straight-line cam; translating roller follower.

Solution for straight-line profile: In Chapter 12 it is shown that the acceleration a of the center of the roller follower for the straight line portion of the profile can be determined by the formula:

$$a = R_{min}\, \omega^2 \cdot \frac{1 + \sin^2 \theta}{\cos^3 \theta}$$

where R_{min} is the minimum pitch radius of the cam (to the center of the follower roller), ω is the angular velocity of the cam shaft, and θ is the angle of cam rotation, θ being equal to zero when the roller is at point A' and equal to $33°$ when the roller is at point A''; the equivalent mechanism is shown in Fig. 7-3.

For $\theta = 0°$ $a = 1.5 \left(\dfrac{1000 \cdot 2\pi}{60} \right)^2 = 16{,}400$ in./sec^2

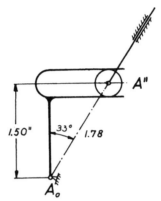

Fig. 7-3. Kinematic equivalent mechanism.

and for $\quad \theta = 33° \quad a = 1.5 \left(\dfrac{1000 \cdot 2\pi}{60}\right)^2 \cdot \dfrac{1 + 0.296}{0.59} = 36{,}000 \text{ in./sec}^2$

A number of values of a are computed in a similar manner and these are plotted in Fig. 7-5.

Solution for circular profile: The equivalent mechanism for the follower moving over the circular profile is in this case a centric slider crank, as shown in Fig. 7-4. First, the velocity and acceleration of the follower when it is at point A'', (Fig. 7-2) where the circular and the straight-line profiles meet, are found. The velocity v of the follower at this point is:

$$v = 1.0 \cdot \dfrac{1000 \cdot 2\pi}{60} = 105 \text{ in./sec}$$

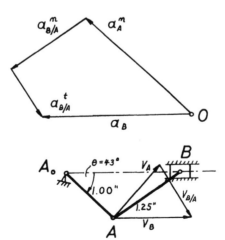

Fig. 7-4. Centric slider crank; determination of acceleration.

Referring to the velocity vector diagram in Fig. 7-4:

$$V_B = V_A \nrightarrow V_{B/A}$$

$$V_{B/A} = 94 \text{ in./sec}$$

Referring to the acceleration vector diagram in Fig. 7-4:

$$a_B^n \nrightarrow a_B^t = a_A^m \nrightarrow a_A^t \nrightarrow a_{B/A}^n \nrightarrow a_{B/A}^t$$

$$(a_B)_n = 0$$

$$a_A^n = 1.0 \left(\frac{1000 \cdot 2\pi}{60} \right)^2 = 11,000 \text{ in./sec}^2$$

$$a_A^t = 0$$

$$a_{B/A}^n = \frac{94^2}{1.25} = 70,500 \text{ in./sec}^2$$

and from the acceleration diagram,

$$a_B = 11,600 \text{ in./sec}^2$$

The acceleration of the translating roller follower when it is at its upper extreme position is calculated from:

$$V_A = 1.0 \left(\frac{1000 \cdot 2\pi}{60} \right) = 105 \text{ in./sec}$$

$$V_{B/A} = 105 \text{ in./sec}$$

$$V_B = 0$$

$$a_B^n \nrightarrow a_B^t = a_A^n \nrightarrow a_A^t \nrightarrow a_{B/A}^n \nrightarrow a_{B/A}^t$$

$$(a_B)_n = 0$$

$$a^n = 1.0 \left(\frac{1000 \cdot 2\pi}{60} \right)^2 = 11,000 \text{ in./sec}^2$$

$$a_A^t = 0$$

$$a_{B/A}^m = \frac{105^2}{1.25} = 8,820 \text{ in./sec}^2$$

$$a_B = a_A^n \nrightarrow a_{B/A}^n$$

$$= 11,000 + 8,820 = 19,820 \text{ in./sec}^2$$

Let us check the acceleration of B in Fig. 7-4, analytically:

$$\sin \beta = - \frac{(1.0)}{(1.25)} \sin 43° = -0.546$$

$$\beta = -33.1°$$

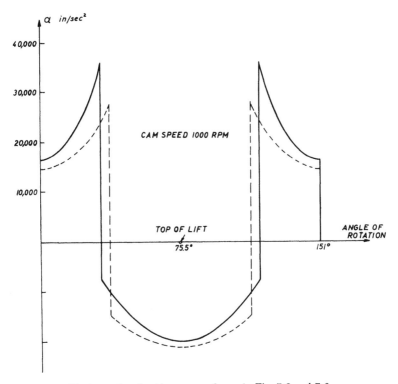

Fig. 7-5.   Acceleration curve of cam in Fig. 7-2 and 7-6.

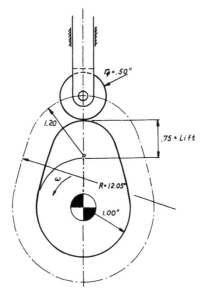

Fig. 7-6.   Circular arc cam; translating roller follower.

$$a_B = (1.0) \left(\frac{1000 \cdot 2\pi}{60} \right)^2 \left[\frac{\cos (43° + 33.1°)}{\cos 33.1°} + \frac{(1.0)}{(1.25)} \cdot \frac{\cos^2 43°}{\cos^3 33.1°} \right]$$

$$= 10,900 \left(\frac{0.24}{0.837} + \frac{1.0}{1.25} \cdot \frac{0.535}{0.59} \right)$$

$$= 10,900 \ (0.287 + 0.726)$$

$$= 11,100 \ \text{in./sec}^2$$

which agrees closely with the result obtained by the graphical method shown in Fig. 7-4.

Another way of solving the problem of profile design is to assume the acceleration characteristic, draw the profile which will give this characteristic and then approximate this profile with circular arcs and straight lines. Assuming a constant acceleration curve and an equal constant deceleration curve, the theoretical profile giving this acceleration characteristic is constructed as outlined in Chapter 4. The actual profile composed only of circular arcs is shown in Fig. 7-6. This profile has improved acceleration characteristics when compared with the profile shown in Fig. 7-2. In Fig. 7-5 the acceleration curve for the former is shown in dashed line and for the latter with full line. This change in cam profile has reduced the maximum acceleration considerably. However, it should be borne in mind that the sudden changes in acceleration makes it unsuitable for high speeds.

A circular cam is a very good cam because of its simplicity in manufacture. A circular cam with a flat-faced follower is shown in Fig. 7-7. The eccentricity e of this cam determines the entire acceleration curve. The size of the cam can be chosen arbitrarily, but if necessary, the cam size can be related to allowable Hertz' pressure (see Chapter 8) because a relatively small increase in cam size will cause a considerably lower contact pressure between the cam and the flat-faced follower. On the other hand, the larger the cam, the larger the sliding velocity between cam and follower.

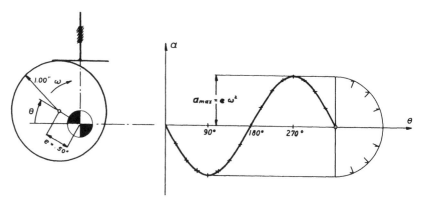

Fig. 7-7. Circular cam with flat-faced follower and corresponding acceleration diagram.

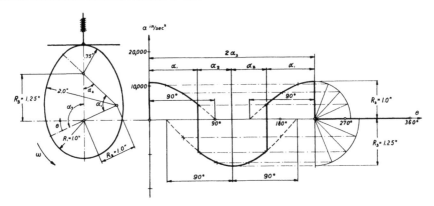

Fig. 7-8. Circular arc cam with flat-faced follower and corresponding acceleration diagram.

Example: Given a flat-faced translating follower in contact with a circular cam profile with dimensions as shown in Fig. 7-7 and with $N = 1000$ rpm. To find the maximum acceleration.

Solution: The equivalent mechanism is a scotch yoke, the acceleration of which follows a simple harmonic curve as shown to the right in the drawing. The maximum amplitude of the acceleration is

$$a = e\omega^2 = 0.5 \left(\frac{1000 \cdot 2\pi}{60}\right)^2 = 5{,}500 \text{ in/sec}^2$$

Example: Given a flat-faced translating follower together with a profile composed of circular arcs (a straight-line profile would cause impact) as shown in Fig. 7-8. The cam rotates counter clockwise with $N = 1000$ rpm. To draw the acceleration curve for one revolution of the cam.

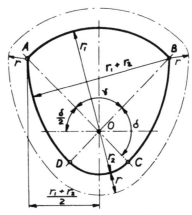

Fig. 7-9. Circular arc-constant diameter cam.

Solution: The work is greatly simplified if each circular arc is considered part of a full circle cam. The full circle cam has a simple harmonic acceleration and this curve can easily be drawn; only the maximum value of the acceleration need be known together with the angle θ. Of course, only that portion of the curve is used where there is an actual cam profile. The acceleration curve is shown at the right in Fig. 7-8.

Constant Diameter Cams

This type of cam is used in motion picture and sewing machine mechanisms. Figure 7-9 shows a general form of a constant diameter cam; it consists of two circular arcs AB and CD, having their center at 0 and radii r_1 and r_2 respectively, and subtending the angle γ, which is the angle of dwell. The distance between A and B, which are both symmetrical about the center line, is $r_1 + r_2$ and the two arcs of dwell are connected with each other by arcs having their centers at A and B and radii $r_1 + r_2$, which are tangentially connected with the small dwell arc at C and D, but make a sharp corner at B and A. In order to avoid early wear at sharp corners A and B, an equidistant curve is drawn in at a distance r outside the original profile which will give exactly the same motion to the follower as the original curve.

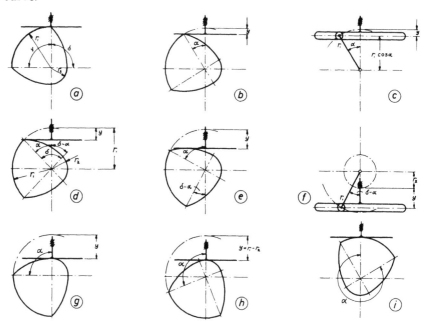

Fig. 7-10. Kinematic equivalent mechanisms.

The cam is shown in different positions in Fig. 7-10 together with the equivalent mechanism for the corresponding position.

From Fig. 7-10c:

$$\cos \alpha = \frac{r_1 - y}{r_1}$$ 7.1

$$y = r_1(1 - \cos \alpha) \qquad 0 \leq \alpha \leq \frac{\delta}{2}$$ 7.2

From Fig. 7-10d:

$$y = \frac{r_1 - r_2}{2} \qquad \alpha = \frac{\delta}{2}$$ 7.3

From Fig. 7-10f:

$$y = r_1 \cos (\delta - \alpha) - r_2 \qquad \frac{\delta}{2} \leq \alpha \leq \delta$$ 7.4

From the above formulas equations for velocities, v, and accelerations, a, are derived:

$$v = r_1 \sin \alpha \frac{d\alpha}{dt} = r_1 \omega \sin \alpha \qquad 0 \leq \alpha \leq \frac{\delta}{2}$$ 7.5

$$a = r_1 \omega^2 \cos \alpha \qquad 0 \leq \alpha \leq \frac{\delta}{2}$$ 7.6

$$v = -r_1 \omega \sin (\delta - \alpha) \qquad \frac{\delta}{2} \leq \alpha \leq \delta$$ 7.7

$$b = -r_1 \omega^2 \cos (\delta - \alpha) \qquad \frac{\delta}{2} \leq \alpha \leq \delta$$ 7.8

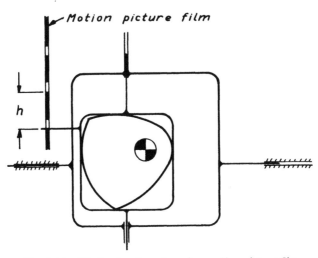

Fig. 7-11. Mechanism for advancing motion picture film.

In the following, formulas are derived for the constant diameter cam, Fig. 7-9, in connection with its use in cameras. The formulas arrived at are general in nature and apply also for other cases. The ratio of motion S is defined by:

$$S = \frac{t_1}{T} = \frac{\delta}{360°}$$

7.9

where t_1 is the time during which the film is moved, T the time for one full revolution, and δ the angle of rotation of cam, for which the film strip is moving; a principal arrangement for moving the film strip is shown in Fig. 7-11.

$$h = r_1 - r_2$$

From Fig. 7-9:

$$\sin \frac{\gamma}{2} = \frac{r_1 + r_2}{2r_1}$$

$$\gamma = 2 \sin^{-1} \frac{r_1 + r_2}{2r_1}$$

$$\delta = 180° - \gamma = 180 - 2 \sin^{-1} \frac{r_1 + r_2}{2r_1}$$

$$\delta = S \cdot 360° = 180 - 2 \sin^{-1} \frac{r_1 + r_2}{2r_1}$$

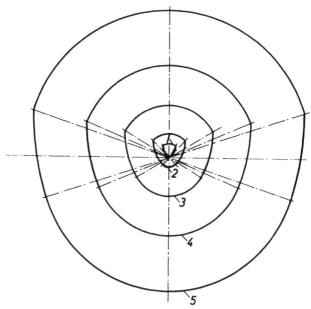

Fig. 7-12. Circular arc-constant diameter cam for different dwell periods.

$$r_1 = \frac{0.5h}{1 - \cos{(S \cdot 180°)}}$$

$$r_2 = r_1 - h$$

r_1 and r_2 can now be calculated for a given stroke h and a given ratio of motion S.

Table 7-1 shows the maximum velocities and accelerations for various ratios S when $h = 0.30$ in. and number of strokes per sec $= 16$.

Table 7-1

S	$\delta[°]$	$r_1[\text{in.}]$	$r_{max2}[\text{in.}]$	$v_{max}[\text{in./sec}]$	$a_{max}[\text{in./sec}^2]$
$1:3$	120	0.30	0.0	26	3,000
$1:4$	90	0.512	0.212	36.2	5,120
$1:6$	60	1.12	0.818	55.8	11,200
$1:8$	45	1.98	1.68	75.5	19,800
$1:10$	36	3.07	2.73	95.0	30,700

For a given stroke h, the motion ratio S has a great influence on the size of the cam. The smallest cam is obtained for $S = 1:3$ and Fig. 7-12 illustrates this. Table 7-2 gives the calculated values of radii r_1 and r_2 for $h = 0.3$ in.

Table 7-2

S	$h[\text{in.}]$	r_1	r_2	Cam Profile
$1:3$	0.30	0.3	0	1
$1:4$	0.30	0.512	0.212	2
$1:6$	0.30	1.12	0.818	3
$1:8$	0.30	1.98	1.68	4
$1:10$	0.30	3.07	2.73	5

Illustrative Problem for Swinging Arm Follower

Given a swinging roller follower and a circular arc—straight line profile as shown in Fig. 7-13a, with a cam speed of 1000 rpm counterclockwise. Find the acceleration of the follower when in contact with a point of the circular arc profile and also when in contact with the straight-line portion of the profile. Suggest ways of improving the profile.

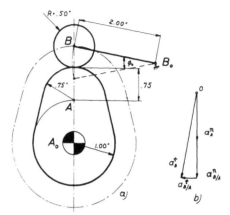

Fig. 7-13. Circular arc-straight line cam with swinging roller follower.

Figure 7-13b shows the determination of the acceleration for a point of the circular arc profile, namely, when B is in its extreme upper position. The following calculations were made:

$$V_A = (1.00) \frac{1000 \cdot 2\pi}{60} = 105 \text{ in./sec}$$

$$V_B = 0 \ (B \text{ is in extreme position})$$

$$V_B = V_A \dotplus V_{B/A} \ \therefore \ V_{B/A} = V_A = 105 \text{ in./sec}$$

$$\underset{4}{\underline{a_B^n}} \dotplus a_B^t = \underline{\underline{a_A^m}} \dotplus \underline{\underline{a_A^t}} \dotplus \underline{\underline{a_{B/A}^m}} \dotplus \underset{4}{a_{B/A}^t}$$

(A term underlined twice means it is a vector that is completely known. A sign $\angle$ designates a vector which is known only in direction. See also Chapter 12.)

$$a_B^n = 0$$

$$a_A^m = 1.00 \left(\frac{1000 \cdot 2\pi}{60} \right)^2 = 11{,}000 \text{ in./sec}^2$$

$$a_A^t = 0$$

$$a_{B/A}^m = \frac{105^2}{1.25} = 8{,}600 \text{ in./sec}^2$$

From the vector diagram: $a_B^t = 20{,}000 \text{ in./sec}^2$

The above result is now checked with the help of equations 12.1 to 12.10 (see Chapter 12). For explanation of symbols see Fig. 12-5.

$a = 1.00 \text{ in.}, \quad b = 1.25 \text{ in.}, \quad c = 2.00 \text{ in.}, \quad d = 2.70 \text{ in.}, \quad \theta_2 = 46.5°, \quad$ and $N = 1000 \text{ rpm counterclockwise.}$

$$l = \sqrt{2.7^2 + 1.0^2 - 2(2.7)(1.0)\cos 46.5°}$$
$$= 2.14 \text{ in.} \tag{12.1}$$

$$\beta = \sin^{-1}\left(\frac{1.0}{2.14} \cdot \sin 46.5°\right)$$
$$= 19.8° \tag{12.2}$$

$$\psi = \cos^{-1}\left(\frac{1.25^2 + 2.14^2 - 2.0^2}{(2)(1.25)(2.14)}\right)$$
$$= 66.7° \tag{12.3}$$

$$\theta_3 = 66.7° - 19.8°$$
$$= 46.9° \tag{12.4}$$

$$\lambda = \sin^{-1}\left(\frac{1.25}{2.00}\sin 66.7°\right)$$
$$= 31.5° \tag{12.5}$$

$$\theta_4 = 360° - 31.5° - 19.8°$$
$$= 308.7° \tag{12.6}$$

$$\delta = 46.5° - 308.7°$$
$$= -262.2° \tag{12.7}$$

$$\epsilon = 46.9° - 308.7°$$
$$= -261.8° \tag{12.7}$$

$$\gamma = 46.5° - 46.9°$$
$$= -0.4° \tag{12.7}$$

$$\omega_b = -\frac{1.0\sin(-262.2°) \cdot 1000 \cdot 2\pi}{1.25\sin(-261.8°) \cdot 60}$$
$$= -83.7 \text{ radians/sec} \tag{12.8}$$

$$\omega_c = \frac{1.0\sin(-0.4°) \cdot 1000 \cdot 2\pi}{2.0\sin(-261.8°) \cdot 60}$$
$$= 0 \tag{12.9}$$

$$\alpha_c = 0 + \frac{1.0 \cdot 11{,}000\cos(-0.4°) + 1.25 \cdot 7000 + 0}{2.0\sin(-261.8°)}$$
$$= 10{,}000 \text{ radians/sec}^2 \tag{12.10}$$

The tangential acceleration of the roller is:

$a_t = 2 \times 10{,}000 = 20{,}000$ in./sec^2, which is very close to the value found for the translating follower in the same position (Fig. 7-5)

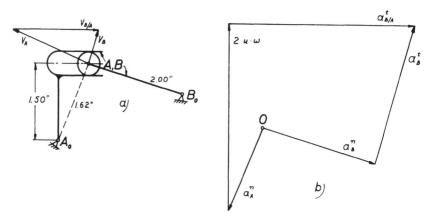

Fig. 7-14. Kinematic equivalent mechanism.

Figure 7-14a shows an equivalent mechanism corresponding to a position where the roller is in contact with the straight line profile. Notice that point A is on the center line of the slotted member and point B is the center of the roller and both points are coincident in this position.

$$A_0A = 1.62 \text{ in.} \quad BB_0 = 2.0 \text{ in.}$$

$$V_A = 1.62 \frac{1000 \cdot 2\pi}{60} = 170 \text{ in./sec}$$

$$V_B = V_A \twoheadrightarrow V_{B/A}; \quad V_B = 69 \text{ in./sec}; \quad V_{B/A} = 175 \text{ in./sec}$$

$$a_B^m \twoheadrightarrow a_B^t = a_A^m \twoheadrightarrow a_A^t \twoheadrightarrow a_{B/A}^m \twoheadrightarrow a_{B/A}^t \twoheadrightarrow 2u\omega$$

$$a_B^n = \frac{69^2}{2.0} = 2{,}380 \text{ in./sec}^2$$

$$a_A^n = \frac{170^2}{1.62} = 17{,}800 \text{ in./sec}^2$$

$$a_B^t = 0$$

$$a_{B/A}^n = 0$$

$$2u\omega = (2.0)(175) \frac{1000 \cdot 2\pi}{60} = 36{,}600 \text{ in./sec}^2$$

and from the vector diagram (Fig. 7-14b)

$$a_A^t = 28{,}200 \text{ in./sec}^2$$

The profile can be improved by constructing the cam contour as was done previously according to the desired acceleration and then approximate with circular arcs and straight lines.

Forces, Contact Stresses, and Materials

The different factors which influence forces acting upon and stresses within cams will now be discussed. The reader must realize, however, that in a book of this type a comprehensive treatment of stress is impossible and for further information the reader should consult any of the excellent treatises on this topic.

The main factors influencing cam forces are:

1. Displacement and cam speed (forces due to acceleration)
2. Dynamic forces due to backlash and flexibility
3. Linkage dimensions which affect weight and weight distribution
4. Pressure angle and friction forces
5. Spring forces.

The main factors influencing stresses in cams are:

1. Radius of curvature for cam and roller
2. Materials.

In Chapter 2 a comparison was made of different displacement diagrams, and of the three commonly used curves it was shown that the parabolic motion has an acceleration factor of 4, simple harmonic motion 4.93, and cycloidal motion 6.28. These factors indicate the relative size of the acceleration forces if the cam and follower system were perfectly rigid and without any backlash. However, no system is perfectly rigid and few systems are without backlash.

During the last two decades an extensive investigation of the increase in forces due to elasticity and backlash has been made, with good practical results.

Dynamic Forces

As shown in Fig. 8-1, let the load W be supported on the two movable stops A and B in such a way that there is a zero upward force in the spring. Next, let the two stops A and B be simultaneously withdrawn permitting the mass to drop onto the spring. The resulting instantaneous deflection of the spring will be *double* that which the spring will have if the mass merely rests quietly on it. It can be proved that the theoretical factor is exactly 2.

For the same reason, the stress acting on the strut, Fig. 8-2, when the load W is transferred instantaneously from the stops A and B, to the strut (without impact) is double the stress which occurs when equilibrium has been restored. It is assumed that when A and B are supporting the mass there is exactly zero clearance between W and the strut. Any positive clearance will result in a stress magnification greater than 2.

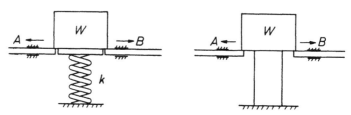

Fig. 8-1. (Left) Sudden removal of support doubles spring force. Fig. 8-2. (Right) Sudden removal of support doubles stress created in strut.

Although parabolic motion seems to be the best with respect to minimizing the calculated maximum acceleration and, therefore, also the maximum acceleration forces, nevertheless in the case of high speed cams, cycloidal motion yields the lower maximum acceleration forces. Thus, it can be shown that due to the sudden change in acceleration (called *jerk* or *pulse*) in the case of parabolic motion, the actual forces acting on the cam are doubled and sometimes even tripled at high speed, whereas with cycloidal motion, due to the gradually changing acceleration, the actual dynamic forces are only slightly higher than the theoretical. Therefore, the force found due to acceleration should be multiplied by at least a factor of 2 for parabolic and 1.05 for cycloidal motion.

Determination of Acceleration

Example: Given a cam with $h = 3.5$ in., $\beta = 180°$, and $N = 100$ rpm. Find its maximum acceleration for parabolic, simple harmonic, and cycloidal motions.

Solution: Use the nomogram in Fig. 8-3 which is for cam speeds of 50 to 500 rpm. Draw line #1 from $\beta = 180°$ to $N = 100$ rpm. This gives $T = 0.3$ sec. Next, draw line #2 from $h = 3.5$ in. through $T = 0.3$ sec to intersect the a_{max}-scale for parabolic and simple harmonic motion.

The values so found are:

Parabolic motion	$a_{max} = 0.4$ g
Simple harmonic motion	$a_{max} = 0.5$ g
Cycloidal motion	$a_{max} = 0.62$ g

(The value of a_{max} for cycloidal motion is found by going from the a_{max}-scale for parabolic motion horizontally to the cycloidal motion scale.)

When computed, the exact values are found as follows:

For parabolic motion:

$$a_{max} = 4\,\frac{h}{T^2} \text{ (formula 2.2), \quad where } \quad T = \frac{60}{N}\cdot\frac{\beta}{360}$$

$$= \frac{(4)(3.5)(100)^2(360)^2}{(60)^2(180)^2} = 155 \text{ in./sec}^2,$$

which can also be written as

$$\frac{155}{386} = 0.402 \text{ g}$$

For simple harmonic motion:

$$a_{max} = \frac{\pi^2}{2}\cdot\frac{h}{T^2}$$

$$= \frac{\pi^2(3.5)(100)^2(360)^2}{(2)(60)^2(180)^2(386)} = 0.495 \text{ g} \tag{2.3}$$

For cycloidal motion:

$$a_{max} = 2\pi\,\frac{h}{T^2}$$

$$= \frac{(2\pi)(3.5)(100)^2(360)^2}{(60)^2(180)^2(386)} = 0.630 \text{ g} \tag{2.4}$$

If N had been 1000 rpm, then to find a_{max}, the nomogram in Fig. 8-4, which is for cam speeds of 500 to 5,000 rpm, would have been used:

Parabolic motion	$a_{max} = 40$ g
Simple harmonic motion	$a_{max} = 50$ g
Cycloidal motion	$a_{max} = 62$ g

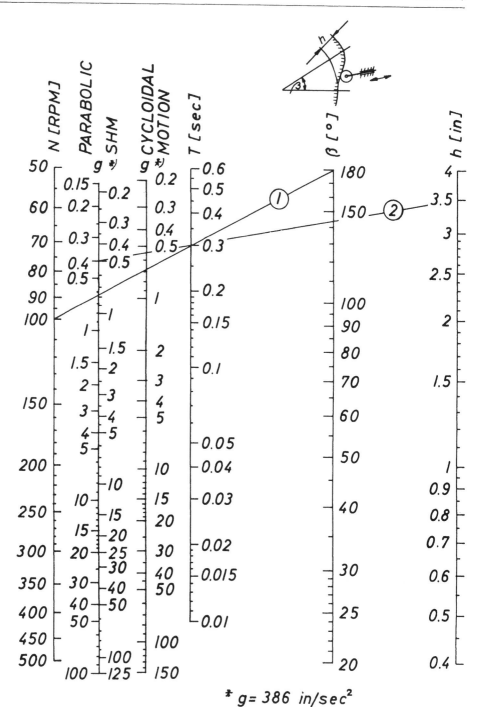

Fig. 8-3. Acceleration nomogram ($N = 50 - 500$ rpm).

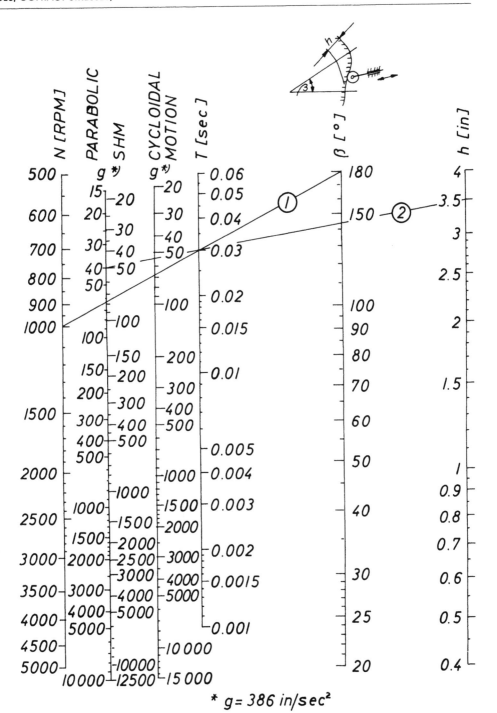

$$* \; g = 386 \; in/sec^2$$

Fig. 8-4. Acceleration nomogram ($N = 500 - 5,000$ rpm).

Linkage Dimensions

The increase in forces due to the flexibility of the linkage system can be partly counteracted: Build parts rigid and light, use materials with a high modulus of elasticity, make the time for rise and return as long as practicable, and finally, make the linkage as small as possible.

Pressure Angle

It has been stated in Chapter 5 that, in general, the maximum pressure angle α_m should be 30 degrees or less for a translating roller follower and 45 degrees or less for a swinging roller follower. It has also been shown how to proportion the cam mechanism for given pressure angles. Let us now analyze the forces acting on a translating roller follower.

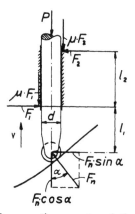

Fig. 8-5. Forces acting on a translating follower.

Figure 8-5 shows a translating roller follower. The pressure angle is α in the position shown and the normal force between the cam and the roller is F_n. This force is broken down into two components, namely, $F_n \sin \alpha$ which is perpendicular to the follower shaft and $F_n \cos \alpha$ which acts in the direction of motion of the follower. The follower shaft is normally guided by a long bearing and because of the component $F_n \sin \alpha$ the two forces F_1 and F_2 are the result. Although F_1 and F_2 are shown as concentrated forces, it is assumed that the actual distributed load on the bearing can be approximated sufficiently closely by these two forces. F_1 and F_2 give rise to the friction forces μF_1 and μF_2 which act on the follower shaft opposite to the direction of motion. Finally, a force P is acting against the direction of motion of the follower and P represents the inertia forces of the rest of the linkage together with friction forces, spring forces, and outer forces.

In a system such as the one described, the summation of forces along the X- and Y-axes has to be zero and summation of the moments of forces around any point must also be zero. Thus,

$$F_1 - F_2 - F_n \sin \alpha = 0 \tag{8.1}$$

$$P + \mu F_2 + \mu F_1 - F_n \cos \alpha = 0 \tag{8.2}$$

Taking the moments about the center of the roller:

$$F_1 l_1 - F_2(l_1 + l_2) - \mu F_1 \frac{d}{2} + \mu F_2 \frac{d}{2} = 0 \tag{8.3}$$

Solving for F_n:

$$F_n = \frac{P l_2}{l_2 \cos \alpha - (2\mu l_1 + \mu l_2 - \mu^2 d) \sin \alpha}$$

or

$$F_n = \frac{P}{\cos \alpha - \left(2\mu \dfrac{l_1}{l_2} + \mu - \mu^2 \dfrac{d}{l_2}\right) \sin \alpha} \tag{8.4}$$

From this F_n can be found for a given system.

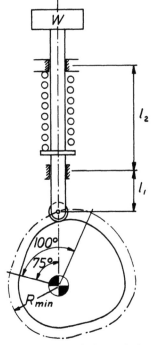

Fig. 8-6. Force analysis of radial translating roller follower.

The torque acting on the cam from a radial translating roller follower is:

$$T_0 = F_n \frac{y'}{\omega} \cos \alpha \qquad\qquad 8.5$$

(See the example later in this chapter).

Example: A radial translating roller follower system is shown in Fig. 8-6. The follower is moved with cycloidal motion over a distance of $h = 1$ in. and an angle of lift of $\beta = 100°$. The cam speed is $N = 900$ rpm. The spring constant is $k = 120$ lb/in. and the spring has an initial compression of 0.35 in. when the roller follower is in its lowest position. The weight of the mass to be moved (including the follower) is $W = 2$ lb., the diameter of the follower stem is $d = 0.75$ in., $l_2 = 4$ in., $l_1 = 1.5$ in. for the position shown, $R_{min} = 2.0$ in., and the coefficient of friction $\mu = 0.05$. Compute F_n and the cam shaft torque when the cam has rotated 75 degrees.

Solution: The equations for cycloidal motion are:

$$\left.\begin{array}{c} y = h\left[\dfrac{t}{T} - \dfrac{1}{2\pi}\sin\left(2\pi\dfrac{t}{T}\right)\right] \\[2mm] v = y' = \dfrac{h}{T}\left[1 - \cos\left(2\pi\dfrac{t}{T}\right)\right] \\[2mm] a = y'' = \dfrac{2\pi h}{T^2}\sin\left(2\pi\dfrac{t}{T}\right) \end{array}\right\} \quad 0 \leq t \leq T$$

where

$$T = \frac{60}{N}\cdot\frac{\beta}{360}$$

$$T = \frac{60}{900}\cdot\frac{100}{360} = \frac{1}{54}\ \text{sec.}$$

$$y_{75} = 1\left[\frac{75}{100} - \frac{1}{2\pi}\sin\left(2\pi\frac{75}{100}\right)\right]$$

$$= 0.75 + 0.159 = 0.909\ \text{in.}$$

$$y'_{75} = \frac{1}{1/54}\left[1 - \cos\left(2\pi\frac{75}{100}\right)\right] = 54\ \text{in./sec}$$

$$y''_{75} = \frac{(2\pi)(1)}{(1/54)^2}\sin\left(2\pi\frac{75}{100}\right) = -18{,}300\ \text{in./sec}^2$$

Because y'' is negative, the inertia force due to this acceleration actually decreases F_n; the inertia force is:

$$\frac{W}{g}y'' = \frac{(2)(18{,}300)}{386} = 94.7\ \text{lb. (upward)}$$

The spring compression at start of lift is 0.30 in. The spring force F_s is therefore:

$$F_s = k(0.30 + y) = 120(0.30 + 0.909) = 145 \text{ lb. (downward)}$$

The pressure angle α in the 75-degree position is found to be approximately 10 degrees by the method described in Chapter 5.

$$P = F_s - \frac{W}{g} y'' = 145 - 94.7 = 50.3 \text{ lb.}$$

$$F_n = \frac{P}{\cos \alpha - \left(2\mu \dfrac{l_1}{l_2} + \mu - \mu^2 \dfrac{d}{l_2}\right) \sin \alpha}$$

$$= \frac{50.3}{0.985 - \left[(2)(0.05)\left(\dfrac{1.5}{4}\right) + 0.05 - (0.05)^2 \left(\dfrac{0.75}{4}\right)\right] 0.174}$$

$$= \frac{50.3}{0.97} = 51.8 \text{ lb.} \tag{8.4}$$

The cam shaft torque is found from equation 10.5:

$$T_0 = F_n \frac{y'}{\omega} \cos \alpha$$

$$= 51.8 \cdot \frac{(54)(60)}{(900)(2\pi)} \cdot 0.985$$

$$T_0 = 29.3 \text{ lb-in.}$$

Recommendations for decreasing F_n:

Make α_m small (low max. pressure angle)
Make μ small (smooth surfaces, low friction materials)

Make the guidance ratio $\dfrac{l_2}{l_1}$ large

Make d large in order to avoid bending of the shaft; however, this will cause some additional frictional resistance to the motion of the shaft.

Let us now make a comparison of the forces for parabolic and cycloidal motion when the effect of pressure angles are considered, too. Although the comparison is made for translating follower cams, the results are, with a good approximation, also valid for cams with swinging followers.

In Fig. 8-7a the variation of the pressure angle for parabolic and cycloidal motion is shown. The maximum pressure angle for given rise h and time T is the same for both, because both curves have the same maximum velocity, and therefore they also have the same maximum slope; the slope was chosen so that $\alpha_m = 45°$.

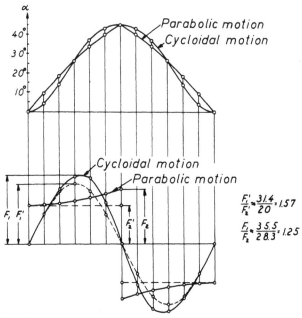

Fig. 8-7a. Variation of pressure angle over stroke for parabolic and cycloidal motion.
Fig. 8-7b. Variation of forces on roller over stroke for parabolic and cycloidal motion.

In Fig. 8-7b the dashed line and curve show the variation of the force due to acceleration for parabolic and cycloidal motion respectively. The theoretical maximum force ratio is:

$$\frac{F'_1}{F'_2} = 1.57$$

If, however, the effect of the pressure angle is considered, as shown by the full lines, the actual ratio becomes:

$$\frac{F_1}{F_2} = 1.25$$

Spring Forces

The follower is kept in contact with the cam by either form-connection (for instance, closed-track cams) or by force-connection, namely, gravity or spring forces. Gravity forces can only keep the follower in contact with the cam at low speeds.

When springs are used they have to be strong enough to counteract dynamic forces, friction forces, spring surge and keep the roller in firm contact with the cam so that the roller rolls and does not slide. On the

other hand, spring forces should be kept as small as possible in order that
the allowable Hertz's pressure is not exceeded. (See later section in this
chapter). Due allowances for variations in spring manufacture must be
made. To a large degree this can be mitigated by proper selection of springs
after the springs have been tested in a suitable manner. In many cases
closed-track cams are used because they will positively move the machine
parts; if springs were used they could fail and thereby cause considerable
damage to the machine.

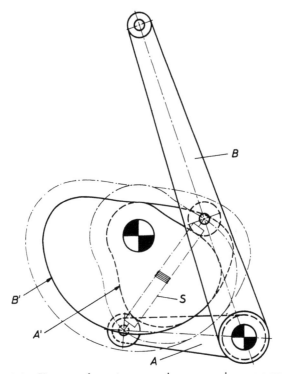

Fig. 8-8. Two complementary cams improve spring arrangement.

In some cases a positive drive is wanted in one direction but not necessarily in both directions. In such cases, the principle of Fig. 8-8 is often
worth considering. The two arms A and B are pulled toward each other
with the help of the spring S. Arm A is moved by cam A' and arm B by B'.
Cam B' imparts the desired output motion to arm B and arm A controlled
by A' is moved so that the spring S does not change its length. Only in
case that arm B is prevented from moving is the spring stressed. Arm B is
moved positively to the right by cam B' but to the left it is pulled back by
spring action only.

Radius of Curvature

This subject was treated in Chapter 5.

Table 8-1. Cam Materials

Cam Material for use with Roller of Hardened Steel	Maximum permissible compressive stress, psi
Gray-iron casting, ASTM A 48-48, Class 20, 160-190 Bhn, phosphate coated	58,000
Gray-ron coating, ASTM A 339-51T, Grade 20, 140-160 Bhn	51,000
Nodular-iron casting, ASTM A 339-51T Grade 80-60-03, 207-241 Bhn	72,000
Gray-iron casting, ASTM A 48-48, Class 30, 200-220 Bhn	65,000
Gray-iron casting, ASTM A 48-48, Class 35, 225-255 Bhn	78,000
Gray-iron casting ASTM A 48-48 Class 30 heat treated (Austempered) 255-300 Bhn	90,000
SAE 1020 steel, 130-150 Bhn	82,000
SAE 4150 steel, heat treated to 270-300 Bhn, phosphate coated	220,000
SAE 4150 steel, heat treated to 270-300 Bhn	190,000
SAE 1020 steel, carburised to 0.045 in. depth of case, 50-58 Rc	226,000
SAE 1340 steel, induction hardened to 45-55 Rc	200,000
SAE 4340 steel, induction hardened to 50-55 Rc	226,000

Based on United Shoe Machinery Corp. data by Guy J. Talbourdet.

Choice of Cam Material

In considering materials for cams it is difficult to select any single material as being the best for a practical application. Often the choice is based on custom and on the machineability of the material rather than its strength. However, the failure of a cam or roller is commonly due to fatigue, so that an important factor to be considered is the limiting wear load which depends upon the surface endurance limits of the materials used and the relative hardnesses of the mating surfaces.

In Table 8-1 are given maximum permissible compressive stresses (surface endurance limits) for various cam materials when in contact with a roller of hardened steel. The stress values shown are based on 100,000,000 cycles or repetitions of stress.

Where the repetitions of stress are considerably greater than 100,000,000, where there is appreciable misalignment, or where there is sliding, more conservative stress figures must be used.

Calculation of Stresses

When a roller follower is loaded against a cam, the compressive stress, developed at the surface of contact may be calculated from

$$S_c = 2290 \sqrt{\frac{F_n}{b} \left(\frac{1}{r_f} \pm \frac{1}{R_c} \right)}$$

for a steel roller against a steel cam. For a steel roller on a cast iron cam, use 1850 instead of 2290 in equation 8.6.

S_c = maximum calculated compressive strength, psi
F_n = normal load, lb
b = width of contact, in.
R_c = radius of curvature of cam surface, in.
r_f = radius of roller follower, in.
E_1 = modulus of elasticity of the cam material
E_2 = modulus of elasticity of the follower roller.

For steel on steel ($E_1 = E_2 = 30 \times 10^6$ psi) and what may be called the load factor equals 19.05×10^{-8}. The equation then becomes:

$$S_c = \left(\frac{F_n}{b} \left(\frac{1}{r_f} \pm \frac{1}{R_c} \right) \frac{1}{19.05 \cdot 10^{-8}} \right)^{\frac{1}{2}}$$

The plus sign between R_c and r_b is used in calculating the maximum compressive stress when the roller is in contact with the convex portion of the cam profile and the minus sign is used when the roller is in contact with the·concave portion. When the roller is in contact with the straight (flat) portion of the cam profile, $R_c = \infty$ and $1/R_c = 0$. The above equations can be solved with the help of the nomograms in Figs. 8-9, 8-10 and 8-11. Figure 8-9 is used when the roller is in contact with a convex portion of the cam profile; Fig. 8-10 when it is in contact with a concave portion; and Fig. 8-11 when in contact with a straight (flat) portion. Actually, the greatest compressive stress is most apt to occur when the roller is in contact with that part of the cam profile which is convex and has the smallest radius of curvature.

Example 1: Given a cam in which the radius of the roller $r_f = 0.25$ in., the minimum convex radius of the cam $R_c = 0.35$ in., the width of contact

estimated realizing that from top to bottom the guide lines between the K_1 and K_2 axes become progressively more slanting). The value obtained using the dimensions and force given is $S_c = 40,000$ psi.

All of the steels listed in Table 8-1 have maximum permissible compressive stress values that are greater than 40,000 psi. From the standpoint of surface endurance limits any of the steels given in this table can be used. However, if unfavorable conditions such as misalignment, a higher degree of sliding, shock loads, etc. are present, a safety factor must be applied.

Example 2: To illustrate the use of the nomograms in Figs. 8-10 and 8-11 to find the allowable normal load, assume that the maximum permissible compressive stress is 40,000 psi and the roller and width-of-contact dimensions are the same as in Example 1. What would be the permissible normal load for contact of the roller with the concave part of the cam profile if it has the same radius of curvature as given for the convex portion in Example 1? What would be the permissible normal load for contact of the roller with the straight part of the cam?

Solution: Using the nomogram in Fig. 8-10 we find that the permissible normal load for contact with the concave part of the profile would be $F_n = 45$ lbs. Using Fig. 8-11 we find that for contact with a straight portion of the cam profile the permissible normal load would be $F_n = 15$ lbs.

Factors Affecting Wear Life

It is pertinent at this point to discuss the effect insofar as wear life is concerned of the kind of contact between the cam and the follower. When a cylindrical roller is used on a plate cam it is assumed that line contact exists between the roller and the cam and that the roller rolls without sliding. Often, misalignment will exist so that there is point rather than line contact. This will increase wear. But misalignment can also take place in such a way that the axis of the roller is parallel to the cam surface but no longer parallel with the shaft; this will cause the roller to eventually roll on the cam at a certain point but slide on it at all other points where contact occurs.

Another factor which can also cause sliding is the varying circumferential speed of the cam as it passes under the roller. If the variation is too large or rapid, the tangential force between the roller and the cam may not be large enough to rotate the roller with the angular acceleration required and sliding occurs. One remedy is, of course, to make the cam smaller (see Chapter 5.)

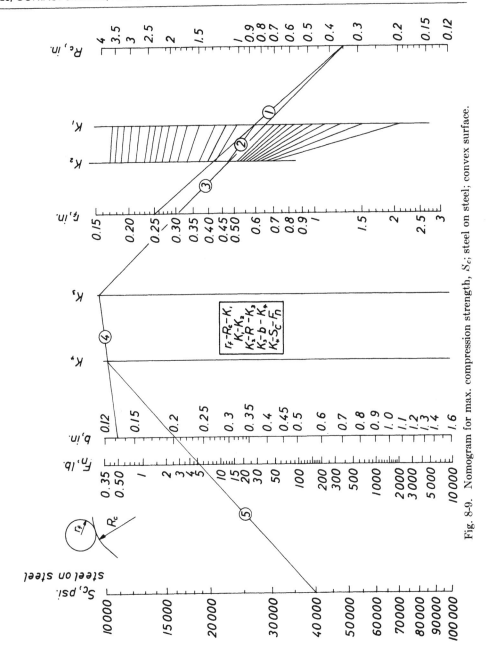

Fig. 8-9. Nomogram for max. compression strength, S_c; steel on steel; convex surface.

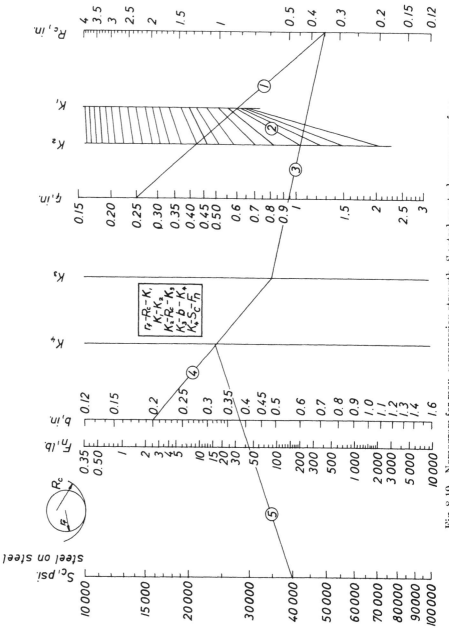

Fig. 8-10. Nomogram for max. compression strength, S_c; steel on steel; concave surface.

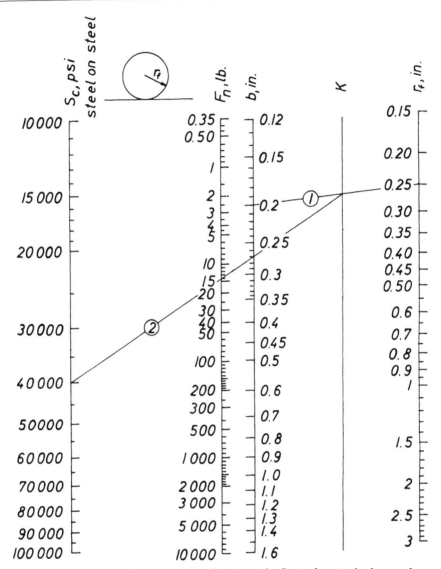

Fig. 8-11. Nomogram for max. compression strength, S_c; steel on steel; plane surface.

$b = 0.132$ in. and the normal load $F_n = 5.5$ lbs. Find the maximum surface compressive stress. Both cam and roller are made of steel.

Solution: Use the nomogram in Fig. 8-9. (Note that in drawing line No. 2 from the intersection of line No. 1 with the K_1 axis to the K_2 axis, its direction or slant will be between the slant of the guide line directly above it and the slant of the guide line next below it. This direction has to be

When a roller follower is used, the use of lubricant has been found to cause the roller to slide. Although this is an undesirable factor, nevertheless, a modest amount of lubricant is recommended because this will protect the cam surface from galling. If the follower is of a sliding type, lubrication of the surface is recommended.

Because misalignment almost always exists, a roller with a slightly spherical surface is a good choice. The roller followers are commercially available from a number of companies and are usually in the form of a roller or needle bearing on a stud.

For given cam proportions the lowest max. compressive stress occurs if, at the point of max. compressive stress at the convex portion of the cam, the radius of curvature of the cam equals the roller radius r_f.

Methods of Cam Manufacture

It has been shown in Chapter 4 how to obtain the cam profile by calculation, i.e., in numerical form, or by drawing it on paper. The problem we are now faced with is how can the desired curve be cut in metal? It is obvious that if the curve can be cut into metal without any intermediate steps, i.e., no master cam is used, then an accuracy is obtained which is dependent only on the device necessary to cut the cam.

Actually, there is only one kind of curve that can be cut directly with high accuracy and that is a circular cam.

Circular Cams (See also Chapter 6)

Figure 9-1 shows a four-bar linkage, A_0ABB_0, where the crank $A_0A = a$, the coupler $AB = b$, the rocker $B_0B = c$, and the frame $A_0B_0 = d$. Figure 9-2 shows a circular cam, derived from this four-bar linkage, where

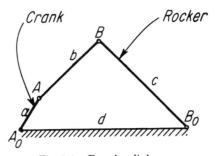

Fig. 9-1. Four-bar linkage.

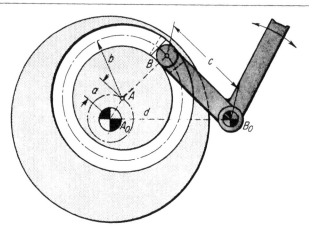

Fig. 9-2. Kinematic equivalent mechanism.

the center A of the circular groove is displaced a distance a from A_0, the center of cam rotation, and where the radius of the center line of the circular groove is b, and where $B_0B = c$. This cam will impart exactly the same motion to the output member B_0B as the four-bar linkage to the rocker B_0B. The circular groove can be cut on a turret lathe to close tolerances with an excellent surface finish. For low speed operation, higher forces can be transmitted by replacing the roller with a ring sector shaped sliding piece as indicated by B in Fig. 9-2.

The reason why this kind of cam is employed so little is that very few are aware of the possibility of using it and besides the dimensioning has ɔeen rather time consuming. However, Chapter 6 shows how, without too much difficulty, this kind of mechanism can be proportioned for optimum performance.

Transferring the Drawing Directly to the Steel Plate

In Chapter 4 it was shown how the cam contour could be drawn. A method frequently used is to transfer the drawing of the cam contour directly onto a steel plate. This is done by placing the drawing on the steel plate and pricking holes along the cam contour so that when the drawing is taken away the contour of the cam appears clearly outlined on the steel plate. Now by combined sawing and filing a master cam is made which can then be used to manufacture the desired number of cams by profile copying.

For packaging and similar machines running at 150–200 rpm this method of making cams has proven successful, even though parabolic curves with their inferior dynamic characteristics as compared with cycloidal motion, for instance, were used. It should be noted that machines running at 150–200 rpm often make a great deal of noise and without any doubt the use of cams with cycloidal profiles could improve this condition, even if the method of manufacturing were the same. The only advantage in designing for parabolic motion is that the points can be easily calculated with the help of a slide-rule. (See Chapter 3).

Drawing of the Profile Directly on the Steel Plate

This method is used with advantage when the cam profile is composed of circular arcs and straight lines (Chapter 7). The cam profile is laid out exactly as if it were drawn on a piece of paper and then by sawing and filing, the master cam is made.

Special Manufacturing Technique for Cam Profiles Composed of Circular Arcs and Straight Lines

For comparatively small cams the technique described here has proven quite successful. Figure 9-3 shows a small cam, the contour of which is composed of circular arcs and straight lines, namely, a circular arc $1–2$ with center at O_2, arc $3–4$ with center at O_3, arc $4–5$ with center at O_1, arc $5–6$ with center at O_4, arc $7–1$ with center at O_1, and the tangents $2–3$ and $6–7$.

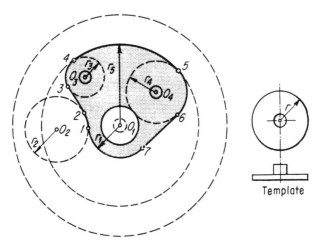

Fig. 9-3. Cam contour composed of circular arcs and straight lines.

First, small holes of about 0.1 in. diameter are drilled at points O_1, O_3, and O_4, and then a hole of diameter $2r_2$ is drilled with its center at O_2. Next, the cam is fixed in a turret lathe with the center of rotation at O_1, and the steel plate is cut until it has a diameter of $2r_5$. Next, hardened pieces are placed at O_1, O_3, and O_4. One such hardened piece is shown to the right in Figure 9-3; it has a small pin, which fits into a hole (O_1, O_3, or O_4) and has a radius r (r_1, r_3, or r_4). One such piece is placed in O_1 and has a radius of r_1, another piece in O_3 with a radius of r_3, etc. Now the arcs *7–1*, *3–4*, and *5–6* can be filed using the hardened pieces as a guide. Then the straight lines *6–7* and *2–3* are cut or machined and the final operation is to drill the hole at O_1, to such a size that a hub can be fastened to the cam. There is no doubt that this method compares favorably with the method of tracing from a drawing. It is also possible to use a compound rotary table to offset centers and then mill part of the contour by turning the cam around the offset center.

Precision-point Cam Manufacture

A combination of drilling and filing is often used to produce precision-point cams. Points on the pitch curve are accurately calculated and at these points holes are milled with diameters equal to that of the follower roller. Obviously the accuracy of the resulting cam depends on the close-ness of the center-to-center distance between the holes. After the hole milling operation, the resulting serrated profile, Fig. 9-4, must be filed away until the desired smoothness is achieved.

The method of calculating points on a cam curve is shown in Chapter 4. The points can be calculated using either rectangular or polar coordinates. Rectangular coordinates are used when a coordinate miller is at hand and if a dividing head is available, polar coordinates can be used.

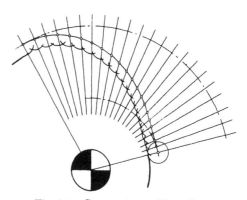

Fig. 9-4. Cam contour with scallops.

In general, a larger number of precision points are used for large cams and fewer points on small cams. On large cams the requirement is often that the follower be at a predetermined spot at a certain time. In some cases, the speed of small cams is often most important and therefore a large number of calculated points may be required.

The following is an excerpt of an article by Raymond E. Cheney, "High-speed Master Cams Generated Mechanically", *Machinery*, October 1961, pp. 93–99, 148. In this article a cutting technique for producing a soft master cam with high accuracy is described together with a grinding device to grind a hardened master cam from the soft cam.

"After careful experimentation and process refinement, the following procedure was adopted as a standard method for cutting the original active and complementary master cams:

"Special shop-design cam-blanks are machined on a conventional lathe. For complementary systems, two similar blanks are prepared for eventual mounting as a pair, to be successively contoured in a single setup. The blanks usually have a face width of approximately ⅛ to ³⁄₁₆ inches. A high-grade AISI 4150 steel, toughened to Rockwell C 28 to 33, was found to produce the best end result. Preparation of the mounting and timing index holes in the blanks is directly reflected in the final accuracy of the master cam. Therefore, the use of honing techniques to prepare the mounting is absolutely essential to obtaining maximum precision and control of the process.

"Both the active and complementary blanks are mounted together on a precision nut arbor carried vertically in an indexing device capable of accuracy within 2 seconds of arc. The indexing device is mounted in an optical jig borer, with its nut arbor zeroed in with the jig-borer spindle prior to mounting the cam

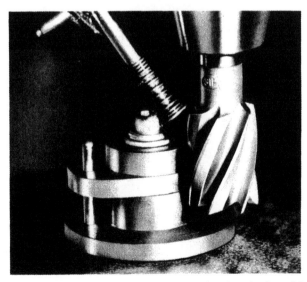

Fig. 9-5. Cam-and-cutter relationship, front view, showing the honed indexing hole arrangement. Sperm oil mist is the cutting lubricant.

master blanks, Fig. 9-5. Optical equipment is recommended for this operation because the table direction can be reversed without backlash compensation, and can also be incrementally set rapidly.

"A multiple-fluted (six-flute preferred), side-cutting stub end mill, concentrically ground to the cam-follower size diameter and carefully mounted in the jig-borer spindle, is used to machine the cam master, Fig. 9-6. The heavier the cutter and the shorter the tool flutes, the more accurate it will cut. If the cam-follower size is less than $\frac{1}{2}$ inch in diameter, it is advisable to have the cam calculated to be cut with a larger-diameter cutter. In such cases, two sets of figures (radial and index) can be provided by a computer. One group of figures is used to cut the cam with an oversize cutter; the other set is used to inspect the completed cam with a regular follower-size probe.

"In operation, and in reference to a complementary cam pair (this also applies in part to a single cam master), the table is moved clear of the cutter and the spindle lowered past the face of the top blank and to a point just above the face of the lower blank, as in Fig. 9-5. The blanks to be cut must be at least 0.040 inch in diameter greater than the finished diameter of the cam at its highest point.

"With the tool turning at approximately 300 rpm (for a $\frac{3}{4}$-inch diameter cutter), the work is advanced into the side of the cutter to a depth of about 0.010 inch. Sperm oil is the lubricant, delivered in a vapor spray to the point of cutting. To determine the actual cutting size of the spiral mill, a complete revolution of the blank is cut by slowly and steadily hand-feeding the index-table. This trial cut is made at the beginning of each set-up to insure actual cutter orientation. Measuring the blank diameter thus cut, and relating it to the zero setting on the jig-borer table, indicates any required correction of the radial zero reference.

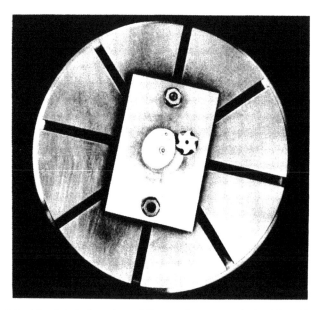

Fig. 9-6. Simulated plan view of the cam-to-cutter relationship. This is a "spindle-eye" view, looking down the axis of the cutter, from above.

"The top cam-blank can now be cut. It should be roughed about 0.010 inch oversize and then finished to final size by an additional rotation of the indexing table, Fig. 9-7. The roughing cut may require some rough milling, depending on the final cam shape. Plunge cutting will remove large areas of the blank stock if desired. The final roughing cut should start at the index position of the greatest radial distance. With the cutter turning and lowered to its original setup position, the jig-borer table should advance the blank into the cutter 0.010 inch less than the finish position for that index point. This will leave 0.010 inch of material at this point to be removed by the finishing cut. With computer data properly arranged to match the rotation of the indexing table, the blank should be rotated to the next full-degree index position and stopped. The jig-borer table is then advanced into the cutter to the radial dimension for that index position with the 0.010 plus material factor included. Then, with the jig-borer table resting at this position, the dividing head is indexed to the next full-degree position and rested. Again the jig-borer table is advanced into the cutter to radial distance for the index position with a 0.010 plus material factor withheld. This process is repeated for each full degree of descending radial values. At any point on the cam where the radial values do not change (as in the case of a dwell) rotation of the index-table continues until a change in radius is required.

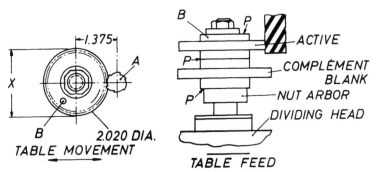

Fig. 9-7. Blank setup and machine orientation procedure involves taking a preliminary cut (broken line) on the actual blank in order to establish the precise cutting size of the end mill. Diameter X should measure 2.0000 inches after the trial cut. If it does not, the zero radial reference dial should be corrected by one-half the difference between 2.0000 and the actual measurement of diameter X. In the sketches, A is the expected 0.750-inch cutting diameter, B is the working pin, and P signifies surfaces that are ground and parallel.

"At any position where the next radial distance is increased, or at the start of a "rise" in the cam curve, it is essential that the sequence of tool and work feed be reversed. For example, if at the end of a dwell the next radial distance is greater, the dwell-distance feeding of the indexing table should be stopped at the last index point in the dwell. The jig-borer table is withdrawn from the cutter a distance sufficient to meet the next radial point. Then the blank is index-advanced to that point. This, of course, requires a machine that does not have to compensate for table lead-screw backlash. This process is continued around the cam until the entire roughing cut is finished.

"The finish-size contour is accomplished in the same manner, using all of the index positions supplied, while exercising even greater care to insure smooth feed and precise setting of the jig-borer table and the index-table. Actual results have proved it is more accurate and more economical to make this sequence a two-man operation. One man indexes the blank while the other man advances the table and reads the script aloud.

"After machining of the top blank has been completed, the timing index hole is carefully drilled and jig-bored through both blanks and into a shoulder of the nut arbor. The top blank can then be removed. The lower cam-blank is dowel-pinned into the arbor to prevent angular slippage during both the removal of the top cam and subsequent rough machining. Now the lower cam-blank can be cut in the same manner as the top cam, starting from the same zero orientation.

"Prototype cams generated in this manner can be held to tolerances well within 0.0001 inch in point-to-point radial variation, and well within 0.0005-inch total variation. Such cam masters will have a surface finish between 8 and 11 micro-inches, just as they come off the jig borer. To rough and finish the average single cam by this procedure, and cutting every $\frac{1}{2}$ degree of rotation, using two men as described, requires about four hours' elapsed time, for a total of eight man-hours. By other methods that would achieve the same degree of accuracy, the required time would be at least forty man-hours.

"The third phase in the cam-making program is to produce the hardened, finished prototype cam for use in the model machine. The narrow-faced, relatively soft, high-precision master cam produced on the jig borer must control, without appreciable loss, the finishing of the final hard cam.

"At the same time this method was being developed, no suitable equipment could be purchased to carry out this operation in a satisfactory manner. Most commercially available equipment used pantographic linkage to control the grinding wheel, or else was controlled by a master and simple follower system too remote from the grinding wheel to have the sensitive response desired. Therefore, the laboratory developed a cam duplicator that not only meets specifications, but produces a blended cam surface actually better than the original jig-bored master.

"A survey of data concerning high-speed cam designs that had met acceleration requirements revealed no case where the radius of curvature was less than 1 inch. This meant that rarely, if ever, would a grinding wheel of less than 2 inches in diameter be required to finish-grind a cam if the grinding axis were held parallel to the cam axis during the grinding operation.

"On this basis, a toolroom type cam-master duplicating machine was designed and built incorporating these functional specifications:

1. One-to-one duplicating ratio;
2. Duplication follower diameter and corresponding grinding wheel diameter between 2 and 4 inches;
3. Direct coupling between the follower and the wheel, with their axes parallel;
4. Common counterbalancing and adjustable gravity loading of grinding wheel and follower;
5. Common shaft for adjacent mounting of master and part;
6. An instrumentation system to dynamically compare the master with the part during the grinding operation.

"Incorporating high precision and a fundamentally stable design resulted in a cam grinder that meets these six specifications, while providing other desirable work features. The unit is a highly sensitive machine capable of exacting performance with excellent repeatability. The oversize follower, Fig. 9-8, by bridging the control points on the master with a larger circumferential surface, controls the finish-grinding operation with blended continuity better than the original master. The result is a smoother contour and better surface finish.

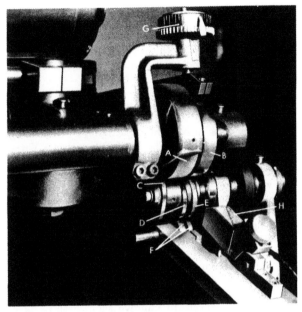

Fig. 9-8. Close-up of the high-precision cam-grinding arrangement showing the electronic comparator probe operating from the reeds touching the tracking rollers.

Fig. 9-9. Side view of the cam grinder in operating alignment showing the two drives and control adjustment. The large knurled knob at the rear is the vernier for fine grinding.

"Of crucial importance is absolute control of the timing hole location with angular reference to the cam contour phase. This is essential in any complementary system using separate cams. Angular precision is achieved during grinding by mounting the original cam master and the blank on a common shaft, with the timing holes in each oriented by a common pin or dowel. Shaft and dowel holes are carefully honed to fit. This setup is shown, wheel raised, in Fig. 9-9.

"With the constant comparing instrument, the effect of wheel wear, loading, and other cutting variables can be monitored and, therefore, closely controlled. Experience has proved this type of device to be remarkable insensitive to minor diameter variations in wheel-to-follower sizes. Variations of 0.020 to 0.030 inch in wheel diameter can usually be tolerated without detectable loss of accuracy. It is true, however, that cams having severe pressure angles call for closer control of wheel size. The company's experience to date indicates that in all cases the available control of wheel size on this machine is satisfactory.

"After inspection, the cams thus made are used in the engineering of prototype calculating machines and computers. Hardened cams produced in this phase may also be used as production masters for cam manufacture and also as inspection masters to provide production quality control for production cams.

"This procedure for making superlative prototype and master cams is still being refined and improved. Engineers desire cams accurate to the fifth decimal place, because such parts, it is hoped, will minimize fretting and noise. IBM engineering expects to improve shop methods of cam fabrication to such an extent that this goal will be reached."

Cam Manufacture Using Numerically Controlled Machines

With the advent of numerically controlled machines cam manufacture is no longer an art but only a question of programming the right profiles and the correct dimensions.

In this book we have emphasized the proportioning of cam mechanisms from a geometric point of view namely to proportion the cams so that the lowest possible pressure angles are obtained. We have also shown how to calculate radius of curvature in order to check for undercutting and too large surface stresses.

Once these conditions are fulfilled the next thing is to cut the profile in metal and if the profiles are given in mathematical terms (which is most often the case) then there are several cam manufacturers in the USA who can do the job from there. See for instance Thomas Register where there are more than 200 companies specializing in the manufacture of cams.

The production time and the accuracy obtainable are excellent: for instance an accuracy of $\pm .0001$ and a production time of only 15 minutes with a minicomputer coupled to machine tools. Data was typed directly into terminal and cam is produced directly from data.

Cam Copying Machines

When a cam has been made by one of the previously mentioned methods it is possible to install this cam directly in the machine where it is to be used, or it can be used as a master for the duplication of the actual working cams.

It is also possible to use a drawing as a master. This is done with optically controlled devices, but the method is not recommended for high speed cams. For low speed cams the method is economical.

Checking of Cams

Comparison: The master cam and the production cam are mounted in a device, rotated together, and an indicator arrangement records the actual deviation of the two profiles. Such an arrangement was shown in Fig. 9-9.

Measuring (Indicator): The profile of the cam to be checked is mounted on a shaft and the polar coordinates are found by rotating the cam a certain angle and for each angle reading the indicator. This method is not suitable for production inspection and does not tell anything about the dynamic behavior of a profile which deviates from the master profile.

Use of the method of finite differences will provide information about the acceleration characteristics (See Chapter 12).

The electronic method uses a velocity type pick-up as the cam is being rotated at a constant speed. By means of differentiating circuits the acceleration curve can be obtained.

Dynamics of Cam Mechanisms

In the case of cams rotating at high speed the factors of elasticity and backlash must be taken into consideration if vibration and impact loads are to be avoided or minimized. One way of doing this is to use the polydyne cam design method. This method was originally applied to automotive cam systems, and consists essentially of two parts: (1) The use of polynomials to obtain a wide variety of curves having certain characteristics, and (2) modification of the polynomials to take into account any elasticity of the system which causes a discrepancy between the cam "command" and its follower mass response.

As was shown in Chapter 3, the basic curves such as parabolic, cycloidal, etc., can be combined to obtain certain more desirable cam characteristics. These characteristics can also be obtained with curves based on polynomial equations. The kind of polynomial equations to be used is dependent upon whether the cam curve is of the type: Dwell-Rise-Dwell-Return, Dwell-Rise-Return, or Rise-Return.

Dwell-Rise-Dwell-Return (DRDR)

Unity displacement in unity time is used in the following, but to obtain follower displacement y at a given angle of rotation θ or time t for a cam with maximum displacment h and corresponding angle of rotation β or time T, multiply the right hand side of the equation by h and replace x by

$$\frac{\theta}{\beta} \quad \text{or} \quad \frac{t}{T}$$

In Fig. 10-1 it is desired to connect points A and B with a cam displacement curve. Let us start with two boundary conditions:

when:
$$x = 0; \quad y = 0$$
$$x = 1; \quad y = 1$$

Because two boundary conditions are given, a polynomial with two terms is employed:

$$y = C_0 + C_1 x \qquad 10.1$$

Substituting the boundary condition into equation 10.1:

$$0 = C_0 + C_1(0); \quad C_0 = 0$$
$$1 = C_0 + C_1(1); \quad C_1 = 1$$

or $y = x$, which is the equation for a straight line, Fig. 10-1, and represents constant velocity motion.

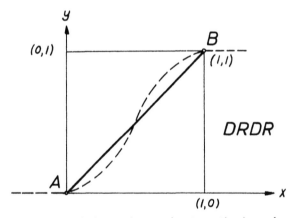

Fig. 10-1. Constant velocity motion; resultant equation is a polynomial of low order and inferior dynamic characteristics.

As has already been shown, this is a dynamically poor curve for a cam contour. Let us therefore impose two further boundary conditions, thus:

when
$$x = 0; \quad y = 0, \quad y' = 0$$
$$x = 1; \quad y = 1, \quad y' = 0$$

Because four boundary conditions are used, a polynomial with four terms is employed:

$$y = C_0 + C_1 x + C_2 x^2 + C_3 x^3$$
$$y' = C_1 + 2C_2 x + 3C_3 x^2 \qquad 10.2$$

Substituting the boundary conditions into equation 10.2:

$$0 = C_0 + C_1(0) + C_2(0)^2 + C_3(0)^3; \quad C_0 = 0$$

$$0 = C_1 + 2C_2(0) + 3C_3(0)^2; \quad C_1 = 0$$

$$1 = C_2(1)^2 + C_3(1)^3 \quad C_2 = 3$$

$$0 = 2C_2(1) + 3C_3(1)^2 \; ; \quad C_3 = -2$$

or
$$y = 3x^2 - 2x^3$$

$$y' = 6x - 6x^2$$

$$y'' = 6 - 12x = \begin{cases} 6 \\ -6 \end{cases} \quad \text{for} \quad x = \begin{cases} 0 \\ 1 \end{cases}$$

Although the resulting curve, Fig. 10-1 (dashed line), is much better for cam design than the constant velocity curve, there is a sudden change in acceleration at the beginning and at the end of the motion. In order to provide further modification, two more boundary conditions are added:

when
$$x = 0; \quad y = 0, \quad y' = 0, \quad y'' = 0$$

$$x = 1; \quad y = 1, \quad y' = 0, \quad y'' = 0$$

For this, a polynomial with six terms is used:

$$y = C_0 + C_1 x + C_2 x^2 + C_3 x^3 + C_4 x^4 + C_5 x^5$$

$$y' = C_1 + 2C_2 x + 3C_3 x^2 + 4C_4 x^3 + 5C_5 x^4$$

$$y'' = 2C_2 + 6C_3 x + 12C_4 x^2 + 20C_5 x^3$$

and substituting the boundary conditions we get:

$$y = 10x^3 - 15x^4 + 6x^5 \tag{2.10}$$

This is the equation for a 3–4–5 polynomial curve (2-10) and takes its name from the powers of x that are involved (See Fig. 2-9).

So far we have controlled the *acceleration* at the end points and one could ask whether there is any value in going further; i.e., controlling y''', y'''', etc. As we will see later in this chapter, it is important in certain cases.

A simplified procedure for calculating the coefficient will now be applied to the following boundary conditions:

$$x = 0; \quad y = 0, \quad y' = 0, \quad y'' = 0, \quad y''' = 0$$

$$x = 1; \quad y = 1, \quad y' = 0, \quad y'' = 0, \quad y''' = 0$$

and the corresponding polynomial

$$y = C_0 + C_1 x + C_2 x^2 + C_3 x^3 + C_4 x^4 + C_5 x^5 + C_6 x^6 + C_7 x^7 \quad 10.3$$

If $\dfrac{d^n y}{dx^n} = 0$ when $x = 0$, then $C_n = 0$, and the following tabulation can be made:

Table 10-1

Initial Powers	Zero Derivatives (at $x = 0$)	Final Powers
0 to 1	. . . .	1
0 to 3	y'	2–3
0 to 5	y', y''	3–4–5
0 to 7	y', y'', y'''	4–5–6–7

Therefore, the polynomial equation 10.3 can now be written:

$$y = C_4 x^4 + C_5 x^5 + C_6 x^6 + C_7 x^7$$

The coefficients of this polynomial can be evaluated with the help of the following general formulas:

when

$$y = C_p x^p + C_q x^q + C_r x^r + C_s x^s \ldots$$

then

$$C_p = \frac{qrs \ldots}{(q-p)(r-p)(s-p) \ldots}$$

$$C_q = \frac{prs \ldots}{(p-q)(r-q)(s-q) \ldots}$$

$$C_r = \frac{pqs \ldots}{(p-r)(q-r)(s-r) \ldots}$$

$$C_s = \frac{pqr \ldots}{(p-s)(q-s)(r-s) \ldots}$$

Example:

If

$$y = C_4 x^4 + C_5 x^5 + C_6 x^6 + C_7 x^7$$

then

$$C_4 = \frac{(5)(6)(7)}{(5-4)(6-4)(7-4)} = 35$$

$$C_5 = \frac{(4)(6)(7)}{(4-5)(6-5)(7-5)} = -84$$

$$C_6 = \frac{(4)(5)(7)}{(4-6)(5-6)(7-6)} = 70$$

$$C_7 = \frac{(4)(5)(6)}{(4-7)(5-7)(6-7)} = -20$$

and
$$y = 35x^4 - 84x^5 + 70x^6 - 20x^7$$

This is the equation for a 4–5–6–7 polynomial curve. A comparison of the displacements for this curve with those of other often used curves is shown in Fig. 10-2.

The foregoing process can be easily varied to obtain a desired characteristic of velocity or displacement. A simple method of progressively varying the powers and plotting the displacement or velocity curve is used until the

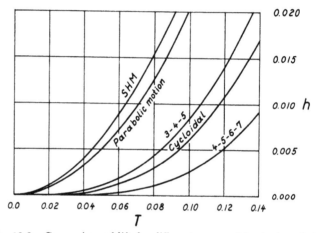

Fig. 10-2. Comparison of lift for different curves at beginning of rise. Range shown is $0.14T$ and $0.02h_1$ where T = time for total life and h = maximum displacement of follower.

desired curve is approached with sufficient accuracy. An example illustrates the technique. Assume that the same boundary conditions set up for the 4–5–6–7 polynomial are again employed. That is, when $x = 0$; $y = 0, y' = 0, y'' = 0, y''' = 0$. When $x = 1$; $y = 1, y' = 0, y'' = 0$. Then, with the known simplification of the 4–5–6–7 polynomial expressed in generalized form.

$$y = C_p x^p + C_q x^q + C_r x^r + C_s x^s$$

But, let p, q, r, and s have different sets of values, such as:

Table 10-2

Curve	p	q	r	s
1	4	5	6	7
2	4	6	8	10
3	4	7	10	13
4	4	8	12	16

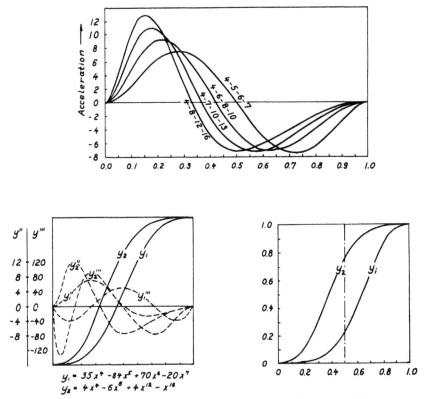

Fig. 10-3a. Acceleration curves for different polynomials. Fig. 10-3b. Displacement, acceleration, and 2nd acceleration curves for the 4-5-6-7 and the 4-8-12-16 polynomials. Fig. 10-3c. A new curve y_i can be obtained by rotating the curve y_2 180°.

The evaluation of coefficients proceeds exactly as before and the resulting displacement, acceleration, and jerk curves are plotted in Figs. 10-3a and b. The effect of higher powers are easily observed. Thus, the higher the powers, the faster the rise (or return) and the greater the maximum acceleration and jerk, both of which shift towards the start of the rise (or end of return).

Curves of the same powers can be built up on the other side, i.e., to the right of the 4–5–6–7 curve by inverting the calculated points or by turning the y_2-curve 180 degrees so that the start and end of rise are inverted. That is, $x_i = 1 - x$ and $y_i = 1 - y$ where x and y are calculated coordinates and x_i and y_i are the inverted coordinates. Such inversion is shown for the 4–8–12–16 curve in Fig. 10-3c. By inverting the y_2 curve to the y_i curve a slower motion at the beginning of the rise is obtained. The process of power selection can continue ad infinitum, of course, and in doing this there need be no fixed increments between the powers. However, the application of powers according to some systematic plan facilitates interpolation

towards the powers which will most closely yield the desired displacement curve.

Just as the foregoing example demonstrates variations of the 4–5–6–7
polynomial, so can similar methods be used with the 3–4–5 or any basic
polynomial of more terms.

Dwell-Rise-Return (DRR)

It is required again that displacement, velocity, and acceleration must
be zero at the boundaries, but at the end of the rise, velocity should be
zero, displacement 1, and acceleration unspecified:

$$x = 0; \qquad y = 0, \qquad y' = 0, \qquad y'' = 0$$

$$x = 1; \qquad y = 1, \qquad y' = 0, \qquad y''' = 0$$

(No abrupt change of acceleration at the end of rise.)   Also the curve for
the return is mirrored about $x = 1$.

$$y = C_0 + C_1 x + C_2 x^2 + C_3 x^3 + C_4 x^4 + C_5 x^5$$

which reduces to:

$$y = C_2 x^2 + C_4 x^4 + C_5 x^5$$

Using the general formulas previously given, the coefficients can be
evaluated, and:

$$y = \frac{10}{3} x^2 - 5x^4 + \frac{8}{3} x^5 \qquad\qquad 10.4$$

Equation 10.4 is shown as a curve in Fig. 10-4.

It is seen that maximum negative acceleration is much greater than
maximum positive acceleration.   To change this, the powers must be
altered.   The second power should not be altered, for the term containing
it defines acceleration at $x = 0$.   Neither should any terms whose coefficients were found to be zero from the given conditions be re-instated.
Hence, in this case, power changes must be confined to the last two terms.
As a trial, which proves to be satisfactory, let the powers 2–4–5 be replaced
by 2–5–6; that is,

$$y = C_2 x^2 + C_5 x^5 + C_6 x^6, \text{ Fig. 10-5}; \qquad a_{max} = \begin{cases} 5.24 \\ -5 \end{cases}$$

Figure 10-6 shows the curves resulting from progressive changes in
powers for a group of higher-order polynomials.   The conditions for each
of these curves are:

$$x = 0; \qquad y = 0, \qquad y' = 0, \qquad y''' = 0$$

$$x = 1; \qquad y = 0.35, \qquad y' = 0, \qquad y'' = 0, \qquad y''' = 0, \qquad y'''' = 0$$

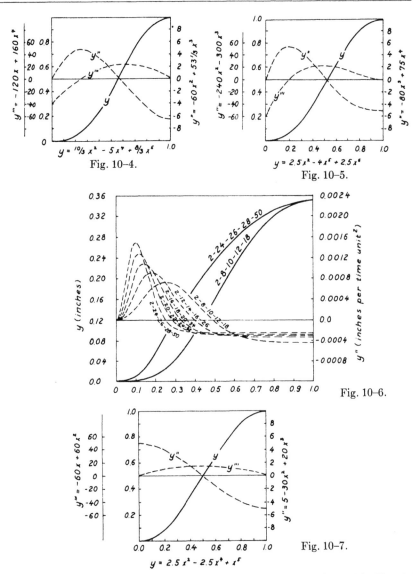

Fig. 10-4.

Fig. 10-5.

Fig. 10-6.

Fig. 10-7.

Fig. 10-4. Acceleration and 2nd accleration curves for the 2-4-5 polynomial. Fig. 10-5. Acceleration and 2nd acceleration curves for the 2-5-6 polynomial. Fig. 10-6. Displacement and acceleration curves for different polynomials. Fig. 10-7. Acceleration and 2nd acceleration curves for a 2-4-5 polynomial.

The solid line curves represent displacement and the dashed curves, acceleration. These curves characterize the type successfully adapted to automotive cam design and can be adapted to other special cases; for instance, textile machine where the cams are operated at high speed.

Rise-Return (RR)

In Chapter 6 it was shown how a cam, having a circle as profile, can be proportioned and it was stated that this was the best profile of all, resulting in better contour accuracy, surface smoothness, and also a profile that could be produced economically.

When it is not possible to use this profile—and the charts in Chapter 6 quickly give information about this—other cam contours can be used and polynomials are again convenient for determining these. Thus, given the boundary conditions:

$$x = 0; \qquad y = 0, \qquad y' = 0, \qquad y''' = 0$$
$$x = 1; \qquad y = 1, \qquad y' = 0, \qquad y''' = 0$$

Substituting the boundary conditions into the following equations:

$$y = C_0 + C_1x + C_2x^2 + C_3x^3 + C_4x^4 + C_5x^5$$
$$y' = C_1 + 2C_2x + 3C_3x^2 + 4C_4x^3 + 5C_5x^4$$
$$y'' = 2C_2 + 6C_3x + 12C_4x^2 + 20C_5x^3$$
$$y''' = 6C_3 + 24C_4x + 60C_5x^2$$

yields: $y = 2.5x^2 - 2.5x^4 + x^5$ 10.5a

The displacement y, acceleration y'' and jerk y''' curves for these equations are shown in Fig. 10-7.

Example: Use equation 10.5a to find the equations for rise and return if $h = 2$ in., $\beta_1 = 150°$, and $\beta_2 = 210°$.

Solution: Equation 10.5a is the equation for rise in the interval $0 \leq x \leq 1$. However, the return takes place in the interval $-1 \leq x \leq 0$. By subtracting the right-hand side of equation 10.5a from 1, it is transformed into:

$$y = 1 - 2.5x^2 + 2.5x^4 - x^5 \quad \text{for} \quad -1 \leq x \leq 0 \qquad 10.5b$$

The equation for rise is:

$$y = 2\left[2.5\left(\frac{\theta}{150}\right)^2 - 2.5\left(\frac{\theta}{150}\right)^4 + \left(\frac{\theta}{150}\right)^5\right] \quad 0 \leq \theta \leq 150°$$

The equation for return is:

$$y = 2\left[1 - 2.5\left(\frac{\theta}{210}\right)^2 + 2.5\left(\frac{\theta}{210}\right)^4 - \left(\frac{\theta}{210}\right)^5\right] \quad \text{for} \quad -210 \leq \theta \leq 0$$

It should be noted that the combination of the curves for rise and return will cause a sudden change of the second derivative because $\beta_1 \pm \beta_2$; the third derivative however will be continuous.

Polydyne Cam Design

In general, when constructing the time-displacement diagram for a cam, it is assumed that the motion of the end mass of the follower will be exactly as desired. This is most often not the case because clearances and flexibility of the system cause deviation, which is also dependent on cam speed. The polydyne cam design method takes these deviations into account by:

(1) Determining clearances and flexibility of the system
(2) Developing the actual cam profile to provide the desired motion of the follower end mass.

Often, control of the motion of the end element is vital, as for instance in the valve motion in a gasoline engine. In other cases, control of the dynamic properties of the system is essential for good performance. Thus, the end element often has a high mass, and therefore controlling the motion of this end mass has a great influence on the behavior of the follower system.

A cam designed by the polydyne method has two features:

(1) It is a good cam except when operating close to the critical speed (speed when excessive vibration occurs) of the follower system
(2) It is designed for a certain speed and its performance is sensitive to speed-changes.

The following equation expresses the various factors to be taken into account, (Figs. 10-8 and 10-9):

$$y_0 = y + y_c + y_d + y_i + y_p \qquad\qquad 10.6$$

where $\qquad\qquad y_0 = $ Equivalent cam lift $= R_l Z_0$

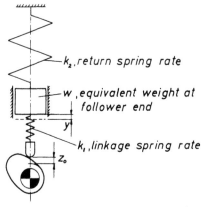

k_2, return spring rate

w, equivalent weight at follower end

y

k_1, linkage spring rate

Z_0

Fig. 10-8. Dynamically equivalent system to actual cam mechanism.

R_l is the mechanical linkage ratio (constant or varying) between the cam follower and the end mass.

Z_0 is the lift at the cam itself.

y = Actual lift of end element of follower linkage.

y_c = Clearance in the system that must be taken up before force is transmitted to the end element.

y_d = Dynamic deflection of the system, a function of speed.

y_i = y-increment deflection, a function of the spring rates of the linkage system and the return spring.

y_p = Initial deflection required to overcome the combined preload of the linkage system and the return spring before any end element lift can occur.

Equation 10.6 can also be written as:

$$y_0 = y_r + C_y + y_d \qquad\qquad 10.7$$

where

$$y_r = y_c + y_p$$

$$C_y = y + y_i$$

$$y_r = \text{ramp height}$$

The terms in equation 10.7 will now be discussed in more detail.

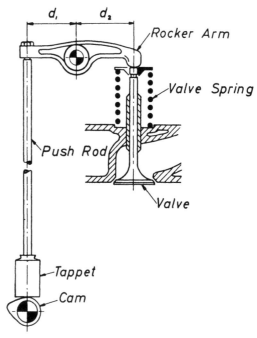

Fig. 10-9. Automotive cam linkage system.

The Ramp

In the case of internal combustion engine cams the ramp or pre-lift is designed to take up the clearance in the system and to overcome the preload to such a point that the end element is on the verge of moving.

The total ramp height can be found in a relatively simple manner by measuring the displacement of the follower just before the valve starts moving. It is, of course, also possible to calculate y_r but an exact value cannot be obtained because the amount of clearance differs from one engine to another and only approximations can be made. However, if it is not possible to measure the necessary ramp height, then calculating the ramp height is better than having no ramp at all. The ramp can be of the inflected* type, a 3–4–5 polynomial, or a cycloidal motion, for example.

Another curve often used for ramp design is the straight line which gives constant velocity motion. With this type of curve the same contact velocity occurs in each linkage, no matter what the differences in clearances and flexibility may be. This kind of ramp curve is also the simplest to design and its height may be changed without affecting the main cam profile. Often a parabolic curve is employed for making a smooth transition between the dwell and ramp.

Increment Deflection

The difference between end mass displacement and follower displacement is:

$$y_i = y_c - y \qquad 10.8$$

If the force applied by the cam is F, then

$$F = k_1 y_i = k_1 (y_c - y) \qquad 10.9$$

But in the static condition:

$$F = k_2 y \qquad 10.10$$

Combining equations 10.8–10.10:

$$y + y_i = y + \frac{F}{k_1} = y + \frac{k_2}{k_1} y = \frac{k_1 + k_2}{k_1} y$$

or $\qquad y + y_i = Cy \qquad\qquad\qquad 10.11$

where $\qquad C = \dfrac{k_1 + k_2}{k_1} \qquad\qquad 10.11a$

* Inflected type of ramp has a curve which causes acceleration of the follower over the first part of the curve and deceleration over the rest of the curve.

The spring rates k_1 and k_2 can either be measured or they can be calculated from design specifications. k_1 is the spring rate of the linkage and k_2 is the return or compression spring rate.

Dynamic Deflection

The dynamic deflection y_d of the linkage system arises from the acceleration of the equivalent mass and the relationship is:

$$y_d k_1 = \frac{W}{g} a \qquad 10.12$$

where

$$W = \text{effective weight, lb}$$

$$g = \text{gravitational constant} = 386 \text{ in./sec}^2$$

$$a = \text{acceleration, in./sec}^2$$

If we wish to express a in terms of degrees rather than seconds, then:

$$a = 36 N^2 y'' \qquad 10.13$$

where

$$N = \text{cam shaft speed, rpm}$$

$$t = \frac{60}{N} \frac{\theta}{360} \qquad \text{or} \qquad \theta = 6Nt$$

Whence*

$$y_d = \frac{36 W N^2 y''}{386 k_1} = 0.093 \frac{W N^2 y''}{k_1} \qquad 10.14$$

or

$$y_d = D y'' \qquad 10.14a$$

where

$$D = 0.093 \frac{W N^2}{k_1} \qquad 10.14b$$

and equation 10.7 can now be written as:

$$y_0 = y_r + Cy + D y'' \qquad 10.15$$

and differentiating:

$$y_0' = Cy' + D y''' \qquad 10.16a$$

$$y_0'' = Cy'' + D y'''' \qquad 10.16b$$

$$* \frac{d^2 y}{ds^2} = \left(\frac{d\theta}{ds}\right)^2 \frac{d^2 y}{d\theta^2} \quad \text{and} \quad \frac{d\theta}{ds} = 6N$$

$$\text{or} \quad \frac{d^2 y}{ds^2} = 36 N^2 \frac{d^2 y}{d\theta^2}$$

Effective Weight

It will be noted from equation 10.14 that the dynamic deflection y_d of the linkage system is a function of its total effective weight.

In a gasoline engine valve cam train the total effective weight of the follower system is made up of the effective weights of the push rod, the rocker arm, and the valve unit, Fig. 10-9, or:

$$W = W_p + W_r + W_v \qquad\qquad 10.17$$

where
W_p = effective weight of push rod, lb

W = effective weight of rocker arm, lb

W_v = effective weight of valve unit, lb

The total effective weight is considered to be concentrated in the end element, or in this case, in the valve unit.

In the case of a spring, the effective weight is considered to be one-third of the weight of the spring itself.

In the case of a lever, the moment of inertia I_0 with respect to the center of rotation is first found:

$$I_0 = mk_0^2$$

where I_0 = moment of inertia with respect to center of rotation, lb-in.-sec^2

m = mass of lever, lb sec^2 in.

k_0 = radius of gyration, in.

The effective weight of a lever can be considered as the weight of the lever concentrated in one point which is at a distance k_0 from the center of rotation.

Tappet. The weight of the tappet is not contained in the total effective weight because from the standpoint of dynamic deflection no matter how heavy the tappet is, the motion of the valve will not be changed unless bouncing occurs. The mass of the tappet will, however, affect the total forces acting on the cam.

Push rod. Since the push rod is relatively long and thin, it is considered to be a spring and its effective weight is:

$$W_p' = \tfrac{1}{3}W_1$$

where
W_1 = weight of push rod, lb

The effective weight of the push rod when concentrated at the valve unit is:

$$W_p = W_p' \left(\frac{d_1}{d_2}\right)^2 \qquad \text{or} \qquad W_p = \tfrac{1}{3}W_1 \left(\frac{d_1}{d_2}\right)^2 \qquad\qquad 10.18$$

Rocker arm. The rocker arm is a lever and hence the effective weight of the rocker arm is considered to be its total weight W_a concentrated at a distance k_0 from the center of rotation. The effective weight W_r when concentrated at the valve unit is then:

$$W_r = W_a \frac{k_0^2}{d_2^2} \qquad\qquad 10.19$$

Valve unit. The valve unit consists of two parts, the valve and the valve spring. The effective weight of the valve is the weight of the valve itself. The effective weight of the valve spring is one-third the weight of this spring.

The values in the following example have been chosen arbitrarily.

Example: A cam rotates with $N = 750$ rpm, the weight of the push rod is 4 lb, and the weight of the rocker arm is 3 lb, $k_0 = 2.4$ in., the weight of the valve $\frac{3}{4}$ lb, the weight of the valve spring $\frac{1}{2}$ lb, end element lift is $\frac{3}{4}$ in. over a 60-degree cam rotation, $d_1 = 3.5$ in., $d_2 = 5.5$ in., $k_1 = 30,000$ lb/in., $k_2 = 350$ lb/in., and the initial spring load S_1 is 150 lb. There is zero clearance and the external load L acting on valve is 100 lb. Using the 3–4–5 polynomial, establish the cam profile.

Solution: The equation for the 3–4–5 polynomial is:

$$y = 10x^3 - 15x^4 + 6x^5$$

The above equation gives the desired motion of the valve for unity rise and unity time and with an end element lift of $\frac{3}{4}$ in. over a 60-degree cam rotation:

$$y = \frac{3}{4}\left[10\left(\frac{\theta}{60}\right)^3 - 15\left(\frac{\theta}{60}\right)^4 + 6\left(\frac{\theta}{60}\right)^5\right] \qquad 0 \le \theta \le 60°$$

$$y' = \frac{3}{4}\left[\frac{(10)(3)}{60}\left(\frac{\theta}{60}\right)^2 - \frac{(15)(4)}{60}\left(\frac{\theta}{60}\right)^3 + \frac{(6)(5)}{60}\left(\frac{\theta}{60}\right)^4\right]$$

$$y'' = \frac{3}{4}\left[\frac{1}{60}\left(\frac{\theta}{60}\right) - \frac{3}{60}\left(\frac{\theta}{60}\right)^2 + \frac{(1)(4)}{(2)(60)}\left(\frac{\theta}{60}\right)^3\right]$$

Now the equivalent cam lift is equal to:

$$y_0 = y_r + Cy + Dy'' \qquad\qquad (10.15)$$

Because we only want to find the cam profile for actual lift, we substitute the total ramp height for y_r in the above equation, and

$$(y_r)_{\max} = r_k + \frac{S_1 + L}{k_1} \qquad\qquad 10.20$$

where r_k = clearance

S_1 = initial compression spring force with valve at starting position, lb

L = external load acting on valve, lb

$$(y_r)_{max} = 0 + \frac{150 + 100}{30,000} = 0.0083 \text{ in.}$$

$$C = \frac{k_1 + k_2}{k_1}$$

$$= \frac{30,000 + 350}{30,000}$$

$$= 1.012 \qquad\qquad (10.11a)$$

The total effective weight:

$$W = W_p + W_r + W_v \qquad\qquad (10.17)$$

$$W_p = (\tfrac{1}{3})(4) = 1.33 \text{ lb}$$

$$W_r = 3\left(\frac{2.4}{5.5}\right)^2$$

$$= 0.57 \text{ lb} \qquad\qquad (10.19)$$

$$W_v = \tfrac{3}{4} + \tfrac{1}{3}(\tfrac{1}{2}) = 0.92 \text{ lb}$$

$$W = 1.33 + 0.57 + 0.92$$

$$= 2.82 \text{ lb}$$

$$D = 0.093\,\frac{(2.82)(750)^2}{30,000} = 4.92 \qquad\qquad (10.14b)$$

If $R_l = \dfrac{d_2}{d_1}$ and Z_0 = lift at the cam itself

then $y_0 = R_l Z_0 = (y_r)_{max} + Cy + Dy''$

$$Z_0 = \frac{3.5}{5.5}(0.0083 + 1.012y + 4.92y'') \qquad\qquad 10.21$$

where $y = \dfrac{3}{4}\left[10\left(\dfrac{\theta}{60}\right)^3 - 15\left(\dfrac{\theta}{60}\right)^4 + 6\left(\dfrac{\theta}{60}\right)^5\right]$

and $y'' = \dfrac{3}{4}\left[\dfrac{\theta}{3,600} - \dfrac{\theta^2}{72,000} + \dfrac{\theta^3}{6,480,000}\right]$

Cam Accuracy

For the cam profiles found by this method, the cutting accuracy necessary for good valve train dynamics is in the order of ± 0.00020 inch (point to point for 1-degree cam rotation) with an overall tolerance of ± 0.0020 inch.

Minimizing Vibration

There are a number of factors which cause vibrations in the cam mechanism but it is possible to reduce them by practical means.

Shape of acceleration curve. It has been shown by J. A. Hrones* that a suddenly applied acceleration (jerk $= \infty$) causes forces that are double and triple the calculated values. Therefore, taking into consideration other factors, a cam curve should be selected which will minimize this. As has been previously shown, a cycloidal motion is a good choice of cam curve causing acceleration forces which are only about 5 per cent above those calculated.

Surface irregularities. Even small irregularities can cause unwanted accelerations, the size of which are proportional to the square of the distance from the cam shaft center. Make the cam surface smooth and to specifications in order to approach the mathematical curve as closely as possible. It is also of definite advantage to make the cam as small as practicable (see Chapter 5).

Cam imbalance. Most plate cams are unbalanced and can cause vibrations. Make the cam shaft diameter sufficiently large enough to keep dynamic flexure to a minimal value or balance the cam.

Cam speed and natural frequency of the cam-follower linkage. At certain cam speeds the vibration of the cam will be of the same or multiple frequency as the cam-follower linkage and may result in excessive vibrations which could be damaging to the system. The solution is to make the natural frequency of the cam-follower linkage high by choosing light materials and making all parts as rigid as practicable.

Crossover Shock

At any point where acceleration changes from positive to negative, or vice-versa, there will be some degree of impact or crossover shock. To minimize this use a rigid follower system and remove backlash from the system by using the arrangement shown in Chapter 1, Figure 1-2a, b or c. Also use of cycloidal motion or other curves with a finite pulse is recommended.

*"Key Factors in Cam Design and Application," *Machine Design*, 1949, April, pp. 127–132, May pp. 107–111, 178, June pp. 124–126.

Vibrations of Cams, Designed by the Polydyne Method

Theoretically at least, a polydyne cam run at design speed will cause no vibration. However, in actual practice, small vibrations will exist at design speed. Figure 10-10 gives an approximate picture of the effect of speed vari-

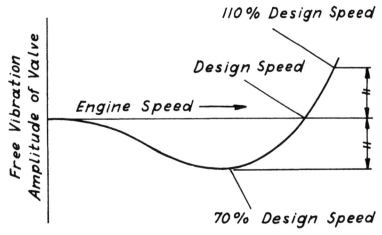

Fig. 10-10. Relationship between engine speed and free vibration amplitude of valve.

ation. It is seen that if the speed is about 70 percent of design speed there is a maximum of vibration amplitude. At about 110 per cent of the design speed the amplitude of free vibration equals the amplitudes at 70 percent of the design speed.

Cam Mechanisms

It should be borne in mind that cams, despite their advantages, have severe limitations, namely that they cannot transmit large forces unless the disc is made very thick and that significant additional acceleration forces are generated when the cam is run with high speed.

In the following is shown the versatility of the cam mechanism and also some examples showing how to neutralize the disadvantages.

Engraving Mechanism

Figure 11-1 shows a cam A with two grooves. One groove translates member B and the other groove member C. In B and C are slots and in the slots is a sliding piece D. B controls the motion of D in vertical direction and C controls the motion in horizontal direction. By the combined motion the center of D will write the letters JK.

Two Movements Obtained from One Cam

For a special purpose it was desired to move the head C along the right-angle path shown in Fig. 11-2. This was accomplished by using one cam and two followers, 180 degrees apart. The swinging follower D moves the head C in a horizontal direction. The head C is guided by a frame attached to the translating follower B. When follower B is moved in a vertical direction it carries head C in a vertical direction. The cam groove is cut so that when follower B is in a dwell position, follower D is moving the head horizontally; when follower D is in a dwell position, follower B is moving the head vertically.

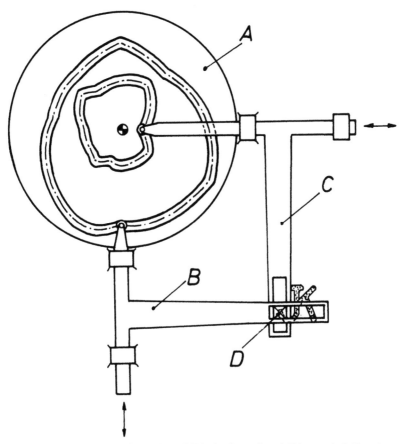

Fig. 11-1. Cam controls motion of B in horizontal and C in vertical direction, point D thereby describing the letters JK (J K).

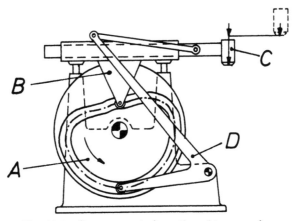

Fig. 11-2. One cam controls motion of two members.

Improved Spring Arrangement

An often employed spring arrangement is shown in Fig. 11-3a. The follower K is kept in contact with the open track cam at all times with the help of the spring L, fastened to the machine frame with the help of bracket S. It can be seen that for each revolution of the cam the spring is stretched and then permitted to contract.

Another arrangement is shown in Fig. 11-3b. The arms K and G are pulled toward each other by the spring L, but kept apart with the help of the clutch shown in the pictorial view. This clutch permits the two levers to come as close to each other as indicated in Fig. 11-3b, but also allows them to be pulled away from each other. Normally the arms K and G pivot together as the cam rotates and the spring L does not change its length. No bracket similar to S in Fig. 11-3a is necessary.

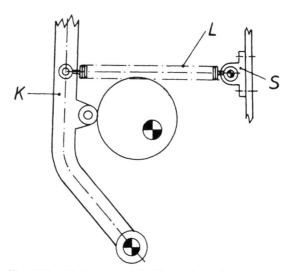

Fig. 11-3a. Spring is stretched for each revolution of cam.

Cam-Actuated Intermittent Worm-Drive Mechanism

A worm drive used on a wire-forming machine is illustrated in Fig. 11-4. The mechanism comprising this drive converts a continuous rotary motion into an intermittent rotary motion at a reduced rotative speed. The object of employing the worm and worm-wheel is to give a compact high-ratio speed reduction in combination with the intermittent motion. Referring to Fig. 11-4, shaft A, mounted in bearings C, carries the single-thread worm B (and rotates in the direction indicated by the arrow). Shaft A receives its motion from the driving shaft E through the splined sleeve D, which

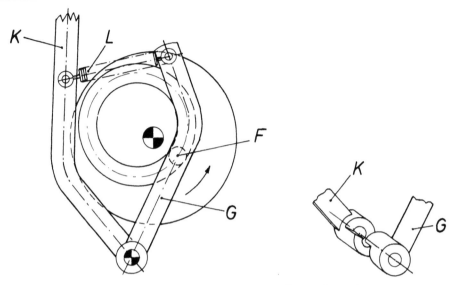

Fig. 11-3b. Length of spring is normally not changed.

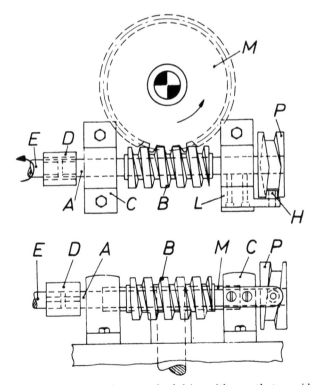

Fig. 11-4. Worm and worm-wheel drive with cam that provides
intermittent rotary movement.

permits axial movement of shaft A. Worm B meshes with worm-gear M, to which it transmits motion in the direction indicated. Shaft A carries cam P, which rotates with it. Bracket L, attached to bearing C, carries roller H, which operates in the groove of cam P. It can be seen that, owing to the fixed position of roller H, the rotation of cam P will cause shaft A to be moved axially. The groove in cam P is shaped to produce a uniform axial motion in one direction during one-half revolution, and in the reverse direction during the other half revolution. The lead of cam P is equal to the lead of worm B.

If worm B were fixed against axial movement, one revolution of worm B would produce a movement of gear M equivalent to the lead of the worm. However, in addition to rotative motion, worm B is also given an axial motion by cam P acting against roller H, as mentioned; thus the rotation of gear M is affected by both the rotative and axial movements of worm B. As the lead of worm B and cam P are equal during one-half of the cycle of cam rotation the motion of gear M equals that which would be produced by an axially fixed worm of double the lead of worm B.

As shaft A continues to rotate, the high point of cam P passes by roller H and the axial movement of shaft A is reversed. When this occurs, there is no movement of gear M, as the axial movement of shaft A produced by cam P is equal to the lead of worm B, but is in the reverse direction. In this manner the axial movement of shaft A neutralizes the lead of worm B, the worm merely turning or threading itself back to its original position without imparting any motion to gear M. The effect is to produce a series of partial revolutions of gear M with equal rest periods between the movements.

Intermittent Motion from Two Synchronized Cams

Packaging machines often require mechanisms to transmit a particular motion during each fifth revolution of the main camshaft. Such a need might arise where five packages are to be grouped, then pushed from the machine at the same time. A mechanism that has been arranged to satisfy these particular requirements is shown in Fig. 11-5.

The principal operating elements of this mechanism are two synchronized cams and one follower-lever. The upper end of the single bell-crank A drives the package-ejector unit (not shown). At the opposite end of the lever are two follower-rollers B_1 and B_2 which are held in contact with cams C_1 and C_2, respectively, by a spring D (attached to the upright lever arm).

Cam C_1 is pinned directly to the constantly rotating camshaft E, while the motion for cam C_2 is obtained indirectly from a gear F, also pinned to the camshaft. By means of the gear train F, G, I, H—the latter gear being

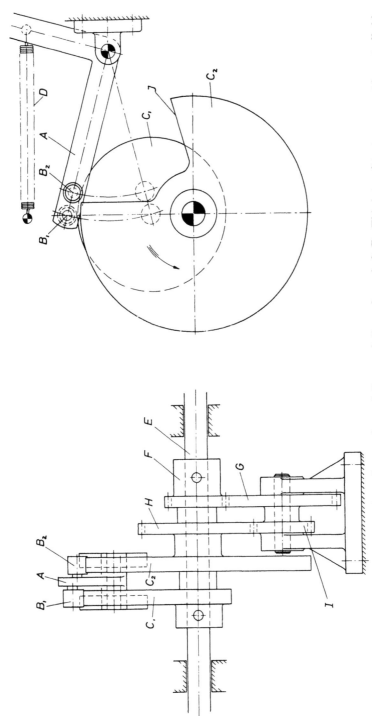

Fig. 11-5. Lever A is permitted to function only once for each five revolutions of camshaft E. This intermittent movement is controlled by the action of two synchronized cams C_1 and C_2.

keyed to cam C_2—movement of gear H is reduced to one-fifth that of gear F. Thus the speed of Cam C_2 is only one-fifth that of the camshaft and C_1, although the rotational movement of both of the cams is in the same direction.

Bearing this in mind, and noting the cam configurations and positions in the right-hand view, it can be seen that during one revolution of cam C_1, cam C_2 will rotate a distance equal to the width of its cutout J. This cutout occupies approximately one-fifth of the otherwise circular cam.

During this rotation of the camshaft, follower-roller B_2 is disengaged from the surface of cam C_2. This permits roller B_1 to track along the entire surface of cam C_1, thus causing lever A to pivot. For the next four rotations of the camshaft, follower-roller B_2 will ride along cam C_2, thereby preventing roller B_1 from being affected by the contour of cam C_1, and causing lever A to remain motionless. Of course, many variations of this mechanism are possible.

Varying the Cam Dwell with Two Adjustable Follower Rollers

It is often necessary to alter the motions in wire-forming machines, and especially to vary the time for dwell. To do this, two adjustable rollers are used as shown in Fig. 11-6. The two dwelling periods can be varied to suit requirements. The cam, indicated at A, is secured to the driving shaft and

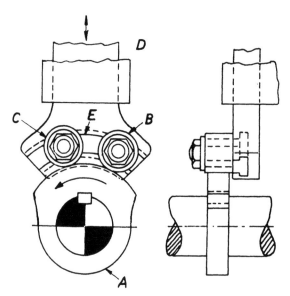

Fig. 11-6. The dwell period and timing of this cam can be varied by simply changing the positions of the two roller followers.

engages both rollers B and C. The rollers are mounted on flanged bushings and secured to the follower D by studs. They can be adjusted to any position along the curved T-slot E.

The amount of dwell and the timing of the rise and fall of the follower depend upon the distance between the two rollers and their locations along the T-slot. For instance, if the follower were required to dwell longer in its upper position, the distance between the rollers would be increased. On the other hand, if the dwelling time in the upper position was to be decreased, the rollers would be brought closer together. The time at which rise and fall of the follower occurs may be varied by adjusting the rollers along the T-slot without changing their center distance.

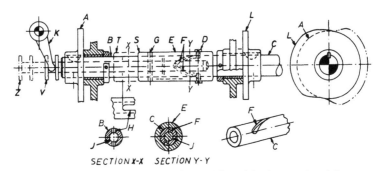

SECTION X-X   SECTION Y-Y

Fig. 11-7. Mechanism for changing angular position of feed-cams A and L to vary rate of tool-feeding movements as required for different machinery operations.

Mechanism for Making Quick Change in Angular Positions of Feed-Cams

The staggered production requirements and the available tool equipment for rough-turning several parts of similar design necessitated changing the angular relationship of two principal feed-cams on one shaft for each tool set-up. The arrangement provided to permit the positions of the cams to be changed quickly to suit the machining requirements of the different parts is shown in Fig. 11-7. Cam A on shaft B is driven by shaft C through keys D in sleeve E. Keys D operate in spiral slots F. Pin G fits in sleeve E and extends through slots H of hollow shaft B and also through the shifter shaft J within shaft B. Axial movement of shifter shaft J, by means of lever K, from position V to position Z causes sleeve E to move from position S to position T and causes keys D to operate in slots F of shaft C. This results in cam A advancing clockwise in relationship to cam L. Shaft J is piloted in shaft C to maintain the alignment of shafts B and C. The follower on cam A is released when changing cam positions. Shifter lever K is provided with conventional means (not shown) for locking in any of the required positions.

Mechanism for Changing Stroke while Machine is Running

In automatic machinery it is often desired to change the stroke of a machine member when the machine is still running. A mechanism that can accomplish this is shown in Fig. 11-8. Shaft A drives cam B which moves the translating follower C. The double lever D rotates around shaft E and transfers the motion of C to F with the help of the sliding pieces K and L,

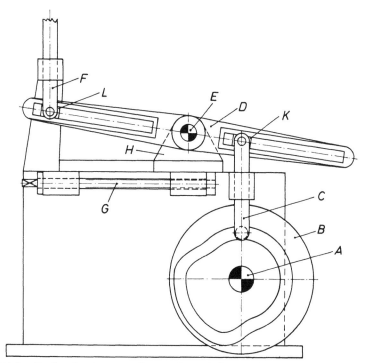

Fig. 11-8. Stroke of output member can be varied even when machine is running.

which slide is the slots of lever D. The bracket H which carries shaft E can be moved backward and forward with the help of the screw G, and the lever ratio between follower C and F can be varied within a large interval.

Slide Motion Differential

Packaging machines are often equipped with automatic inspection devices that discard cartons which, for weight or other reasons, are above or below standard. The mechanism illustrated in Fig. 11-9 operates in the following manner:

The object is to change the motion of the slide E not only with respect to stroke, but also in timing. This is done by selection of latch M or B (see illustration). The drive is rotating shaft Z. Levers A and L follow open track cams D and O. This view shows lever L blocked out by latch M. Lever A follows cam D under tension from spring C. When the left end of connecting-rod F is driven to point E, it carries the intermediate lever to position G–H–J. The standing lever is thus carried to position I and the slide E moves to point K.

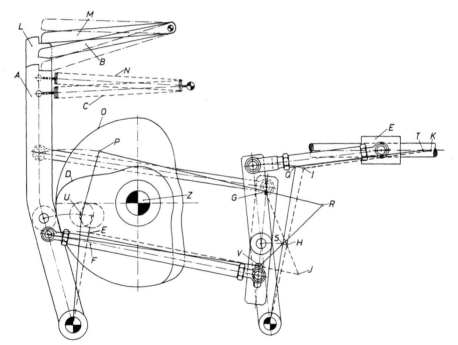

Fig. 11-9. The output motion of the slide E, right, is changed timewise and lengthwise by alternating the use of latches B and M.

When lever A is latched out by B and lever L is allowed to follow cam O, the connecting rod F is again driven to the right, carrying the intermediate lever to position V–S–R. The standing lever moves to point Q and the slide E moves to point T.

Because lever A is now the driven member and follows cam D, whereby an entirely different timing of the output motion of the slide E is obtained, this motion will now discard a carton rather than bring it to the next station of the machine for further processing.

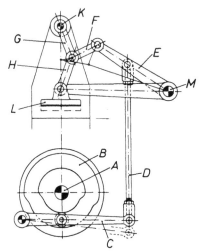

Fig. 11-10.   Cam actuated toggle links for transmission of large forces.

Cam Actuated Toggle Mechanism
for Operating Pressure Pad

In general, cams are not suited to transmitting large forces, so that when large forces are involved toggle links may be used to solve the problem. In Fig. 11-10 cam B is rotated by shaft A and acts on link E through the follower C and the connecting rod D.   Link E is supported by the shaft M and E and F are one pair of toggle links which in turn act on another pair of toggle links H and G. G is supported by the shaft K and H carries the pressure pad L.   Through the use of two pair of toggle links in series large forces can be transmitted because the output force of the one toggle link is greatly increased and when acting as an input force to the second toggle link, the input force is further increased.

A series of devices designed to lower the maximum pressure angle are now described.

Sliding Cam

In Fig. 11-11 the input shaft A is keyed to the disk B, which carries guides C for the double-edged cam D.   The roller R is stationary and when the shaft A rotates, the roller R will lift the cam D the distance h_1 due to the one edge, and the other edge will cause the translating roller follower to move the distance h_2, the total distance travelled by E being therefore $h_1 + h_2$.

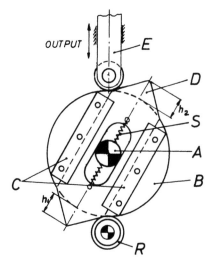

Fig. 11-11. Sliding cam.

The torque on the input shaft may be considered to be the sum of two torques; one of which raises the follower E by the amount h_2 and one which serves to lift the cam B by the amount h_1. Therefore the pressure angle must be considered as occurring at two places, and is approximately one-half of that which would occur if the cam were fixed on the shaft and the follower E were, alone, raised the whole amount $h_1 + h_2$.

The double spring S will push the cam D back to its neutral position.

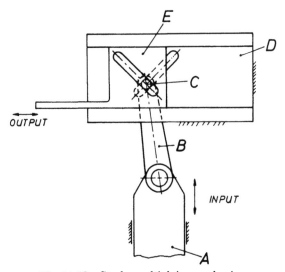

Fig. 11-12. Stroke-multiplying mechanism.

Stroke-multiplying Mechanism

In a punch press the vertical stroke of the translating member A, Fig. 11-12, was limited to a certain value, but a horizontal stroke of about 1.4 times as large was desired. This was accomplished by adding a lever B, carrying the pin C which moves in the two slots shown. The one slot is in D, which is stationary, and the other slot is in E, that is movable and guided by D.

The stroke of the output member E will be about 1.4 times that of A, the ratio being determined by the angle of the slots.

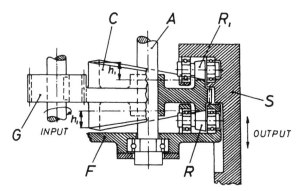

Fig. 11-13. Double-faced cam for double the stroke by approximately half the pressure angle.

Double-faced Cam for Double the Stroke at Approximately Half the Pressure Angle

This mechanism, Fig. 11-13, is based on somewhat the same idea as the foregoing. The input gear G drives the double-faced cam C ,which is supported by the shaft A, but otherwise is free to slide on this shaft. The shaft A is supported by the frame F, which supports the roller R. When the cam C rotates it will be lifted a certain distance h_1 due to R, but because the other face of the cam will move the roller R_1 a distance h_1 the total stroke of the translating member S will be $2h_1$.

Cam-and-rack Movement for Increasing the Throw of a Lever

Here, as shown in Fig. 11-14, the added stroke is obtained by giving the output member C a combined translation and rotation.

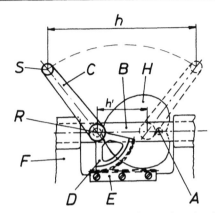

Fig. 11-14. Cam-and-rack movement for increasing the throw of a lever.

The input shaft A carries the well-known heart-shaped cam H, which will move the roller R with uniform velocity. The lever C which is made to rotate about the center of roller R has attached to it the gear segment D, which is in mesh with the rack E, attached to the frame F. Due to the combined translation and rotation of C, the output stroke will be h instead of h'.

Fall and Rise of Follower Over Only 72-degree Cam Shaft Rotation

The device shown in Fig. 11-15 makes use of the fact that if a cam is rotated faster but still moves the output member exactly as before; i.e., the time displacement diagram remains unchanged, then the maximum pressure angle for the faster rotating cam must be smaller than the one for the slower rotating cam. Suppose a cam rotates with 1 rpm and the follower rises over 60 degrees of cam shaft rotation, then obviously the time for the rise is $\frac{1}{6}$ second. If the cam is now rotated with 2 rpm the rise takes place over $\frac{1}{12}$ of a second and if it is desired to keep the same time-displacement diagram as before then the angle of rotation for the rise must be increased from 60 degrees to 120 degrees, and this results in a lower pressure angle.

The cut-out cam A, which has a constant radius except for the cut-out portion, is fastened to the shaft A_0 and between the shaft A_0 and the cam C there is a gear-train to step up the speed of the cam C by a ratio of $5:1$ over that of A_0. The roller follower B has a double roller R on it; the one roller rides on A, the other on C. The cut-out portion of A subtends an angle of approximately 72 degrees and when the cut-out part of A reaches R, the follower B is allowed to move and will follow the shape of cam C. Cam C will make one complete revolution during the interval that the cut-out

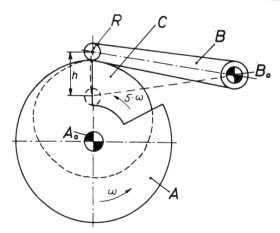

Fig. 11-15. Fall and rise of follower over 72° cam shaft rotation and stroke h. Cut-out cam A keyed to shaft A_o; gear ratio from A to C is 5:1 (not shown).

portion of A is passing under roller R. Thus, the follower B will move from its upper position to its lowest and back again when the input shaft A_0 has rotated 72 degrees only. Other variations are possible.

Up to this point when rotation of cams was dealt with, the rotation was uniform. It is, however, also possible to let the cam rotate with varying velocity, and if the cam is rotated faster at points of maximum slope, the value of its pressure angle must be lowered, if the time displacement diagram is to remain unchanged.

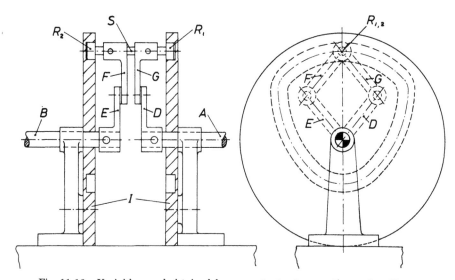

Fig. 11-16. Variable speed obtained from constant rotary motion; axis collinear.

Mechanisms for Varying Angular Velocity

First, let us show three mechanisms for imparting a varying angular velocity to the driven shaft.

In Fig. 11-16 the driving shaft A carries the arm D, which is pin-connected to G, which in turn carries the shaft S with the two rollers R_1 and R_2. E and F are two arms similar to D and G respectively, but arranged symmetrically as shown in the right-hand view. B is the output shaft. The rollers R_1 and R_2 are guided by the two cams I. As long as R_1 and R_2 run on an arc concentric with A, A and B will move with the same velocity, but as soon as the distance between the rollers and the center A varies, the motion of B will be non-uniform and, of course, dependent on the form of the two cams I.

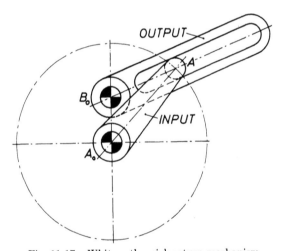

Fig. 11-17. Whitworth quick-return mechanism.

The device has the advantage that shafts A and B are collinear, but it has the disadvantage of being rather complex and costly.

A relatively simple method of imparting a varying motion to the output shaft B_0 is shown in Fig. 11-17 and is known under the name of Whitworth quick-return motion. The input shaft A_0 carries the arm with roller A. The roller A slides in the slot of another arm, which is fastened to the output shaft B. The mechanism is simpler than that of Fig. 11-16 but the axes are not collinear. Also, the contact stress between the roller and the slot is high because of line contact. The slot is costly to machine.

Another way of obtaining a varying output motion is with the help of a drag-link mechanism, Fig. 11-18. A_0A rotates with uniform velocity and imparts with the help of the link AB a varying rotary motion to link B_0B.

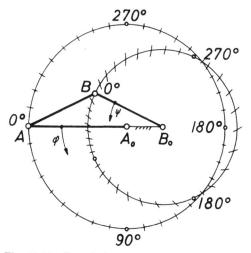

Fig. 11-18.   Drag-link mechanism for variable output.

Using Mechanism to Reduce Size of Cam

Let us now use this mechanism to reduce the size of a cam, and let the displacement diagram, Fig. 11-19, be given.   The solid line curve indicates the desired displacement diagram.   With the help of the drag-link mechanism, the objective is to reduce the pressure angle τ_2.

First, the angular displacement ϕ of the input link A_0A is plotted against the angular displacement ψ of the output link B_0B, Fig. 11-19.   In this diagram, find the point V_2', which has the lowest slope.   When the position of the drag-link mechanism corresponds to point V_2', the output link B_0B is obviously rotating much faster than the input link; i.e., a given increase in ϕ produces the maximum increase in ψ.

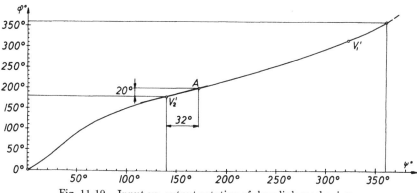

Fig. 11-19.   Input vs. output rotation of drag-link mechanism.

Next go to some point A in the ψ-ϕ diagram. As shown in Fig. 11-19, this point is situated 20 degrees from V_2' in the $+\phi$ direction and 32 degrees from V_2' in the $+\psi$ direction.

In the diagram, Fig. 11-20, locate point A' which is on the actual displacement curve and go horizontally to the right to A'' so that the abscissa difference between A'' and V_2 is 32 degrees. Repeat the procedure.

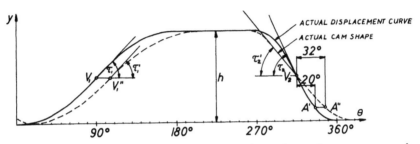

Fig. 11-20. Modifying the displacement diagram to lower the maximum pressure angle.

The curve so obtained (dashed line) determines the actual cam shape and when the drag-link mechanism is placed between the input shaft and the cam, the actual motion of the output member will follow the solid line curve of Fig. 11-20. Note that the new pressure angle τ_2' on the actual cam contour at point V_2 is less than the original pressure angle τ_2.

Velocities, Accelerations, and Dynamic Forces in Linkages and Cam Mechanisms

As has been previously shown, certain forms of four-bar linkages can be substituted in place of certain cams for the purpose of analysis and synthesis. This chapter will cover the analysis of these mechanisms with respect to velocities, accelerations and forces. A graphical and analytical procedure will be used. A purely analytical method requires many time-consuming calculations; and the purely graphical method requires much drawing and often the lines fall outside the paper. Analytical expressions are also given, and if computers are available these expressions can be programmed. The methods are applicable for all types of mechanisms.

Determination of Velocities in the Four-bar Linkage

Relative velocity method: This method makes use of the vector equation

$$V_B = V_A + V_{B/A} \qquad (12.1)$$

which reads: the velocity of B is equal to the velocity of A plus vectorially the velocity of B relative to A, Fig. 12-1. The velocity of A is given by $V_A = A_0A \times \omega$ and is represented by a vector V_A perpendicular to A_0A and having the same sense with respect to A_0 as ω. The vector V_A can be given any length but conveniently is given the same length as A_0A. The procedure is as follows.

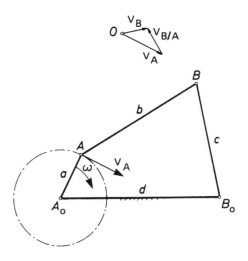

Fig. 12-1. Relative velocity method.

1. Through the head of V_A is drawn a line perpendicular to AB and somewhere on this line must be the head of the vector $V_{B/A}$ because the velocity of B relative to A can only be perpendicular to AB. Otherwise B would either come closer to or farther away from A and this is not possible since A and B are fixed points on the same body, namely, the coupler.

2. Through O is drawn a perpendicular to $B_0 B$ because V_B must be perpendicular to $B_0 B$.

3. Where the perpendicular to AB intersects the perpendicular to $B_0 B$, is the terminal point of $V_{B/A}$ and V_B, respectively.

Determination of Acceleration in the Four-bar Linkage

The acceleration of point B, Fig. 12-2, is determined with the help of the vector equation

$$a_B = a_A + a_{B/A}$$

which reads: the acceleration of B is equal to the acceleration of A plus vectorially the acceleration of B relative to A.

The three terms in this equation are broken down into normal and tangential components as follows:

$$a_B^n + a_B^t = a_A^n + a_A^t + a_{B/A}^n + a_{B/A}^t \qquad (12.2)$$

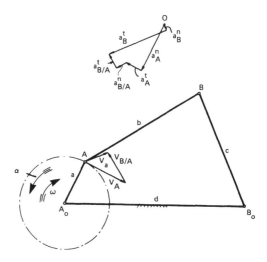

Fig. 12-2. Determination of acceleration.

Next, each of these terms is investigated. Let the crank A_0A have the angular velocity ω clockwise and the angular acceleration α_a counterclockwise.

Now
$$V_A = A_0A \cdot \omega$$

The velocities V_B and $V_{B/A}$ are determined with the help of the vector polygon drawn from point A.

$$a_B^n = \frac{V_B^2}{B_0B}$$

a_B^t is known only in direction

$$a_A^n = A_0A \cdot \omega^2$$

$$a_A^t = A_0A \cdot \alpha$$

$$a_{B/A}^n = \frac{(V_{B/A})^2}{AB}$$

$a_{B/A}^t$ is known only in direction

Indicating a known vector by underlining it with two lines and a vector by which only the direction is known by placing the sign ∡ under it, the vector equation takes on the form:

$$a_B^n + \underset{4}{a_B^t} = a_A^n + a_A^t + a_{B/A}^n + \underset{4}{a_{B/A}^t}$$

Because this equation has only two terms with a ∡ whereas the other terms are fully known, the equation can be solved with the help of the vector diagram in Fig. 12-2.

Example: Let $a = 1$ in., $b = 2.9$ in., $c = 2.58$ in., $d = 4$ in., $\theta = 50°$, $N = 1000$ rpm clockwise, and no angular acceleration of A_0A. Find angular acceleration of B_0B.

Solution: First, the four-bar linkage is drawn to scale, placing A_0A so that $\theta = 50°$, Fig. 12-3.

$$V_A = \overline{A_0A} \cdot \omega = (1) \cdot \frac{(1000) \cdot (2\pi)}{(60)} = 105 \text{ in./sec}$$

The vector diagram drawn at point A, Fig. 12-3, yields

$$V_B = V_A \cdot \frac{\overline{V_B}}{\overline{V_A}} = 105 \cdot \frac{(0.59)}{(2.10)} = 29.5 \text{ in./sec}$$

$$V_{B/A} = V_A \cdot \frac{\overline{V_{B/A}}}{\overline{V_A}} = 105 \cdot \frac{1.89}{2.10} = 94.5 \text{ in./sec}$$

where $\overline{V_A}$, $\overline{V_{B/A}}$, and $\overline{V_B}$ are the lengths of the vectors V_A, $V_{B/A}$, and V_B, respectively, as measured on the drawing.

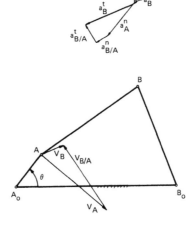

Fig. 12-3.   Determination of acceleration.

$$\underline{\underline{a_B^n}} \rightarrowtail a_B^t = \underline{\underline{a_A^n}} \rightarrowtail \underline{\underline{a_A^t}} \rightarrowtail \underline{\underline{a_{B/A}^n}} \rightarrowtail a_{B/A}^t$$
$$\qquad\quad 4 \qquad\qquad\qquad\qquad\qquad\qquad 4$$

$$a_B^n = \frac{V_B^2}{B_0 B} = \frac{(29.5)^2}{2.58} = 337 \text{ in./sec}^2$$

$$a_A^n = \frac{V_A^2}{A_0 A} = \frac{(105)^2}{(1)} = 11,000 \text{ in./sec}^2$$

$$a_A^t = A_0 A \alpha_A = 0$$

$$a_{B/A}^n = \frac{(V_{B/A})^2}{AB} = \frac{(94.5)^2}{(2.9)} = 3,080 \text{ in./sec}^2$$

The acceleration vector diagram is next drawn as shown in Fig. 12-3 and the following results are obtained:

$$a_B^t = 13,700 \text{ in./sec}^2$$

$$\alpha_c = \frac{a_B^t}{B_0 B} = \frac{13,700}{2.58} = 5,310 \text{ radians/sec}^2 \text{ counterclockwise}$$

Analytical Expressions for Velocity and Acceleration in the Four-bar Linkage

The procedure described above will now be applied to the four-bar linkage (Fig. 12-4).
The usual vector equation

$$V_B = V_A \rightarrowtail V_{B/A}$$

can be written in complex form as

$$-ic\omega_c e^{i\theta_4} = ia\omega e^{i\theta_2} + ib\omega_b e^{i\theta_3}$$

where the minus sign on the left hand side of the equation is used because θ_4 indicates that the position of B_0 relative to B will be used and that therefore

$$ic\omega_c e^{i\theta_4} = V_{B_0/B} = -V_{B/B_0} = -V_B$$

Hence

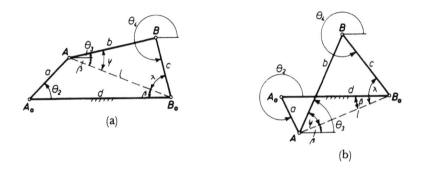

Fig. 12-4a. Designation of links and angles; $0 < \theta_2 < 180°$.

Fig. 12-4b. Designation of links and angles; $180° < \theta_2 < 360°$.

$$a\omega e^{i\theta_2} + b\omega_b e^{i\theta_3} + c\omega_c e^{i\theta_4} = 0 \qquad (12.3a)$$

Equation 12.3a can be written in complex rectangular coordinates as

$$a\omega(\cos\Theta_2 + i\sin\Theta_2) + b\omega_b(\cos\Theta_3 + i\sin\Theta_3) + c\omega_c(\cos\Theta_4 + i\sin\Theta_4) = 0 \qquad (12.3b)$$

In the above equation it is necessary that the sum of real and imaginary parts taken separately is equal to zero, i.e. the sum of all horizontal and vertical components taken separately must be equal to zero or:

$$\Sigma HC: \; a\omega \cos\Theta_2 + b\omega_b \cos\Theta_3 + c\omega_c \cos\Theta_4 = 0 \qquad (12.3c)$$

$$\Sigma VC: \; a\omega \sin\Theta_2 + b\omega_b \sin\Theta_3 + c\omega_c \sin\Theta_4 = 0 \qquad (12.3d)$$

Solving for ω_b and ω_c,

$$\omega_b = -\frac{a \sin(\Theta_2 - \Theta_4)}{b \sin(\Theta_3 - \Theta_4)}\omega \qquad (12.3e)$$

$$\omega_c = \frac{a \sin(\Theta_2 - \Theta_3)}{c \sin(\Theta_3 - \Theta_4)}\omega \qquad (12.3f)$$

To find the accelerations:

$$a_B = a_A \mapsto a_{B/A}$$

$$a_B^n \mapsto a_B^t = a_A^n \mapsto a_A^t \mapsto a_{B/A}^n \mapsto a_{B/A}^t \tag{12.4}$$

and in complex form:

$$c\omega_c^2 e^{i\theta_4} - ic\alpha_c e^{i\theta_4} = -a\omega^2 e^{i\theta_2} + ia\alpha e^{i\theta_2} - b\omega_b^2 e^{i\theta_3} + ib\alpha_b e^{i\theta_3}$$

As was done with velocities this equation is broken down into horizontal and vertical components (using that $e^{i\theta} = \cos \Theta + i \sin \theta$)

$$b\alpha_b \cos \Theta_3 + c\alpha_c \cos \Theta_4 = a\omega^2 \sin \Theta_2 + b\omega_b^2 \sin \Theta_3 + c\omega_c^2 \sin \Theta_4$$

$$- a\alpha \cos \Theta_2 - b\alpha_b \sin \Theta_3 - c\alpha_c \sin \Theta_4 = a\omega^2 \cos \Theta_2 + b\omega_b^2 \cos \Theta_3$$

$$+ c\omega_c^2 \cos \Theta_4 + a\alpha \sin \Theta_2$$

Solving for α_b:

$$\alpha_b = \frac{a\omega^2 \cos (\Theta_2 - \Theta_4) + b\omega_b^2 \cos (\Theta_3 - \Theta_4) + c\omega_c^2}{b \sin (\Theta_4 - \Theta_3)} - \frac{a\alpha \sin (\Theta_2 - \Theta_4)}{b \sin (\Theta_4 - \Theta_3)} \tag{12.5}$$

Solving for α_c:

$$\alpha_c = \frac{a\omega^2 \cos (\Theta_2 - \Theta_3) + b\omega_b^2 + c\omega_c^2 \cos (\Theta_3 - \Theta_4)}{c \sin (\Theta_3 - \Theta_4)} + \frac{a\alpha \sin (\Theta_2 - \Theta_3)}{c \sin (\Theta_3 - \Theta_4)} \tag{12.6}$$

To determine ω_b, ω_c, α_b and α_c it is necessary to find the angles θ_3 and θ_4. The following formulas are needed, Fig. 12-4a and b:

From triangle $A_0 AB_0$: $l = \sqrt{d^2 + a^2 - 2\,ad \cos \Theta_2}$ (12.7)

$$\beta = \cos^{-1}\left[\frac{l^2 + d^2 - a^2}{2\,ld}\right] \tag{12.8}$$

From triangle ABB_0: $\psi = \cos^{-1}\left[\frac{b^2 + l^2 - c^2}{2\,bl}\right]$ (12.9)

There are two cases:

I: $0 \leqslant \theta_2 \leqslant 180°$
II: $180° < \theta_2 < 360°$

Case I:	$\Theta_3 = \psi - \beta$	(12.10)

Case II:	$\Theta_3 = \psi + \beta$	(12.11)

From triangle ABB_0:	$\lambda = \cos^{-1}\left[\dfrac{l^2 + c^2 - b^2}{2\,lc}\right]$	(12.11)

Case I:	$\Theta_4 = 360° - \lambda - \beta$	(12.12)

Case II:	$\Theta_4 = 360° - \lambda + \beta$	(12.12)

A numerical example is given later.

Expressions for the accelerations of the centers of gravity G_3 and G_4 will now be developed.

Using the designations from Fig. 12-5:

$$a_{G_3} = a_A \nrightarrow a_{G_3/A}$$

$$a_{G_3} = a_A^n \nrightarrow a_A^t \nrightarrow a_{G_3/A}^n \nrightarrow a_{G_3/A}^t$$

$$a_A^n = i^2 \cdot A_0A \cdot \omega^2 e^{i\theta_2} = -A_0A \cdot \omega^2 e^{i\theta_2}$$

$$a_A^t = i \cdot A_0A \cdot \alpha e^{i\theta_2}$$

$$a_{G_3/A}^n = i^2 \cdot AG_3 \cdot \omega_b^2 e^{i(\theta_3 + \gamma_3)} = -AG_3 \cdot \omega_b^2 e^{i(\theta_3 + \gamma_3)}$$

$$a_{G_3/A}^t = i \cdot AG_3 \cdot \alpha_b e^{i(\theta_3 + \gamma_3)},$$

where ω and α are given and ω_b and α_b are determined by eqs. (12.3e) and (12.5).

Writing in form of complex vectors:

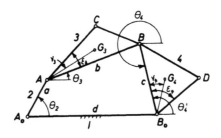

Fig. 12-5. Designation of links and angles.

$$a_{G_3} = -A_0A \cdot \omega^2(\cos \Theta_2 + i \sin \Theta_2) + i \cdot A_0A \cdot \alpha(\cos \Theta_2 + i \sin \Theta_2)$$

$$-AG_3 \cdot \omega_b^2(\cos (\Theta_3 + \gamma_3) + i \sin (\Theta_3 + \gamma_3)) + i \cdot AG_3 \cdot \alpha_b(\cos (\Theta_3 + \gamma_3)$$
$$+ i \sin (\Theta_3 + \gamma_3))$$

$$a_{G_3} = -A_0A \cdot (\omega^2 \cos \Theta_2 - \alpha \sin \Theta_2) - AG_3 \cdot (\omega_b^2 \cos (\Theta_3 + \gamma_3)$$
$$+ \alpha_b \sin (\Theta_3 + \gamma_3)) + i[-A_0A \cdot (\omega^2 \sin \Theta_2 - \alpha \cos \Theta_2)$$
$$- AG_3 \cdot (\omega_b^2 \sin (\Theta_3 + \gamma_3) - \alpha_b \cos (\Theta_3 + \gamma_3))] \qquad (12.13)$$

Similarly for the acceleration of G_4:

$$a_{G_4} = i^2 \cdot B_0G_4 \cdot \omega_c^2 e^{i(\theta'_4 - \gamma_4)} + i \cdot B_0G_4 \cdot \alpha_c e^{i(\theta'_4 - \gamma_4)}$$

$$\Theta'_4 = \pi - (2\pi - \Theta_4) = \Theta_4 - \pi$$

$$a_{G_4} = -B_0G_4 \cdot \omega_c^2 e^{i(\theta_4 - \gamma_4)} e^{i(-\pi)} + i \cdot B_0G_4 \cdot \alpha_c e^{i(\theta_4 - \gamma_4)} \cdot e^{i(\pi)}$$

$$= B_0G_4 \cdot \omega_c^2 e^{i(\theta_4 - \gamma_4)} - i \cdot B_0G_4 \cdot \alpha_c e^{i(\theta_4 - \gamma_4)}$$

$$= B_0G_4 \cdot \omega_c^2 (\cos (\Theta_4 - \gamma_4) + i \sin (\Theta_4 - \gamma_4))$$

$$-i \cdot B_0G_4 \cdot \alpha_c (\cos (\Theta_4 - \gamma_4) + i \sin (\Theta_4 - \gamma_4))$$

$$a_{G_4} = B_0G_4 \cdot (\omega_c^2 \cos (\Theta_4 - \gamma_4) + \alpha_c \sin (\Theta_4 - \gamma_4))$$

$$+ i[B_0G_4(\omega_c^2 \sin (\Theta_4 - \gamma_4) - \alpha_c \cos (\Theta_4 - \gamma_4))] \qquad (12.14)$$

The acceleration of point C, Fig. 12-5, can now be found by writing AC instead of AG_3, and ε_3 instead of γ_3; and similarly for point D:

$$a_C = -A_0A \cdot (\omega_a^2 \cos \Theta_2 - \alpha_a \sin \Theta_2) - AC \cdot (\omega_b^2 \cos (\Theta_3 + \varepsilon_3)$$
$$- \alpha_b \sin (\Theta_3 + \varepsilon_3)) + i[-A_0A \cdot (\omega_a^2 \sin \Theta_2 - \alpha_a \cos \Theta_2)$$
$$- AC \cdot (\omega_b^2 \sin (\Theta_3 + \varepsilon_3) - \alpha_b \cos (\Theta_3 + \varepsilon_3))] \qquad (12.15)$$

$$a_D = B_0D \cdot (\omega_c^2 \cos (\Theta_4 - \varepsilon_4) + \alpha_c \sin (\Theta_4 - \varepsilon_4))$$
$$+ i[B_0D(\omega_c^2 \sin (\Theta_4 - \gamma_4) - \alpha_c \cos (\Theta_4 - \gamma_4))] \qquad (12.16)$$

Example: Let us again choose $a = 1$ in., $b = 2.9$ in., $c = 2.58$ in., $d = 4$ in., $\theta_2 = 50°$, $N = 1000$ rpm clockwise (constant velocity), i.e., $\alpha = 0$.
Solution: Substituting into equations 12.1 to 12.10 (Case 1):

$$l = \sqrt{4^2 + 1^2 - 2 \cdot 4 \cdot 1 \cdot \cos(50°)}$$

$$= 3.44 \text{ in.} \tag{12.7}$$

$$\beta = \cos^{-1}\left(\frac{3.44^2 + 4^2 - 1^2}{(2)(3.44)(4)}\right)$$

$$= 12.8° \tag{12.8}$$

$$\psi = \cos^{-1}\left(\frac{2.9^2 + 3.44^2 - 2.58^2}{(2)(2.9)(3.44)}\right)$$

$$= 47.3° \tag{12.9}$$

$$\theta_3 = 47.3° - 12.8°$$

$$= 34.5° \tag{12.10}$$

$$\lambda = \sin^{-1}\left(\frac{2.9}{2.58} \cdot \sin(47.3°)\right)$$

$$= 55.4° \tag{12.11}$$

$$\theta_4 = 360° - 55.4 - 12.8°$$

$$= 291.8° \tag{12.12}$$

$$\omega_b = -\frac{(1.0)\sin(-241.7°)}{(2.9)\sin(-257.2°)}\left(-\frac{(1000)(2\pi)}{(60)}\right)$$

$$= 32.6 \text{ radians/sec (counterclockwise)} \tag{12.3e}$$

$$\omega_c = \frac{(1.0)\sin(15.5°)}{(2.58)\sin(-257.2°)}\left(-\frac{(1000)(2\pi)}{(60)}\right)$$

$$= -11.2 \text{ radians/sec (clockwise)} \tag{12.3f}$$

$$\alpha_c = +\frac{(1.0)(105)^2 \cos 15.5° + (2.9)(32.6)^2 + (2.58)(11.2)^2 \cos(-257.2°)}{(2.58)\sin(-257.2°)}$$

$$= \frac{10,600 + 3,080 - 71.5}{2.52}$$

$$= 5{,}390 \text{ radians/sec}^2 \text{ counterclockwise} \tag{12.6}$$

Determination of Acceleration in the Slider-Crank

Let us develop analytical expressions first, Fig. 12-6:

$$s = a + h - R \cos \theta - L \cos \beta \tag{12.17}$$

$$e = R \sin \theta + L \sin \beta \tag{12.18}$$

From (12.17):
$$v = R\omega \sin \theta + (L)(\sin \beta)\left(\frac{d\beta}{dt}\right) \tag{12.19}$$

From (12.18):
$$R\omega \cos \theta + (L)(\cos \beta)\left(\frac{d\beta}{dt}\right) = 0$$

$$\beta' = \frac{d\beta}{dt} = -\frac{\omega R \cos \theta}{L \cos \beta} \tag{12.20}$$

(12.19 and 12.20):
$$v_B = R\omega\left[\sin \theta - \frac{\sin \beta \cos \theta}{\cos \beta}\right]$$

$$= R\omega \frac{\sin (\theta - \beta)}{\cos \beta} \tag{12.21}$$

$$a_B = R\omega \frac{\cos \beta \cos (\theta - \beta)(\omega - \beta) + \sin \beta \sin (\theta - \beta)}{\cos^2 \beta}$$

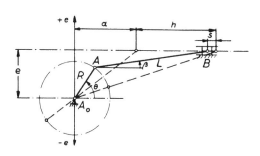

Fig. 12-6.   Eccentric slider-crank.

$$= R\omega\left[\frac{\cos\beta\,\cos\,(\theta-\beta)\omega}{\cos^2\beta} + \frac{\omega R\,\cos\theta\,\cos\beta\,\cos\,(\theta-\beta)}{L\,\cos^3\beta}\right.$$

$$\left. - \frac{\omega R\,\cos\theta\,\sin\beta\,\sin\,(\theta-\beta)}{L\,\cos^3\beta}\right]$$

$$= R\omega^2\left[\frac{\cos\,(\theta-\beta)}{\cos\beta} + \frac{\lambda\cos\theta\,(\cos\beta\,\cos\,(\theta-\beta) - \sin\beta\,\sin\,(\theta-\beta))}{\cos^3\beta}\right]$$

$$= R\omega^2\left[\frac{\cos\,(\theta-\beta)}{\cos\beta} + \lambda\,\frac{\cos^2\theta}{\cos^3\beta}\right] \qquad\qquad (12.22)$$

where $\lambda = \dfrac{R}{L}$

For $e = 0$:

$$R\sin\theta + L\sin\beta = 0 \qquad\qquad (12.18a)$$

$$\sin\beta = -\frac{R}{L}\sin\theta = -\lambda\sin\theta \qquad\qquad (12.23)$$

Example: Let $A_0A = R = 1.00$ in., $AB = L = 3.00$ in., $\theta = 45°$, offset or eccentricity $e = -0.35$ in., and $N = 1000$ rpm clockwise, Fig. 12-7. Find the acceleration of the slider.

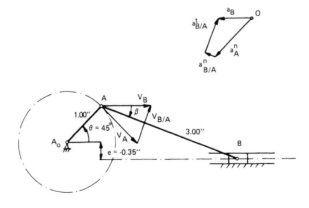

Fig. 12-7. Eccentric slider-crank; determination of acceleration.

Solution: $-0.35 = (1.0) \sin (45°) + (3.0) \sin \beta$ (12-18)

$$\sin \beta = -\frac{1.057}{3.0}$$

$$\beta = -20.6°$$

$$a = (1.00)\left(\frac{(1000) \cdot (2\pi)}{(60)}\right)^2\left[\frac{\cos (45° + 20.6°)}{\cos (-20.6°)} + \left(\frac{1}{3}\right)\left(\frac{\cos^2 (45°)}{\cos^3 (-20.6°)}\right)\right]$$

$$= 10,900\left[\frac{0.413}{0.936} + \left(\frac{1}{3}\right)\left(\frac{0.5}{0.82}\right)\right]$$

$$= 10,900 (0.442 + 0.203)$$

$$= 7,025 \text{ in./sec}^2$$ (12.22)

Next, let us find the acceleration graphically. Velocities and accelerations in the slider crank are found by exactly the same method that was used for the four-bar linkage. Let us use the same values as in previous example:

$$V_B = V_A \rightarrowtail V_{B/A}$$

$$V_A = (1.0)\frac{(1000) \cdot (2\pi)}{(60)} = 105 \text{ in./sec}$$

The velocity vector diagram in Fig. 12-7 gives:

$$V_B = 102 \text{ in./sec}$$
$$V_{B/A} = 80 \text{ in./sec}$$

The vector equation for the acceleration of B is written in the usual way.

$$a_B = a_A \rightarrowtail a_{B/A}$$

$$a_B^n \rightarrowtail a_B^t = a_A^n \rightarrowtail a_A^t \rightarrowtail a_{B/A}^n \rightarrowtail a_{B/A}^t$$

$a_B^n = 0$ because B moves in a straight line

$$a_A^n = (1.0)\left(\frac{1000 \cdot 2\pi}{60}\right)^2 = 10,900 \text{ in./sec}^2$$

$$a_A^t = 0$$

$$a^n_{B/A} = \frac{(80)^2}{3} = 2{,}135 \text{ in./sec}^2$$

$a^t_{B/A} =$ is perpendicular to AB

The vector equation takes therefore the form:

$$\underline{a^n_B} \nrightarrow \underset{4}{a^t_B} = \underline{a^n_A} \nrightarrow a^t_A \nrightarrow \underline{a^n_{B/A}} \nrightarrow \underset{4}{a^t_{B/A}}$$

and is solved with the help of the vector diagram, Fig. 12-7.

$a_B = 7{,}000$ in. sec^2, which is in close agreement with the analytically obtained result of 7,025 in./sec^2.

Static and Dynamic Force Analysis of Linkages

In this section we shall discuss how to determine the forces created in the joints of a four-bar linkage due to external static forces (a static analysis) and internal dynamic forces (a dynamic analysis) created by the irregular movement of the links. The methods described are immediately applicable to cam mechanisms and most other complex mechanisms. A graphical solution as well as an analytical solution is presented. The formulas developed can be programmed on a computer.

Static Force Analysis

First, the case of a single force F_4 acting on the rocker 4 in Fig. 12.8a is considered and the turning moment which must act at the crank to keep the four-bar linkage in static equilibrium is determined together with the forces acting in the joints. Next a single force F_3 acting on the coupler 3 in Fig. 12.9a is considered and again the turning moment and the joint forces are found.

Superposition of the two forces F_3 and F_4 yields the joint forces created when these forces act simultaneous. The necessary turning moment at the crank to balance these forces is found as the vector sum of the two moments created by F_3 and F_4.

For mechanisms with turning joints only—linkages—it is customary and sufficiently accurate from a practical point of view to disregard friction and this is done in the following.

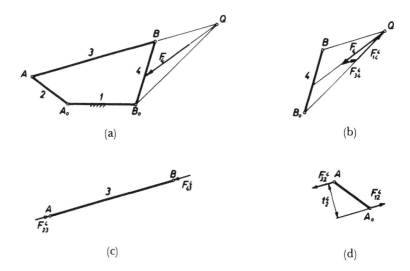

Fig. 12-8a.   Four-bar linkage; a force F_4 acts on the rocker 4.

Fig. 12-8b.   Free-body diagram of the rocker.

Fig. 12-8c.   Free-body diagram of the coupler 3.

Fig. 12-8d.   Free-body diagram of the crank 2.

Graphical Solution: The four-bar linkage in Fig. 12-8a is shown in a position where a static external force is acting on the rocker 4 and it it desired to find the turning moment acting on the crank to counteract this force and also the joint forces are to be found.

To find the external forces acting on bar 4 a free body diagram is drawn, Fig. 12-8b. Bar 4 is acted upon by three forces, namely F_4 which is given, a force F_{34}^4 where the upper "4" designates that force F_{34}^4 is created by F_4 and a force F_{14}^4, which is also created by the force F_4. F_{34}^4 means a force created by the force F_4 and is the force with which body 3 acts on body 4. Similarly F_{14}^4 is created by F_4 and is the force with which body 1 acts on body 4. Force F_4 is given in magnitude and direction (i.e. a vector) and, assuming no friction in joint B, the direction of F_{34}^4 on B is known. The point of concurrency Q† is found as the intersection of F_4 and F_{34}^4 and since body 4 is in equilibrium the direction of F_{14}^4 must be in the direction of B_0Q.

The vector polygon for the three forces acting on body 4 can now be drawn, Fig. 12-8b, and since for equilibrium $\Sigma F = 0$, $F_4 + F_{34}^4 + F_{14}^4 = 0$.

†Q is the point through which the resultant of all external forces pass.

The force F^4_{34} acting from 3 on 4 must be equal to but opposite in direction to the force F^4_{43} acting from 4 on 3 or,

$$F^4_{34} = -F^4_{43}$$

There are only two forces acting on 3, namely F^4_{43} and F^4_{23} acting at the points B and A, respectively. These two forces have the same direction as AB. A free body diagram is shown in Fig. 12-8c and

$$F^4_{23} = -F^4_{43} = F^4_{34}$$

Fig. 12-8d is a free body diagram of crank 2. The only forces acting on 2 are F^4_{32} and F^4_{12} and

$$F^4_{23} = -F^4_{32}$$

and

$$F^4_{12} = -F^4_{32}$$

Summing the forces acting on 2,

$$\Sigma F = F^4_{32} + F^4_{12} = 0$$

but because these two forces are not collinear, an external moment M^4_2 acting on the crank must be added to maintain equilibrium and

$$\Sigma M = 0 \qquad M^4_2 + F^4_{32} \cdot t^4_2 = 0$$

$$M^4_2 = -F^4_{32} \cdot t^4_2$$

where t^4_2 is the moment arm.

If more than one force acts on bar 4 they can always be combined to one single force F_4 and the procedure described used.

The next case to be considered is an external force F_3 acting on coupler 3, Fig. 12-9a. This force gives rise to two other forces, namely F^3_{23} and F^3_{43} at point A and B, respectively. The direction of F^3_{43} is the same as the direction B_0B. The procedure is now exactly as described in connection with the rocker 4. Point Q is found as the intersection of F_3 and B_0B. AQ determines the direction of F^3_{23} and the vector force polygon can now be drawn so that

$$F = F_3 + F^3_{23} + F^3_{34} = 0$$

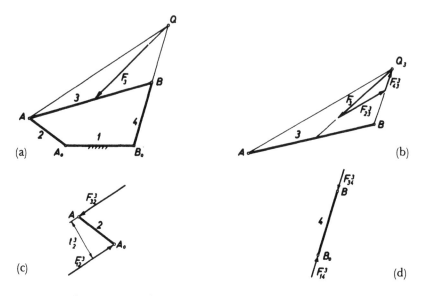

Fig. 12-9a.   Four-bar linkage; a force F_3 acts on the coupler 3.

Fig. 12-9b.   Free-body diagram of the coupler 3.

Fig. 12-9c.   Free-body diagram of the crank 2.

Fig. 12-9d.   Free-body diagram of the rocker 4.

The resultant moment M_2^3 can be found from

$$M = M_2^3 + F_{32}^3 \cdot t_2^3$$

$$M_2^3 = -F_{32}^3 t_2^3$$

Besides $\qquad\qquad F_{12}^3 = -F_{32}^3 \qquad$ (Figs. 12-9b and c)

A free body diagram of 4, Fig. 12-9d, shows the force $F_{14}^3 = -F_{34}^3$.

   Again if a number of forces are acting on the coupler 3 they can be combined into a single force F_3 and the force analysis carried out.

   Fig. 12-10 shows two forces F_3 and F_4 acting on bar 3 and 4, respectively. Using the principle of superposition which says that the combined action of forces F_3 and F_4 can be found considering each force acting separately and then adding the resultant forces vectorially, all joint forces and the moment acting on crank 2 can be found. The analytical solution will be shown later.

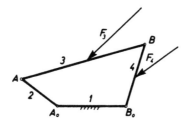

Fig. 12-10.   Four-bar linkage; coupler and rocker are both acted upon by forces F_3 and F_4, simultaneously.

Dynamic Force Analysis

Before proceeding with the analysis of linkages we will examine briefly the forces and the resultant accelerations of a body having plane motion.

A number of external forces F_1, F_2, F_3, ... act on a body free to move in a plane, Fig. 12-11. The center of gravity is designated G. The center of gravity has an acceleration which is determined by

$$\Sigma F = ma_G \qquad (12.24)$$

where

$$\Sigma F = F_1 \rightarrowtail F_2 \rightarrowtail F_3 \rightarrowtail \dots$$

m is the mass of the body, and a_G is the acceleration of G. This acceleration has the same direction as the resultant ΣF of all external forces. The line of action of the resultant ΣF of the external forces will pass through G if the body has no angular acceleration. If there is an angular acceleration then the line of action is determined by

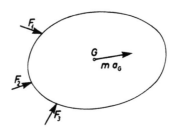

Fig. 12-11.   A number of external forces act on the body; the resultant passes through the center of gravity G.

$$hma_G = I_0\alpha \text{ or } h = \frac{I_0\alpha}{ma_G} \qquad (12.25)$$

where h is the distance between ma_G and the line of action, I_0 is the mass moment of inertia around the center of gravity, and α the angular acceleration of the body. If α is CW then the line of action is determined as in Fig. 12-12a, and if α is CCW, as in Fig. 12-12b. The line of action is parallel with ma_G.

D'Alembert's Principle

The equation

$$\Sigma F = ma_G \qquad (12.24)$$

expresses that the summation of all external forces ΣF acting on a body is equal to the sum of all the external forces acting on the particles of a body this sum being equal to ma_G.

Equation 12.24 can be written as

$$F + (-ma_G) = 0 \qquad (12.24a)$$

and this equation interpreted as saying that the summation of all external forces added vectorially to the reversed effective forces $-ma_G$ equals zero. In effect then, 12.24a represents a condition of static equilibrium and the equations of statics, $\Sigma F = 0$ and $\Sigma M = 0$ apply.

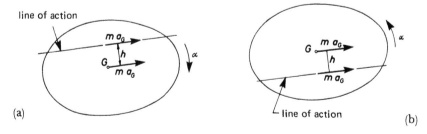

Fig. 12-12a.   A number of external forces act on the body; the resultant does not pass through the center of gravity G.

Fig. 12-12b.   A number of external forces act in the body; the resultant does not pass through the center of gravity G.

It is obvious that ma_G may be found by means of eq. 12.24 if the external forces are known or, if the mass of the body as well as the acceleration of the center of gravity is known.

Dynamic Force Analysis of a Four-bar Linkage

A dynamic force analysis of a linkage is always started with a complete velocity and acceleration analysis.

Example: The proportions of a four-bar linkage, Fig. 12-13, are: $A_0A = 1$ in., $AB = BB_0 = 3$ in., $A_0B_0 = 4$ in., $N = 3500$ rpm CCW. The bars are made of steel $\rho = 0.28$ lbs/in^3, and all bars have the same cross sectional area of 0.5 by 5/32 in^2. Using the distance between the joints of a bar + 0.5 in. as the effective length when calculating weight and moment of inertia of the bars find the turning moment necessary to drive the four-bar linkage of Fig. 12-13 in the position shown. Find all joint forces and disregard forces due to gravity.

Graphical solution: To find velocities:

$$V_A = A_0A \cdot \omega = \frac{(1.0)(3500 \cdot 2\pi)}{60} = 367 \text{ in./sec}$$

$$V_B = V_A \mapsto V_{B/A} \tag{12.1}$$

and from the velocity diagram, Fig. 12-14, when the velocity vectors are drawn to scale:

$$V_B = 367 \text{ in./sec}$$

$$V_{B/A} = 367 \text{ in./sec}$$

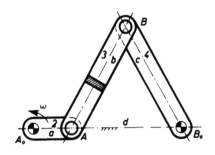

Fig. 12-13.   Four-bar linkage.

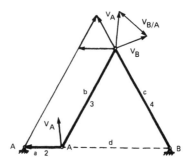

Fig. 12-14.   Four-bar linkage; determination of velocities.

To find accelerations:

$$\underset{4}{\underline{\underline{a_B^n}}} + \underset{}{\underline{a_B^t}} = \underline{\underline{a_A^n}} + \underline{\underline{a_A^t}} + \underline{\underline{a_{B/A}^n}} + \underset{4}{A_{B/A}^n} \tag{12.2}$$

$$a_A^n = A_0A \cdot \omega^2 = (1.0)\frac{(3500 \cdot 2\pi)^2}{(60)} = 134336 \text{ in./sec}^2$$

$$a_{B/A}^n = \frac{V_{B/A}^2}{AB} = \frac{(367)^2}{(3.0)} = 44896 \text{ in./sec}^2$$

$$a_B^n = \frac{V_B^2}{AB} = \frac{(367)^2}{(3.0)} = 44896 \text{ in./sec}^2$$

The above vector equation is solved in Fig. 12-15 by drawing the vectors to scale and measuring on the drawing:

$$a_B^t = 101378 \text{ in./sec}^2$$
$$a_{B/A}^n = 101378 \text{ in./sec}^2$$

and therefore:

$$\alpha_b = \frac{a_{B/A}^t}{AB} = \frac{101378}{3.0} = 33793 \text{ rad/sec}^2 \; CW$$

$$\alpha_c = \frac{a_B^t}{B_0B} = \frac{101378}{3.0} = 33793 \text{ rad/sec}^2 \; CCW$$

The acceleration a_{G_3} of the center of gravity G_3 for bar 3 and a_{G_4} of G_4 of bar 4

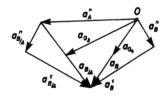

Fig. 12-15. Determination of accelerations.

has been found in Fig. 12-15. Here the head of a_{G_3} must divide $a_{B/A}$ into two equal parts and must have its tail at 0, and a_{G_4} must divide a_B into two equal parts and must also have its tail at 0. By measuring on the drawing:

$$a_{G_3} = 108268 \text{ in./sec}^2$$

$$a_{G_4} = 54134 \text{ in./sec}^2$$

Using an effective bar length of $1 + 0.5$ in. when calculating weights and moments of inertia:

$$W_2 = (1.0 + 0.5)(5/32)(0.5)(0.28) = 0.033 \text{ lb}$$

$$W_3 = W_4 = (3.0 + 0.5)(5/32)(0.5)(0.28) = 0.077 \text{ lb}$$

$$I_3 = I_4 = \frac{1}{12} m(l + 0.5)^2 = \frac{1}{12} \frac{(0.077)}{(386)} (3.0 + 0.5)^2$$

$$= 0.000204 \text{ lb-sec}^2\text{-in}$$

In Fig. 12-16a the four-bar linkage is shown with a_{G_3}, a_{G_4}, α_b, and α_c drawn to scale and direction, and a free body diagram analysis is carried out converting mass forces to external forces, and the analysis is started with bar 4, Fig. 12-16b, where the resultant is calculated from

$$m_4 \cdot a_{G_4} = \frac{(0.077)}{(386)} (54134) = 10.8 \text{ lb}$$

The distance h_4 is found from

$$h_4 = \frac{I_4 \cdot \alpha_4}{m_4 \cdot a_{G_4}} = \frac{(0.000204)(33793)(386)}{0.077 \cdot 54134} = 0.64 \text{ in.}$$

The resultant $m_4 \cdot a_{G_4}$, of the external forces is moved the distance h_4 such that 4 will have angular acceleration in a CCW direction (as explained in Fig. 12-12a and b). $m_4 \cdot a_{G_4}$ is drawn in the opposite direction thus determining$-m_4 \cdot a_{G_4}$. This force is now treated exactly like an external force and drawing the closed vector polygon F_{34}^4 and F_{14}^4 and found. By measuring on the drawing:

$$F_{34}^4 = 8 \text{ lb}$$

$$F_{14}^4 = 2.7 \text{ lb}$$

Next a free body diagram of bar 3, Fig. 12-16c, is drawn and the resultant of the external forces is

$$m_3 \cdot a_{G_3} = \frac{(0.077)(108268)}{(386)} = 21.6 \text{ lb}$$

where it is assumed that bar 4 is not acted upon by any external force other than those due to $m_3 \cdot a_{G_3}$. The distance

$$h_3 = \frac{I_3 \cdot \alpha_3}{m_3 \cdot a_{G_3}} = \frac{(0.000204)(33793)(386)}{(0.077)(108268)} = 0.32 \text{ in.}$$

and repeating the foregoing procedure (Fig. 12-16c):

$$F_{43}^3 = 4.3 \text{ lb}$$

$$F_{23}^3 = 21.0 \text{ lb}$$

It should be noted that the force $F_{43}^4 = -F_{34}^4$ from Fig. 12-16b is not included in the free body diagram of bar 3, Fig. 12-16c, but could of course be added to F_{43}^3 to determine the joint force in B. The same is valid for the joint force in B_0. The force F_{43}^4 is added to the free body diagram of bar 2, Fig. 12-16d, as a force F_{32}^4 where

$$F_{32}^4 = F_{43}^4$$

The combined force at A on bar 2 due to the acceleration forces on bar 2 and 4 is

$$F_{32} = F_{32}^3 + F_{43}^4 = 28.6 \text{ lb}$$

Because of the angular velocity of bar 2 an acceleration force acts in the direction of A_0A the magnitude of which is

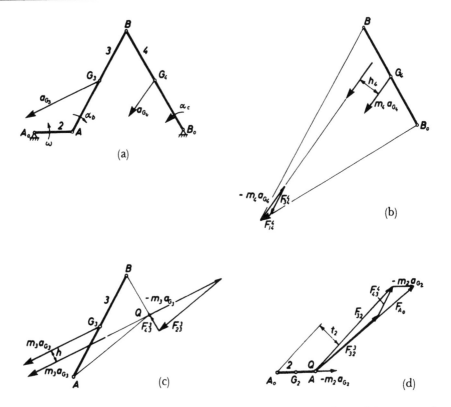

Fig. 12-16a. Four-bar linkage with the acceleration vectors a_{G_3} and a_{G_4} of the centers of gravity G_3 and G_4, respectively.

Fig. 12-16b.   Determination of forces.

Fig. 12-16c.   Determination of forces.

Fig. 12-16d.   Determination of forces.

$$m_2 \cdot a_{G_2} = m_2 \cdot A_0 G_2 \cdot \omega^2 = \frac{(0.033)(0.5)}{(386)} \left(\frac{3500 \cdot 2\pi}{60} \right)^2 = 5.74 \text{ lb}$$

Therefore the joint force in A_0 is

$$^F A_0 = F_{32}^3 + F_{43}^4 + (-m_2 a_{G_2}) = 33 \text{ lb}$$

The turning moment M_2 is found from the equation

$$M_{A_0} = 0 \text{ or } M_2 + F_{32} \cdot t_2 = M_2 + (28.6)(0.7) = M_2 + 20.0$$

$$M_2 = 20.0 \text{ lb-in. } CW$$

since the moment $F_{32} \cdot t_2$ acts CCW.

Analytical Solution: The foregoing example which was solved graphically is solved analytically in the following:

$N = 3500$ rpm CCW; $\omega_a = + 367$ rad/sec (CCW)

$\theta_2 = 0°$

$\theta_3 = 60°$

$\theta_4 = 300°$

$$\omega_b = - \frac{a \sin (\theta_2 - \theta_4)}{b \sin (\theta_3 - \theta_4)} \, \omega_a = \frac{(1.0) \sin (-300°)}{(3.0) \sin (-240°)} .367 = -122 \text{ rad/sec}^2$$

$$\text{(12.3e)}$$

$$\omega_c = \frac{a \sin (\theta_2 - \theta_3)}{c \sin (\theta_3 - \theta_4)} \, \omega_a = \frac{(1.0) \sin (-60°)}{(3.0) \sin (-240°)} .367 = -122 \text{ rad/sec}^2$$

$$\text{(12.3f)}$$

$$\alpha_b = \frac{a \omega_a^2 \cos (\theta_2 - \theta_4) + b \omega_b^2 \cos (\theta_3 - \theta_4) + c \omega_c^2}{b \sin (\theta_4 - \theta_3)} - 0 \qquad \text{(12.5)}$$

$$= \frac{(1.0)(367)^2(0.5) + (3.0)(-122)^2(-0.5) + (3.0)(-122)^2}{(3.0)(-0.866)}$$

$$= -34514 \text{ rad/sec}^2$$

$$\alpha_c = \frac{a \omega_a^2 \cos (\theta_2 - \theta_3) + b \omega_b^2 + c \omega_c^2 \cos (\theta_3 - \theta_4)}{c \sin (\theta_3 - \theta_4)} + 0 \qquad \text{(12.6)}$$

$$= \frac{(1.0)(367)^2(0.5) + (3.0)(-122)^2 + (3.0)(-122)^2(-0.5)}{(3.0)(0.866)}$$

$$= 34514 \text{ rad/sec}^2$$

$$a_{G_3} = [-A_0A(\omega_a^2 \cos\theta_2 - \alpha_a \sin\theta_2) - AG_3(\omega_b^2 \cos\theta_3 + \alpha_b \sin\theta_3)]I$$

$$+ [-A_0A(\omega_a^2 \sin\theta_2 - \alpha_a \cos\theta_2) - AG_3(\omega_b^2 \sin\theta_3 - \alpha_b \cos\theta_3)]J \tag{12.13}$$

$$= [(-1.0)(367)^2 \cos 0° - (1.5)[(-122)^2 \cos(60°) + (-34514)\sin(60°)]I$$

$$+ [(-1.0)(367)^2 \cdot \cos(90°) - 1.5[(-122)^2 \sin(60°)$$

$$- (-34514)\cos(60°)]J = [-134689 - 11163 + 44834]I$$

$$- [19335 + 25886]J$$

(For explanation of I, J, and K components see appendix at the end of the chapter).

$$a_{G_3} = -101018I - 45221J$$

$$a_{G_4} = B_0G_4[\omega_c^2 \cos\theta_4 + \alpha_c \sin\theta_4]I + [B_0G_4(\omega_c^2 \sin\theta_4 - \alpha_c \cos\theta_4)]J \tag{12.14}$$

$$= (1.5)[(-122)^2 \cos(300°) + 34514 \sin(300°)]I$$

$$+ (1.5)[(-122)^2 \sin(300°) - 34514 \cos(300°)]J$$

$$a_{G_4} = -33671I - 45220J$$

$$a_{G_2} = -A_0G_2\omega_a^2 I = (0.5) \cdot (367)^2 = 67345I$$

The results calculated so far and otherwise given and to be used in the following are:

$\alpha_b = -34514K$	$R_A = 1.0 \angle 0° = 1.0I$
$\alpha_c = 34514K$	$R_{G_4} = 1.5 \angle 120° = -0.75I + 1.3QJ$
$a_{G_3} = -101018I - 45220J$	$R_B = 3.0 \angle 120° = -1.5I + 2.6J$
$a_{G_4} = -33671I - 45220J$	$R_{G_3/A} = 1.5 \angle 60° = 0.75I + 1.3J$
	$R_{B/A} = 3.0 \angle 60° = 1.5I + 2.6J$
	$F_{34}^4 \angle 240° = -F_{34}^4 \cdot 0.5I - F_{34}^4 \cdot 0.866J$

$$\Sigma M_{B_0} = 0: R_{G_4} \times (-m_4 a_{G_4}) + (-I_4 \alpha_4) + R_B \times F_{34}^4 = 0$$

$$-m_4 \cdot a_{G_4} = -\frac{0.077}{386}(-33671I - 45220J) = 6.72I + 9.02J$$

$$-I_4 \cdot \alpha_4 = (-0.000204)(34514)K = -7.04K$$

$$R_{G_4} \times (-m_4 \cdot a_{G_4}) = \begin{matrix} I & J & K \\ -0.75 & 1.3 & 0 \\ 6.72 & 9.02 & 0 \end{matrix} = (-6.77 - 8.74)K = -15.51K$$

(For explanation of cross product or moment vector see appendix at the end of the chapter)

$$R_B \times F_{34}^4 = \begin{vmatrix} I & J & K \\ -1.5 & 2.6 & 0 \\ -0.5 \cdot F_{34}^4 & -0.866 \cdot F_{34}^4 & 0 \end{vmatrix} = 2.6\, F_{34}^4 K$$

$$\Sigma M_{B_0} = 0: -15.51K - 7.04K + 2.6K = 0$$

$$F_{34}^4 = \frac{22.54}{2.6} = 8.67 \text{ lb}$$

$$F_{34}^4 = -4.34I - 7.51J$$

$$(\Sigma F)^4 = 0: F_{34}^4 + (-m_4 \cdot a_{G_4}) + F_{14}^4 = 0$$

$$-4.34I - 7.51J + 6.72I + 9.02J + F_{14}^4 = 0$$

$$F_{14}^4 = -2.38I - 1.51J$$

$$\Sigma M_A = 0: R_{G_3/A} \times (-m_3 \cdot a_{63}) + (-I_3 \alpha_3) + R_{B/A} \times F_{43}^3 = 0$$

$$-m_3 \cdot a_{G_3} = -\frac{0.077}{386}(-101018I - 45220J)$$

$$= 20.15I + 9.02J$$

$$-I_3 \alpha_3 = (-0.000204)(-35514K) = 7.04K$$

$$F_{43}^3 = F_{43}^3 \angle 120° = -F_{43}^3 \cdot 0.5I + F_{43}^3 \cdot 0.866J$$

$$R_{G_3/A} \times (-m_3 \cdot a_{G_3}) = \begin{vmatrix} I & J & K \\ 0.75 & 1.3 & 0 \\ 20.15 & 9.02 & 0 \end{vmatrix} = (6.77 - 26.20)K = -19.43K$$

$$R_{B/A} \times F_{43}^3 = \begin{vmatrix} I & J & K \\ 1.5 & 2.6 & 0 \\ -0.5F_{43}^3 & 0.866F_{43}^3 & 0 \end{vmatrix} = (1.3 + 1.3)F_{43}^3K = 2.6F_{43}^3K$$

$$\Sigma M_A = 0: \ -19.43K + 7.04K + 2.6F_{43}^3K$$

$$F_{43}^3 = \frac{12.39}{2.6} = 4.77 \text{ lb}$$

$$F_{43}^3 = -2.39I + 4.13J \qquad (F_{43}^3 = F_{43}^3 \angle 120°)$$

$$(\Sigma F)^3 = 0 \quad F_{23}^3 + (-m_3 \cdot a_{G_3}) + F_{43}^3 = 0$$

$$F_{23}^3 + 20.15I + 9.02J - 2.39I + 4.13J = 0$$

$$F_{23}^3 = -17.76I - 13.15J$$

$$F_{32} = -F_{43}^4 - F_{23}^3 = 4.34I + 7.51J + 17.76I + 13.15J$$

$$= 22.1I + 20.66J$$

$$\Sigma M_{A_0} = 0: \ R_A \times F_{32} + M_2 = 0$$

$$R_A \times F_{32} = \begin{vmatrix} I & J & K \\ 1.0 & 0 & 0 \\ 22.1 & 20.66 & 0 \end{vmatrix} = 20.66K = 20.66 \text{ lb-in } CCW$$

$$M_2 = 20.66 \text{ lb-in } CW$$

$$F_{41} = -F_{14}^4 - F_{43}^3 = 2.38I + 1.51J + 2.39I - 4.13J$$

$$F_{41} = 4.77I - 2.62J$$

The foregoing problem which was solved graphically and analytically for one position may on the basis of the equations used in this connection be programmed on a digital computer. Fig. 12-17 shows ω_b, ω_c, α_b, and α_c as a function of θ for $\theta = 0°[10°]\ 360°$.

APPENDIX

I, J and K Components of a Vector

A vector in space may be written

$$R = R_x + R_y + R_z$$

where R_x, R_y, and R_z are the components of the R-vector in the direction of the x-, y- and z-axis.

The vector R may also be written

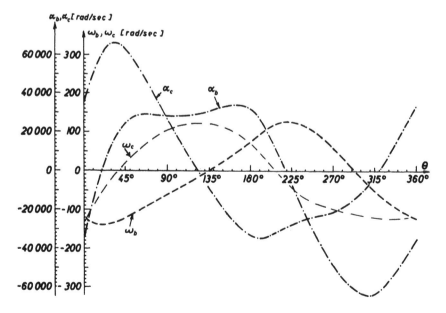

Fig. 12-17. Angular velocities and angular accelerations for one complete revolution of the crank.

$$R = R_x I + R_y J + R_z K$$

where I, J, and K are unit vectors in the direction of the x-, y-, and z-axis, respectively.

The Moment Vector

The moment vector is the cross product of the relative position vector R and the force vector F or,

$$M = R \times F$$

The R and F vector may be written as

$$R = R_x I + R_y J + R_z K$$

and

$$F = F_x I + F_y J + F_z K$$

The cross product M is

$$M = \begin{vmatrix} I & J & K \\ R_x & R_y & R_z \\ F_x & F_y & F_z \end{vmatrix}$$

Notice: It is important to remember to write the cross product as $R \times F$ and not as $F \times R$ because $R \times F = -F \times R$

If $M < 0$ then the moment is acting CW otherwise CCW

Determination of Acceleration in the Inverted Crossed Slide-Crank

Figure 12-18 shows an inverted crossed slide-crank with zero eccentricity; i.e., the center line of the translating follower passes through A_0 and is shown in a position where the input member has rotated the angle θ. By inspection it is seen that

$$y = \frac{R_{min}}{\cos \theta} - R_{min}$$

Differentiating once:

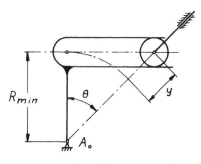

Zero e centricity

Fig. 12-18.   Inverted crossed slide-crank; $c = o$.

$$V = R_{min}\omega \frac{\tan \theta}{\cos \theta}$$

and differentiating once more:

$$a = R_{min}\omega^2 \frac{1 + \sin^2 \theta}{\cos^3 \theta} \qquad\qquad 12.26$$

The formula for the acceleration has been used to make the nomogram, Figs. 12-24 and 12-25.

Example: $\theta = 45°$, $N = 140$ rpm, counterclockwise (+), $R_{min} = 1$ in. Find the acceleration of the translating roller follower.

Solution: A line is drawn from $\theta = 45°$ through $N = 140$ rpm, Fig. 12-24, which gives a point on the pivot line k; a line from the k-axis to $R_{min} = 1$ in. gives $a = 900$ in./sec² in the upward direction.

Calculated values:

$$a = R_{min}\omega^2 \frac{1 = \sin^2 \theta}{\cos^3 \theta}$$

$$= (1)\left(\frac{140 \cdot 2\pi}{60}\right)^2 \frac{1 + 1/2}{1/4\sqrt{2}}$$

$$= 905 \text{ in./sec}^2$$

In Figure 12-19 the center line of the translating follower does not pass through A_0 but has a positive eccentricity e.

The following formula is evident from the geometry of the figure:

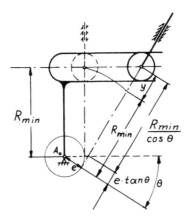

Fig. 12-19.   Positive eccentricity inverted crossed slide-crank; $e > o$.

$$y = \frac{R_{min}}{\cos \theta} - R_{min} + e \cdot \tan \theta$$

from which $$v = \frac{R_{min} \sin \theta}{\cos^2 \theta} + \omega \frac{e}{\cos^2 \theta}$$

and $$a = R_{min}\omega^2 \frac{1 + \sin^2 \theta}{\cos^3 \theta} + e\omega^2 \frac{2 \sin \theta}{\cos^3 \theta} \qquad 12.27$$

In the case of negative eccentricity, Fig. 12-20, the following formulas are similarly obtained:

$$y = \frac{R_{min}}{\cos \theta} - R_{min} - e \tan \theta$$

$$v = \omega \frac{R_{min} \sin \theta}{\cos^2 \theta} - \omega \frac{e}{\cos^2 \theta}$$

$$a = R_{min}\omega^2 \frac{1 + \sin^2 \theta}{\cos^3 \theta} - e\omega^2 \frac{2 \sin \theta}{\cos^3 \theta} \qquad 12.28$$

Formulas 12.27 and 12.28 differ from 12.26 only by the term $e\omega^2 \dfrac{2 \sin \theta}{\cos^3 \theta}$.

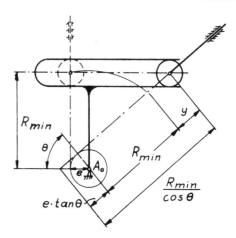

Fig. 12-20.   Negative eccentricity inverted crossed slide-crank; $e > o$.

Therefore, in all three cases the nomogram, Fig. A-1, is used to determine $R_{\min}\omega^2 \dfrac{1 + \sin^2 \theta}{\cos^3 \theta}$ and then the term $e\omega^2 \dfrac{2 \sin \theta}{\cos^3 \theta}$ is added for positive eccentricity and subtracted for negative eccentricity.

The term $\dfrac{2 \sin \theta}{\cos^3 \theta}$ can be obtained from Table 12-1, and the term $c\omega^2$ is found with the help of the $R\omega^2$ nomogram, Fig. A-1. (See Appendix).

Example:

$$\theta = 30°, \ N = 300 \text{ rpm}, \ R_{\min} = 2.5 \text{ in., and } e = +0.5 \text{ in.}$$

Solution:

$$a = 4,750 + (500) \cdot (1.54) = 4,750 + 770 = 5,520 \text{ in./sec}^2 \qquad (12.27)$$
$$\quad\ \ \uparrow \qquad\quad \uparrow \qquad\qquad \uparrow$$

nomogram   nomogram   Table 12-1
Fig. 12-24   Fig. A-1

In the following we will make use of Coriolis acceleration; problems

Table 12-1

θ	0	5	10	15	20	25	30	35	40	45	50
$\dfrac{2 \sin \theta}{\cos^3 \theta}$	0	.176	.358	.576	.825	1.13	1.54	2.08	2.86	4.0	5.77

involving this acceleration are in general considered difficult, but here is given an exact procedure so that no difficulties should arise.

The Coriolis term comes into the picture when the acceleration of one point is measured relative to another point, the two points being coincident.

The vector equation with Coriolis acceleration in it has the following form:

$$a = a_m \leftrightarrow a_p \leftrightarrow 2u\omega,$$

and can be stated in words as follows:

The acceleration of a particle moving on a path which is, itself, rotating, is the sum of three terms. The first term a_m is the absolute acceleration of a point on the path coincident with the particle at the instant considered. The second term a_p is the acceleration of the particle moving along the path considered as fixed. The third term is the Coriolis acceleration.

The velocity u of the particle relative to the path is tangent to the path. ω is the angular velocity of the path.

The direction of $2u \cdot \omega$ is always normal to the path, and the sense corresponds to the direction in which the head of the vector u would move if it were rotated about its tail in the sense of ω.

Example: Given the inverted offset slider-crank, Fig. 12-21; the crank $A_0 A$ rotates at a constant velocity of 500 rpm counterclockwise. Find the angular acceleration of the output link.

Solution: Points A and D are two separate but coincident points; point A is on the link $A_0 A$ and point D is on the link $B_0 D$.

$$V_A = V_D \leftrightarrow V_{A/D}$$

$$V_A = (2.20) \frac{(500)(2\pi)}{60} = 115 \text{ in./sec}$$

and from the velocity vector diagram:

$$V_{A/D} = 117 \text{ in./sec}$$
$$V_D = 47 \text{ in./sec}$$

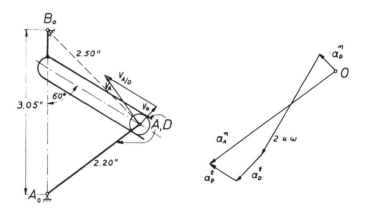

Fig. 12-21.   Inverted offset slider-crank.

Point A is the point moving on the path, the path being the slot, and the vector equation, therefore, is written in the form:

$$a_A = a_D \dashrightarrow a_p \dashrightarrow 2u\omega$$

or $\quad a_A^n \dashrightarrow a_A^t = a_D^n \dashrightarrow a_D^t \dashrightarrow a_p^n \dashrightarrow a_p^t \dashrightarrow 2u\omega \qquad\qquad 12.29$

$$a_A^n = (2.20)\left(\frac{(500)(2\pi)}{(60)}\right)^2 = 6,000 \text{ in./sec}^2$$

$$a_A^t = 0$$

$$a_D^n = \frac{V_D^2}{B_0 D} = \frac{47^2}{2.5} = 883 \text{ in./sec}^2$$

$a_p^n = 0$ (because the path is a straight line and there can therefore be no acceleration perpendicular to the path or the slot.)

$$2u\omega = (2)(V_{A/D})\left(\frac{V_D}{B_0 D}\right) = (2.0)(117)\left(\frac{47}{2.50}\right)$$

$$= 4,400 \text{ in./sec}^2$$

(ω is the angular velocity of the link containing the path, i.e., in this case the angular velocity of link $B_0 D$).

2uω is perpendicular to the path and has a direction indicated by the direction the head of the vector $V_{A/D}$ would move if it were rotated about its tail in the sense of ω.

$$a_A^n \rightarrowtail a_A^t = a_D^n \rightarrowtail \underset{4}{a_D^t} \rightarrowtail a_p^n \rightarrowtail \underset{4}{a_p^t} \rightarrowtail \underline{\underline{2u\omega}}$$

The acceleration diagram is also drawn in Fig. 12-20 and,

$$a_D^t = 1{,}360 \text{ in./sec}^2$$

$$\alpha_{B_0D} = \frac{(1{,}360)}{(2.50)} = 544 \text{ radians/sec}^2$$

In the following example input and output is reversed.

Example: Given the inverted offset slider-crank, Fig. 12-22; the input link B_0D rotates at 500 rpm clockwise. Find the angular acceleration of the output link A_0A.

Solution: The method is the same as in the previous example and,

$$V_A = V_D \rightarrowtail V_{A/D}$$

$$V_D = (2.50)\left(\frac{(500)\cdot(2\pi)}{(60)}\right)$$

$$= 131 \text{ in./sec}$$

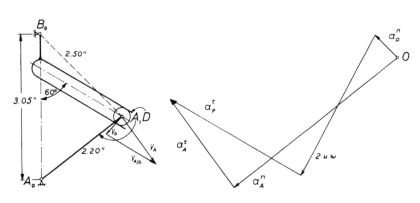

Fig. 12-22.   Inverted offset slider-crank.

and from the velocity vector diagram:

$$V_{A/D} = 322 \text{ in./sec}$$

$$V_A = 315 \text{ in./sec}$$

$$a_A = a_D \nrightarrow a_p \nrightarrow 2u\omega$$

$$a_A^n \nrightarrow a_A^t = a_D^n \nrightarrow a_D^t \nrightarrow a_p^n \nrightarrow a_p^t \nrightarrow 2u\omega$$

$$a_A^n = \frac{315^2}{(2.20)} = 45,100 \text{ in./sec}^2$$

$$a_D^n = (2.50)\left(\frac{(500)(2\pi)}{(60)}\right)^2 = 6,820 \text{ in./sec}^2$$

$$a_D^t = 0$$

$$a_p^n = 0$$

$$2u\omega = (2)(322)\left(\frac{(500)(2\pi)}{(60)}\right) = 33,700 \text{ in./sec}^2$$

$$\underline{\underline{a_A^n}} \nrightarrow \underset{4}{a_A^t} = \underline{\underline{a_D^n}} \nrightarrow \underline{\underline{a_D^t}} \nrightarrow \underline{\underline{a_p^n}} \nrightarrow \underset{4}{a_p^t} \nrightarrow \underline{\underline{2u\omega}}$$

and the vector diagram, Fig. 12-22 yields

$$a_A^t = 23,100 \text{ in./sec}^2$$

$$\alpha_{A_0A} = \frac{(23,100)}{(2.20)} = 10,500 \text{ radians/sec}^2 \text{ counter clockwise}$$

Determination of Acceleration in the Crossed Slide-Crank

Example: Given the crossed slide crank, Fig. 12-23. The input member A_0A turns counterclockwise at 650 rpm. Find the acceleration of the output member.

Solution: A and B are two coincident points; A belongs to the input member and is considered a fixed point on the center line of the slot and B is a point on the output member.

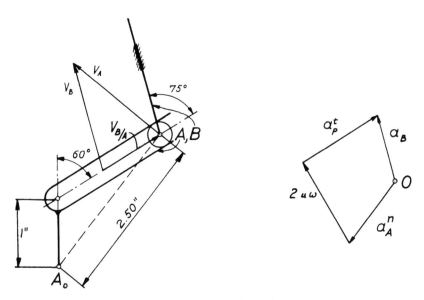

Fig. 12.23. Crossed slide-crank.

$$V_B = V_A \mapsto V_{B/A}$$

$$V_A = (2.50)\frac{(650)(2\pi)}{(60)} = 170 \text{ in./sec}$$

$$V_{B/A}w = 103 \text{ in./sec}$$

$$V_B = 164 \text{ in./sec}$$

$$a_B = a_A \mapsto a_p \mapsto 2u\omega$$

$$a_B^n \mapsto a_B^t = a_A^n \mapsto a_A^t \mapsto a_p^n \mapsto a_p^t \mapsto 2u\omega \qquad (12.29)$$

$$a_B^n = 0$$

$$a_A^n = (2.50)\left(\frac{(650)(2\pi)}{(60)}\right)^2 = 11,600 \text{ in./sec}^2$$

$$a_A^t = 0$$

$$a_p^n = 0$$

$$2u\omega = (2.0)(103)\frac{(650)(2\pi)}{(60)} = 14{,}000 \text{ in./sec}^2$$

$$a_B^n \mapsto a_B^t = a_A^n \mapsto a_A^t \mapsto a_p^n \mapsto a_p^t \mapsto 2u\omega$$
$$\qquad\qquad 4 \qquad\qquad\qquad\qquad 4$$

From the acceleration diagram, Fig. 12-23,

$$a = 10{,}000 \text{ in./sec}^2$$

Method of Finite Difference Calculations in Kinematic Analysis

Graphical differentiation of the time-displacement diagram will give the velocity and one more differentiation will give the acceleration. The graphical method, however, is inaccurate and time-consuming. Often the time-displacement diagram is given in the form of an equation and it is, of course, then possible to differentiate this equation once to get the velocity

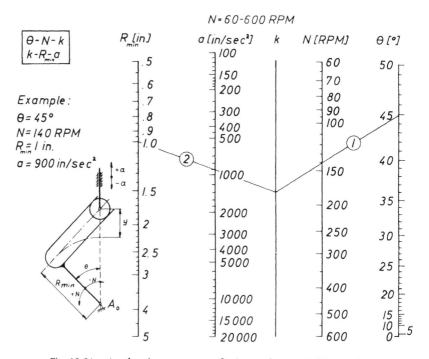

Fig. 12-24.   Acceleration nomogram for inverted crossed slide-crank.

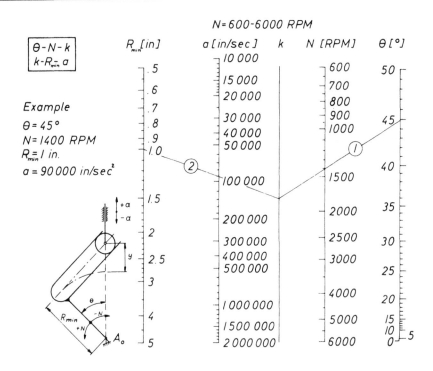

Fig. 12-25. Acceleration nomogram for inverted crossed slide-crank.

and twice to get the acceleration. In many cases, however, the analytical expression is so complex that calculations take much time.

The method of finite differences will enable the velocity and acceleration to be found if a limited number of points on the time-displacement curve are given. The surprising accuracy that can be obtained with the help of this method even if only a comparatively few values of the time-displacement curve are known will be demonstrated by an example later on.

Suppose we know the value of a function y at some pivotal points—y_{i-2}, y_{i-1}, y_i, y_{i+1}, y_{i+2}, evenly spaced at a distance Δt as shown in Fig. 12-26. The derivative y_i' is found from:

$$v = y_i' = \frac{1}{2\Delta t}(y_{i+1} - y_{i-1})$$ (12.30)

with an error of the order of δ^2

and y_i'' from $\quad a = y_i'' = \frac{1}{(\Delta t)^2}(y_{i+1} - 2y_i + y_{i-1})$ (12.31)

This is called the first order method.

By taking more terms into account, a higher accuracy is obtained, for instance:

$$v = y_i' = \frac{1}{12\Delta t}(-y_{i+2} + 8y_{i+1} - 8y_{i-1} + y_{i-2}) \qquad (12.32)$$

$$a = y_i'' = \frac{1}{12(\Delta t)^2}(-y_{i+2} + 16y_{i+1} - 30y_i + 16y_{i-1} - y_{i-2}) \qquad (12.33)$$

with an error of order δ^2

This is called the second order method.

It may also be noted that the close the spacing the more accurate the results.

Example: For a 60-degree cam shaft rotation a translating roller follower rises 1 in. with simple harmonic motion and the cam shaft speed is 1000 rpm. Calculate the displacement for each 10 degrees of cam shaft rotation and using method of finite differences, calculate velocity and acceleration at these points with an error of the order of δ^2 (1st order method) and with an error of δ^4 (2nd order method), Fig. 12-27.

Solution: The time T for rise is found from:

$$T = \frac{60}{N} \cdot \frac{\beta}{360} = \frac{60}{1000} \cdot \frac{60}{360} = \frac{1}{100}\ \text{sec}$$

Whence

$$\Delta t = \frac{T}{6} = \frac{1}{600}\ \text{sec}$$

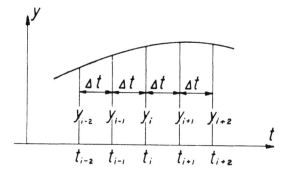

Fig. 12-26. Equal spacing of pivotal points.

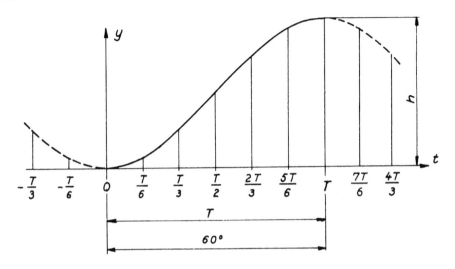

Fig. 12-27.   Displacement curve is extended beyond interval.

Table 12-2 shows the calculated displacement values, which are calculated

from
$$y = \frac{h}{2}\left[1 - \cos\left(\pi \frac{t}{T}\right)\right]$$

It is evident from formulas 12.30–12.33, that since they contain y_{i-1}, y_{i-2}, y_{i+1}, and y_{i+2}, we must have values for two points beyond the beginning and end of the stroke, so that we can find velocities and accelerations at the beginning and end of the stroke. For these, we assume that the displacement curve is extended beyond the given interval in such a way that:

$$y\left(-\frac{T}{6}\right) = y\left(\frac{T}{6}\right) = 0.06699 \text{ in.}$$

Table 12-2

t	0	$\dfrac{T}{6}$	$\dfrac{T}{3}$	$\dfrac{T}{2}$	$\dfrac{2T}{3}$	$\dfrac{5T}{6}$	T
y	0.0	0.06699	0.25000	0.50000	0.75000	0.93301	1.00000

$$y\left(-\frac{T}{3}\right) = y\left(\frac{T}{3}\right) = 0.25000 \text{ in.}$$

$$y\left(\frac{7T}{6}\right) = y\left(\frac{5T}{6}\right) = 0.93302 \text{ in.}$$

$$y\left(\frac{4T}{3}\right) = y\left(\frac{2T}{3}\right) = 0.75000 \text{ in.}$$

(Expressed in mathematical terms one would say: If $y_i' = 0$, then $y_{i-1} = y_{i+1}$, $y_{i-2} = y_{i+2}$ if $\Delta t \ll 1$) ($\ll 1$ means *much less* than 1).

First order method: As an example, values of the velocity and the acceleration at a few pivotal points are calculated:

$$v = y_i' = \frac{1}{2\Delta t}(y_{i+1} - y_{i-1}) \tag{12.30}$$

At $t = 0$: $v = \dfrac{1}{(2)(0.00167)}(0.06699 - 0.06699) = 0$

$t = \dfrac{T}{6}$: $v = \dfrac{1}{(2)(0.00167)}(0.25000 - 0.00000) = 74.850 \text{ in./sec}$

$t = \dfrac{T}{3}$: $v = \dfrac{1}{(2)(0.00167)}(0.50000 - 0.06699) = 129.644 \text{ in./sec}$

$$a = y_i'' = \frac{1}{(\Delta t)^2}(y_{i+1} - 2y_i + y_{i-1}) \tag{12.31}$$

At $t = 0$: $a = \dfrac{1}{(0.00167)^2}(0.06699 - 2(0.0) + 0.06699)$

$$= 48,040 \text{ in./sec}^2$$

$t = \dfrac{T}{6}$: $a = \dfrac{1}{(0.00167)^2}(0.25000 - 2(0.06699) + 0.0)$

$$= 41,601 \text{ in./sec}^2$$

$t = \dfrac{T}{3}$: $a = \dfrac{1}{(0.00167)^2}(0.50000 - 2(0.25000) + 0.06699)$

$$= 24,020 \text{ in./sec}^2$$

Second order method:

$$v = y_i' = \frac{1}{12\Delta t}(-y_{i+2} + 8y_{i+1} - 8y_{i-1} + y_{i-2}) \qquad (12.32)$$

At $t = \dfrac{T}{6}$: $v = \dfrac{1}{12(0.00167)}[-0.5 + (8)(0.25) - (8)(0.0)$

$$+ 0.06699] = 78.193 \text{ in./sec}$$

and for the acceleration:

$$a = y_i'' = \frac{1}{(12)(\Delta t)^2}(-y_{i+2} + 16y_{i+1} - 30y_i + 16y_{i-3} - y_{i-2}) \qquad (12.33)$$

At $t = \dfrac{T}{3}$: $a = \dfrac{1}{(12)(0.00167)^2}[-0.75 + (16)(0.5) - (30)(0.25)$

$$+ (16)(0.06699) - 0.0] = 24{,}557 \text{ in./sec}^2$$

The values in Table 12-3 show that if two neighboring values to y_i are used, namely $y_i - 1$ and y_{i+1}, the accuracy obtained is sufficient for engineering purposes (5 per cent accuracy).

This method can be used to check the acceleration of a cam, the profile of which is found by actually measuring the contours of the cam with a dial indicator. (See Chapter 9).

Table 12-3

t seconds	y inches	v in./sec			a in./sec^2		
		1st Order Method	2nd Order Method	Exact Value	1st Order Method	2nd Order Method	Exact Value
$t = -\dfrac{T}{3} = -0.00333$	0.25000						
$t = -\dfrac{T}{6} = -0.00167$	0.06699						
$t = 0$	0.00000	0.000	0.000	0.000	48040	49114	49348
$t = \dfrac{T}{6} = 0.00167$	0.06699	74.850	78.193	78.540	41601	42529	42737
$t = \dfrac{T}{3} = 0.00333$	0.25000	120.644	135.433	186.035	24020	24557	24674

$t = \dfrac{T}{2} = 0.00500$	0.50000	150.150	156.386	157.080	0.0	0.0	0.0
$t = \dfrac{2T}{3} = 0.00667$	0.75000	130.036	135.437	136.035	−24020	−24552	−24674
$t = \dfrac{5T}{6} = 0.00833$	0.93302	75.075	78.193	78.540	−41601	−42538	−42737
$t = T = 0.01000$	1.00000	0.0	0.0	0.0	−48040	−49104	−49348
$t = \dfrac{7T}{6} = 0.01167$	0.93302						
$t = \dfrac{4T}{3} = 0.01333$	0.75000						

Computer Programs for Analysis and Synthesis of Cam Mechanisms

Six programs are in the disk bound in the back of this book. The programs are written in BASIC for an IBM PC. Before beginning, refer to the directions, *Using the Cam Programs,* at the beginning of the book. In order for you to get acquainted with the programs and what they can do a procedure is worked out for each of the six programs, and examples are given so that you can check the answers the disk programs give. Remember, when you try out values other than those given in the examples, the computer will not always catch "bad data," but if you follow what is said in this book about cam design, this should be no problem. The actual printed values may differ a little, dependent upon the computer model.

13.1. Translating Roller Follower (Fig. 13-1)

Program TRANS1 (Open- and closed track cam)

Insert the disk and get into BASIC(A) mode. Enter **RUN"MAIN"** and press the return key. After a moment you will see the following on the screen:

A—TRANSLATING ROLLER FOLLOWER
ANALYSIS AND SYNTHESIS PROGRAMS FOR PROPOR-
TIONING OF TRANSLATING ROLLER FOLLOWER
MECHANISMS

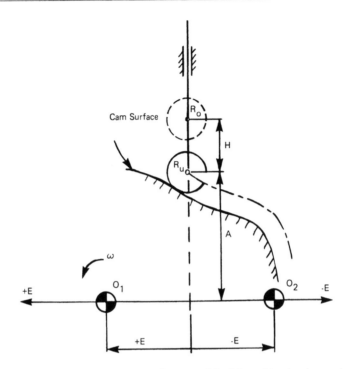

Fig. 13-1. When cam shaft center is at O_1 the eccentricity E is positive, but is negative at O_2. Distance A and E are always defined relative to the path $R_u R_0$ of the translating roller follower. If $E = 0$ then the cam mechanism is a radial translating roller follower.

B—SWINGING ROLLER FOLLOWER
         ANALYSIS AND SYNTHESIS PROGRAMS FOR PROPOR-
         TIONING OF SWINGING ROLLER FOLLOWER MECHANISM.
Z—EXIT PROGRAM AND RETURN TO DOS.
ENTER THE GROUP LABEL YOU WOULD LIKE TO RUN
PROGRAMS FOR (A, B OR Z)?

A—Translating Roller Follower comprises three programs TRANS1, TRANS2, and TRANS3 (see later).
B—Swinging Roller Follower comprises three programs SWING4, SWING5, and SWING6 (see later).

The first program, TRANS1, is for a translating roller follower and will calculate the maximum pressure angle by rise and return for given time-displacement diagrams and cam proportions.

Example 1, Parabolic Motion, Translating Roller Follower (Fig. 13-2): The total

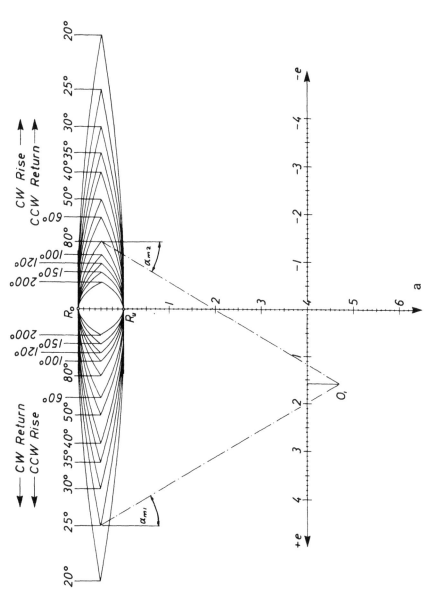

Fig. 13-2.   Pressure angle nomogram; parabolic motion.

stroke of a translating roller follower is $h = 1$ in., the motion is parabolic by rise and return, the angle for rise is $25°$ and the angle for return is $80°$. The eccentricity $E = +1.58$ in., $A = 4.7$ in. and the cam rotates CCW.

Enter **A**. Then enter **1** and the screen will look like this:

RISE		RETURN
1	PARABOLIC MOTION	11
2	SIMPLE HARMONIC MOTION	12
3	CYCLOIDAL MOTION	13
4	3-4 POLYNOMIAL	14
5	3-4-5 POLYNOMIAL	15
6	4-5-6-7 POLYNOMIAL	16
7	MODIFIED TRAPEZOIDAL ACCEL.	17
8	MODIFIED SINUSOIDAL ACCEL.	18
9		19

Parabolic motion was given for both **RISE**, key in **1**, and **RETURN**, key in **11**.

The prompter asks **ECCENTRICITY E?** Key in **+1.58** (or just 1.58). If the eccentricity is negative, don't forget the minus sign.

The prompter asks **DISTANCE A, IN?** Look at the nomogram, Fig. 13-2. The distance A is the distance from the lowest position of the center of the roller to the cam shaft center, measured in the direction of the stroke of the follower. Therefore, key in **4.7**.

TOTAL ANGLE FOR RISE, DEGREES?
Key in **25**.

TOTAL ANGLE FOR RETURN, DEGREES?
Key in **80**.

IF CAM ROTATION IS CW, WRITE "CW"; ELSE WRITE "CCW".
Key in **CCW**.

When you have entered the last value, all values will be displayed so that you can check whether they are correct. If they are correct, **EXIT AND EXECUTE** the program, otherwise enter into correction mode.

The printer, on execution, prints the input values and then prints what is the motion for rise, namely **PARABOLIC MOTION, RISE**, and after some time it prints **MAX. PRESSURE ANGLE BY RISE = 30.01 AT CAM SHAFT ANGLE THETA = 12.5**. This means that the maximum pressure angle by rise is $30.01°$ and that the maximum pressure angle occurs when the cam has rotated $12.5°$ from its lowest position at the start of the rise portion. Then the printer prints **PARABOLIC MOTION; RETURN** meaning that the return of the follower from its highest to its lowest position takes place according to parabolic motion. Finally it prints **MAX. PRESSURE ANGLE BY RETURN = 30.09° AT CAM ANGLE THETA = 40°**, meaning that the maximum pressure angle by return is $-30.09°$ and that the maximum pressure angle occurs when the cam has rotated 40° from its highest position at the start of the return portion.

You are now ready to start another example, still using program TRANS1, which will also be used for Examples 3 through 6.

Example 2, Simple Harmonic Motion, Translating Roller Follower (Fig. 13-3): Given: $h = 1$ in., simple harmonic motion by rise and return, angle for rise = 25°, angle for return = 80°, $E = 1.24$ in., $A = 3.65$ in., CCW rotation of cam. Try this example and the answers should be:

Maximum pressure angle by rise = 29.74° THETA = 12°

Maximum pressure angle by return = −30.06° THETA = 46°.

If you do not get these answers, go back to Example 1 and do that example over again.

Example 3, Cycloidal Motion, Translating Roller Follower (Fig. 13-4): Given: $h = 1$ in., cycloidal motion by rise and return, angle for rise = 25°, angle for return = 80°, $E = 1.58$ in., $A = 4.7$ in., CCW rotation of cam. The answers should be:

Maximum pressure angle by rise = 30.06° THETA = 12°

Maximum pressure angle by return = −30.3° THETA = 43°

If you do not get these answers, go back to Example 1 and do it over again.

Example 4, 3-4-5 Polynomial, Translating Roller Follower (Fig. 13-5): Given: $h = 1$ in., 3-4-5 polynomial by rise and return, angle for rise = 25°, angle for return = 80°, $E = 1.48$ in., $A = 4.45$ in., CCW rotation. The answers should be:

Maximum pressure angle by rise = 29.7° THETA = 12°

Maximum pressure angle by return = −29.9° THETA = 44°

If you do not get these answers, go back to Example 1.

Example 5, Modified Trapezoidal Acceleration, Translating Roller Follower (Fig. 13-6). Given: $h = 1$ in., modified trapezoidal acceleration by rise and return, angle for rise = 25°, angle for return = 80°, $E = 1.59$ in., $A = 4.71$ in., CCW rotation of cam. The answers should be:

Maximum pressure angle by rise = 29.9° THETA = 12.375°

Maximum pressure angle by return = −30.2° THETA = 42°

If you have arrived at the correct answers to the foregoing problems, you are ready for one final example. A definite advantage of this program is, that you can choose any of the given eight time-displacement curves for rise and any of these same eight curves for return, and the program will give the answer. So, as a final demonstration of the program let us choose the following.

Example 6, Translating Roller Follower: Given: $h = 1.57$ in., the time-displacement curve for rise is to be cycloidal motion, the time-displacement curve for return is modified trapezoidal acceleration, angle for rise = 25°, angle for return = 80°, $E = 2.0$ in., $A = 6.0$ in., CCW rotation of cam. Answers:

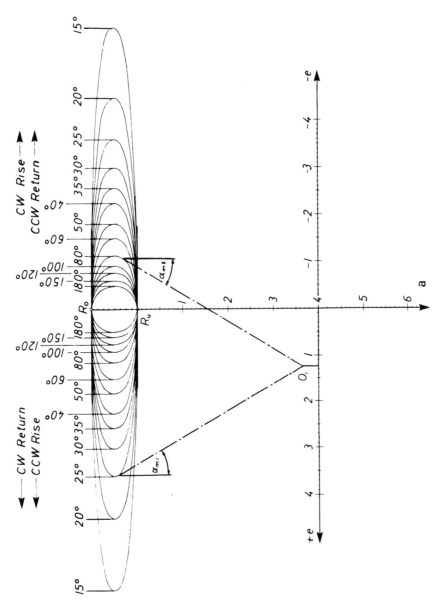

Fig. 13-3.   Pressure angle nomogram; simple harmonic motion.

Maximum pressure angle by rise = 37.6° THETA = 12°
Maximum pressure angle by return = −32.2° THETA = 42°

Program TRANS2 (open- and closed track cam)

In the following Examples 7 through 12, use program TRANS2, which is a synthesis program, i.e., it will calculate the smallest possible cam proportions for given maximum pressure angles by rise and return.

Example 7, Parabolic Motion, Translating Roller Follower: The total stroke of a translating roller follower is $h = 1$ in., the motion is parabolic by rise and return, the angle for rise = 25°, the angle for return = 80°, the maximum pressure angle by rise is 30°, by return 30°. The cam rotates *CCW*. Find the eccentricity E and the distance A. The answers are:

$E = 1.58$ in.

$A = 4.7$ in.

which agrees closely with the values given in Example 1, Fig. 13-2.

Example 8, Simple Harmonic Motion: Given: $h = 1$ in., simple harmonic motion by rise and return, angle for rise = 25°, angle for return = 80°, maximum pressure angle by rise is 30°, by return 30°. The cam rotates *CCW*. Find the eccentricity E and the distance A. The answers are:

$E = 1.22$ in.

$A = 3.63$ in.

The results agree closely with the values in Example 2, Fig. 13-3.

Example 9, Cycloidal Motion, Translating Roller Follower: Given: $h = 1$ in., cycloidal motion by rise and return, translating roller follower, angle for rise = 25°, angle for return = 80°, maximum pressure angle by rise = 30°, by return 30°. The cam rotates *CCW*. Find E and A. The answers are:

$E = 1.57$ in.

$A = 4.73$ in.

The results agree closely with the values given in Example 3, Fig. 13-4.

Example 10, 3-4-5 Polynomial, Translating Roller Follower: Given: $h = 1$ in., 3-4-5 polynomial by rise and return, angle for rise = 25°, angle for return = 80°, maximum pressure angle by rise = 30°, by return 30°. The cam rotates *CCW*. Find E and the distance A. The answers are:

$E = 1.47$ in.

$A = 4.41$ in.

These agree closely with Example 4, Fig. 13-5.

Example 11, Modified Trapezoidal Acceleration, Translating Roller Follower: Given: $h = 1$ in.; modified trapezoidal acceleration by rise and return, translating roller follower, angle for rise = 25°, angle for return = 80°. The maximum pressure angle by rise is 30°, by return 30°. The cam rotates *CCW*. Find E and A. The answers are:

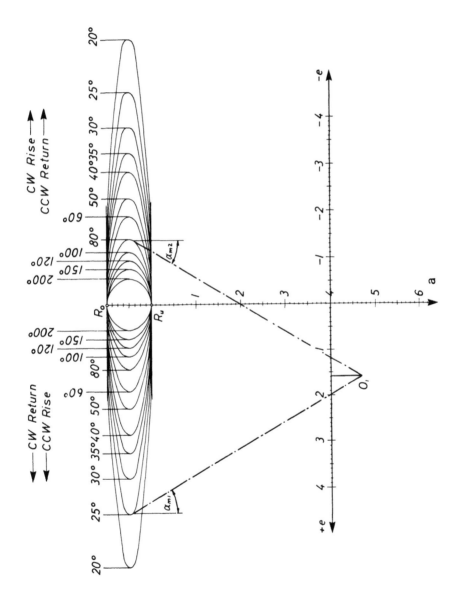

Fig. 13-4. Pressure angle nomogram; cycloidal motion.

$E = 1.57$ in.

$A = 4.72$ in.

and agree closely with example 5, Fig. 13-6.

As a final example, let us take the combination of different motions by rise and return.

Example 12, Translating Roller Follower: Given: $h = 1.59$ in., time-displacement curve for rise is cycloidal motion, by return modified trapezoidal acceleration, angle for rise 25°, for return 80°. The maximum pressure angle by rise is 30°, by return 30°. Find E and A. *CCW* rotation of cam. The answers are:

$E = 2.5$ in.

$A = 7.52$ in.

Let us try to input these values to program TRANS1. For the above values inputted to program TRANS1:

Maximum pressure angle by rise = 29.98°

Maximum pressure angle by return = −30.01°

which is a very close agreement.

Program TRANS3 (Closed-track cam)

The program TRANS3 calculates radius of curvature and the compressive stress for each 1° cam shaft rotation during rise and return for any of the eight time-displacement diagrams and cam proportions for translating roller follower. At the end of rise it prints out the maximum values of the compressive stress during acceleration and deceleration. It then prints out the corresponding values for return. *Do Examples 1 through 12 before doing the following examples.*

Example 13, Parabolic Motion by Rise and Return: Given: $h = 1$ in., $E = 0$, $A = 2.35$ in., angle for rise = 40°, angle for return = 40°, roller radius $R_f = .5$ in., roller width = .625 in., cam rotates with a speed of 300 rpm *CW*, cam material is steel, and weight of mass to be moved = 5 lbs. The answers are:

Maximum compressive stress during 1st half rise = 48408 psi at THETA = 20°

Maximum compressive stress during 2nd half rise = 41678 psi at THETA = 20°

Maximum compressive stress during 1st half return = 41678 psi at THETA = 20°

Maximum compressive stress during 2nd half return = 48408 psi at THETA = 20°

At the end the input values are printed. It is possible in this program to combine any of the eight curves for rise with any of the same eight curves

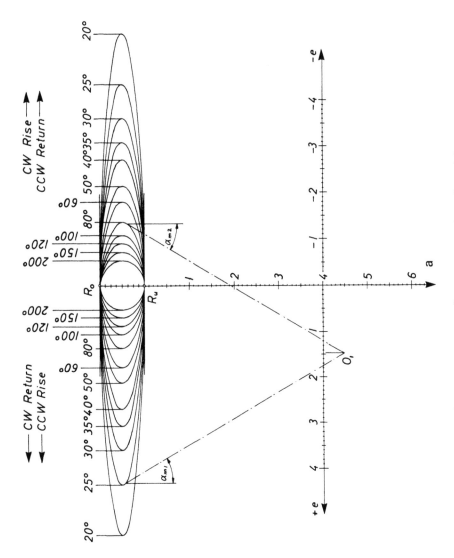

Fig. 13-5. Pressure angle nomogram; 3-4-5 polynomial.

for return. The programs can be used in various other ways than what they are directly intended for. Consider for instance the following.

Example 14, Translating Roller Follower: A radial translating roller follower is driven by an open track cam rotating with 1200 rpm. The stroke is .5 in., the motion is simple harmonic, the angle for rise is 60°, the follower roller is 1¼ in. in diameter, the prime or base circle is 3¼ in. in diameter. Determine whether the grinding wheel, which has a diameter of 4¾ in., is small enough to cut the cam profile.

This problem can actually be solved by means of program TRANS3. If you have been through Example 13 of this chapter, you will have noticed that values of RHO were printed out for each position of the follower roller. RHO is the radius of curvature of the cam, if the roller radius $R_f = 0$. So let us see how this problem is solved.

Solution: From the problem given we can input the following values: Simple harmonic motion for rise and return, $h = .5$ in., angle for rise = 60°, roller radius $R_f = .625$ in. (cam speed of 1200 rpm has no influence on the radius of curvature of the cam).

From the information given we conclude that $E = 0$ (it is a *radial* translating roller follower), we will put the total angle for return = 20° (to make necessary only a few calculations), the roller width $B = .75$ in., and the cam rotates *CCW*. Part of the printout is shown below.

h = .5 in., A = 1.625 in., B1 = 60°, B2 = 20°, R$_f$ = .625 in., B = .75 in., CAM SPEED, RPM = 1200 CCW, MASS TO BE MOVED IS 1 lb, MATERIAL IS STEEL.

SIMPLE HARMONIC MOTION; RISE

THETA	RHO	COMPRESSIVE STRESS
0	4.22	29,951
1	4.26	29,952
2	4.38	29,954
3	4.58	29,955
4	4.89	29,955
5	5.35	29,950
6	6.02	29,936
7	7.03	29,911
8	8.62	29,868
9	11.43	29,803
10	17.47	29,711
11	39.20	29,585
12	−36.53	29,420
13	−24.43	29,208
14	−13.33	28,942

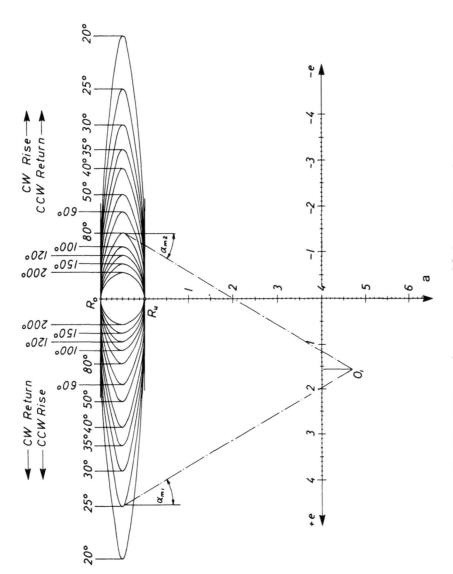

Fig. 13-6. Pressure angle nomogram; modified trapezoidal acceleration.

It is the positive RHO values that are of interest because positive values mean that the cam profile is concave at these points and the only limiting factor for the size of a grinding wheel is the concave part of the profile. It is the smallest positive value of RHO, that is the deciding factor. This value is RHO = 4.22 in., and adding the value of R_f = .625 in give 4.85 in. concave radius, but

$$4.85 > 2.375 \text{ in. (the radius of the grinding wheel)}$$

Inadequate convex curvature of the cam profile is a frequent problem in cam design, in the worst case undercutting may occur (see Fig. 5-29).

Example 15, Translating Roller Follower: A radial translating roller follower is driven by an open track cam rotating with 800 rpm, the stroke h = 1 in., the motion is Simple Harmonic, the angle for rise is 150°, the roller follower is 1¼ in. in diameter, the prime or base circle is 3¼ in. in diameter. Calculate whether there will be undercutting. To solve this problem we again use program TRANS3.

Solution: The inputted values are: h = 1 in., E = 0, A = 1.625 in., total angle for rise = 150°, total angle for return = 150°, roller radius R_f = .625 in., roller width B = .75 in., cam speed 800 rpm *CCW,* cam material is steel, weight to be moved = 1 lb. Part of the printout is shown below.

h = 1 in., E = 0, A = 1.625 in., B1 = 150°, B2 = 150° R_f = .625 in., B = .75 in, CAM SPEED, RPM = 800

SIMPLE HARMONIC MOTION, RISE

THETA	RHO	COMPRESSIVE STRESS
130	−2.0541	10150.9
131	−2.05463	10195.1
132	−2.05515	10237
133	−2.05565	10276.6
134	−2.05612	10313.8
135	−2.05657	10348.7
136	−2.057	10381.3
137	−2.0574	10411.6
138	−2.05777	10439.6
139	−2.05812	10465.4
140	−2.05843	10488.8
141	−2.05872	10510.1
142	−2.05898	10529
143	−2.05921	10545.7
144	−2.05942	10560.2

145	-2.05959	10572.5
146	-2.05973	10582.5
147	-2.05984	10590.3
148	-2.05991	10595.8
149	-2.05996	10599.2
150	-2.05998	10600.3

First of all, when checking for undercutting, it is the smallest negative value of RHO that is decisive. From the printout, the smallest negative value of RHO is −2.05 that occurs for THETA = 130°, i.e., at the end of the rise portion. This value is in close agreement with the value that can be read from the nomogram, Fig. 5-31, namely RHO_{min} = 2.0 in. In order that undercutting does not take place, $R_f \geqslant |RHO|$. The numerical value of RHO is 2.06 in. Substracting R_f = .625 in. yields 1.435 in., which is the actual radius of curvature of the cam at the end of the stroke. Therefore there is no undercutting (and the compressive stresses are OK, too).

The programs SWING4, SWING5, and SWING6 correspond to programs TRANS1, TRANS2, and TRANS3, respectively, with the difference that the roller follower is now on an arm that swings back and forth depending on the shape of the cam.

If you have worked the foregoing examples you should have no difficulties with the following.

13.2. Swinging Roller Follower (Fig. 13-7)

Program SWING4 (Open- and closed-track cam)

This program calculates maximum pressure angles by rise and return for given time-displacement diagrams and cam mechanism proportions. Before you attempt this program, you should have done at least example 1 in the foregoing.

Example 16, Swinging Roller Follower: Parabolic motion by rise and return, ϕ_0 = 28.65°, length of follower arm L_f = 2 in., distance MO = 5 in., R_{min} = 4 in., total angle for rise = 25°, total angle for return = 25°, cam rotates CCW, Fig. 13-7. The answer is:

Maximum pressure angle by rise = 44.3° at cam angle THETA = 12.5°

Maximum pressure angle by return = 46.9° at cam angle THETA = 12.5°

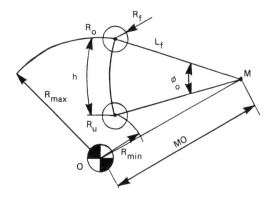

Fig. 13-7.

Program SWING5 (Open- and closed-track cam)

This program determines cam mechanism proportions for prescribed maximum pressure angles by rise and return, swinging roller follower.

Example 17, Swinging Roller Follower: Parabolic motion by rise and return, length of follower arm $L_f = 5$ in., total angle of oscillation of swinging follower arm $= 33°$, total angle for rise $= 100°$, total angle for return $= 180°$, maximum pressure angle by rise $= 45°$, maximum pressure angle by return $= 45°$, cam rotates CCW, Fig. 13-7. The answer is:

Distance $MO = 6.1$ in.

$R_{min} = 1.64$ in.

Program SWING6 (Closed-track cam)

For given time-displacement diagrams and cam mechanism proportions for swinging roller follower, this program will calculate the radius of curvature of the relative path of the center of the roller follower (i.e., the radius of curvature of the cam, if roller radius $R_f = 0$) and the compressive stress for each 1° cam rotation during rise and return.

Example 18, Swinging Roller Follower: Parabolic motion by rise and return, distance $MO = 5.8$ in, length of follower arm $L_f = 5$ in., $R_{min} = 2.35$ in., total angle of oscillation of swinging follower arm $= 11.46°$, total angle for rise $= 40°$, for return $40°$, roller radius $R_f = .5$ in., roller width $= .625$ in, cam speed $= 300$ rpm, CCW, cam material is steel, weight of mass to be moved (concentrated at the follower center) $= 5$ lbs. The answer is:

Maximum compressive stress during 1st half rise = 50,804 psi at THETA = 20°

Maximum compressive stress during 2nd half rise = 41,947 psi at THETA = 20°

Maximum compressive stress during 1st half return = 43,671 psi at THETA = 20°

Maximum compressive stress during 2nd half return = 49,623 psi at THETA = 20°

The allowable stresses for various cam materials are listed below.

Cam Material for use with Roller Hardened Steel	Maximum permissible compressive stress, psi
Gray-iron casting, ASTM A 48-48, Class 20, 160-190 Bhn, phosphate coated	58,000
Gray-iron coating, ASTM A 339-51T, Grade 20, 140-160 Bhn	51,000
Nodular-iron casting, ASTM A 339-51T Grade 80-60-03, 207-241 Bhn	72,000
Gray-iron casting, ASTM A 48-48, Class 30, 200-220 Bhn	65,000
Gray-iron casting ASTM A 48-48, Class 35, 225-255 Bhn	78,000
Gray-iron casting ASTM A 48-48, Class 30 heat treated (austempered) 255-300 Bhn	90,000
SAE 1020 steel, 130-150 Bhn	82,000
SAE 4150 steel, heat treated to 270-300 Bhn, phosphate coated	220,000
SAE 4150 steel, heat treated to 270-300 Bhn	90,000
SAE 1020 steel, carburised to 0.045 in. depth of case, 50-58 Rc	226,000
SAE 1340 steel, induction hardened to 45-55 Rc	200,000
SAE 4340 steel, induction hardened to 50-55 Rc	226,000

Based on United Shoe Machinery Corp. data by Guy J. Talbourdet.

13.3 Program Listing

This section gives a complete printout of the material on the accompanying disk.

```
10 CLS : PRINT : PRINT : PRINT : PRINT
20 PRINT TAB(10)"A- TRANSLATING ROLLER FOLLOWER."
30 PRINT : PRINT TAB(20)"ANALYSIS AND SYNTHESIS PROGRAMS FOR PROPORTIONING OF"
40 PRINT TAB(15)"TRANSLATING ROLLER FOLLOWER"
50 PRINT : PRINT : PRINT TAB(10)"B- SWINGING ROLLER FOLLOWER."
60 PRINT : PRINT TAB(20)"ANALYSIS AND SYNTHESIS PROGRAMS FOR PROPORTIONING OF"
70 PRINT TAB(15)"SWINGING ROLLER FOLLOWER"
80 PRINT : PRINT : PRINT TAB(10)"Z- EXIT PROGRAM AND RETURN TO DOS."
90 PRINT : PRINT : PRINT
100 INPUT"ENTER THE GROUP LABEL YOU WOULD LIKE TO RUN PROGRAMS FOR (A,B OR Z)";G
ROUP$
110 IF GROUP$="Z" OR GROUP$="z" THEN CLS : SYSTEM
120 IF GROUP$="A" OR GROUP$="a" GOTO 400
130 IF GROUP$="B" OR GROUP$="b" GOTO 150
140 GOTO 10
150 CLS: PRINT : PRINT
160 PRINT TAB(25)"SWINGING ROLLER FOLLOWER"
170 PRINT : PRINT TAB(10)"1- SWING4:"
180 PRINT TAB(14)"For a swinging roller follower this program calculates the max
imum"
190 PRINT TAB(12)"pressure angles by rise and return for given time-displacement
"
200 PRINT TAB(12)"diagrams and open track cam proportions"
210 PRINT : PRINT TAB(10)"2- SWING5:"
220 PRINT TAB(14)"Calculates open track cam proportions for given time"
230 PRINT TAB(12)"displacement diagrams and prescribed maximum pressure angles b
y"
235 PRINT TAB(12)"rise and return."
240 PRINT : PRINT TAB(10)"3- SWING6:"
250 PRINT TAB(14)"For given time-displacement diagrams and closed track cam"
260 PRINT TAB(12)"proportions this program calculates the radius of"
270 PRINT TAB(12)"curvature of the relative path of the roller follower and the"

280 PRINT TAB(12)"compressive stress for each 1 degree cam shaft rotation and"
290 PRINT TAB(12)"prints the maximum values by rise and return."
300 PRINT : PRINT TAB(10)"4- EXIT TO MAIN MENU"
310 PRINT : INPUT"ENTER DESIRED PROGRAM NUMBER (1,2,3,or 4)";PROG
320 IF PROG=4 GOTO 10
330 IF PROG>4 GOTO 150
340 CLS : PRINT : PRINT : PRINT : PRINT : PRINT : PRINT : PRINT : PRINT
350 PRINT TAB(25)"LOADING PROGRAM, PLEASE WAIT."
360 IF PROG=1 THEN RUN "SWING4"
370 IF PROG=2 THEN RUN "SWING5"
380 IF PROG=3 THEN RUN "SWING6"
390 GOTO 150
400 CLS : PRINT : PRINT
410 PRINT TAB(25)"TRANSLATING ROLLER FOLLOWER."
420 PRINT : PRINT TAB(10)"1- TRANS1:"
430 PRINT TAB(14)"For a translating  roller follower this program calculates the
"
```

```
440 PRINT TAB(12)"maximum pressure angles by rise and return for given time-"
450 PRINT TAB(12)"displacement diagrams and open track cam proportions"
460 PRINT : PRINT TAB(10)"2- TRANS2:"
470 PRINT TAB(14)"Calculates open track cam proportions for given time-"
480 PRINT TAB(12)"displacement diagrams and prescribed maximum pressure angles b
y"
485 PRINT TAB(12)"rise and return."
490 PRINT : PRINT TAB(10)"3- TRANS3:"
500 PRINT TAB(14)"For given time-displacement diagrams and closed track cam"
510 PRINT TAB(12)"proportions this program calculates the radius of"
520 PRINT TAB(12)"curvature of the relative path of the roller follower and the
"
530 PRINT TAB(12)"compressive stress for each 1 degree cam shaft rotation and pr
ints"
540 PRINT TAB(12)"the maximum values by rise and and return."
550 PRINT : PRINT TAB(10)"4- EXIT TO MAIN MENU"
560 PRINT : INPUT"ENTER DESIRED PROGRAM NUMBER (1,2,3,or 4)";PROG
570 IF PROG=4 GOTO 10
580 IF PROG>4 GOTO 400
590 CLS : PRINT : PRINT : PRINT : PRINT : PRINT : PRINT : PRINT : PRINT
600 PRINT TAB(25)"LOADING PROGRAM, PLEASE WAIT."
610 IF PROG=1 THEN RUN "TRANS1"
620 IF PROG=2 THEN RUN "TRANS2"
630 IF PROG=3 THEN RUN "TRANS3"
640 GOTO 400
650 END




10 'THIS PROGRAM IS FILED UNDER "TRANS1"
20 CLS
30 'TRANSLATING ROLLER FOLLOWER
40 'FINDING MAX. PRESSURE ANGLES BY RISE AND RETURN FOR GIVEN
50 'TIME-DISPLACEMENT DIAGRAMS AND CAM PROPORTIONS
60 PRINT"FINDING MAX. PRESSURE ANGLES BY RISE AND RETURN FOR GIVEN"
70 PRINT"TIME-DISPLACEMENT DIAGRAMS AND CAM PROPORTIONS"
80 PRINT"RISE                          RETURN"
90 PRINT" 1  PARABOLIC MOTION              11"
100 PRINT" 2  SIMPLE HARMONIC MOTION        12"
110 PRINT" 3  CYCLOIDAL MOTION              13"
120 PRINT" 4  3-4 POLYNOMIAL                14"
130 PRINT" 5  3-4-5 POLYNOMIAL              15"
140 PRINT" 6  4-5-6-7 POLYNOMIAL            16"
150 PRINT" 7  MODIFIED TRAPEZOIDAL ACCEL.   17"
160 PRINT" 8  MODIFIED SINUSOIDAL ACCEL.    18"
170 PRINT
180 PRINT
190 INPUT"TIME-DISPLACEMENT CURVE FOR RISE (1-9)";M
```

```
200 INPUT"TIME-DISPLACEMENT CURVE FOR RETURN(11-19)";L
210 INPUT"STROKE H, IN.";H
220 INPUT"EXCENTRICITY E(+ OR -)";E
230 INPUT"DISTANCE A, IN.";A
240 INPUT"TOTAL ANGLE FOR RISE, DEGREES"; B1
250 INPUT"TOTAL ANGLE FOR RETURN, DEGREES";B2
260 INPUT"IF CAM ROTATION IS CW WRITE 'CW':ELSE WRITE 'CCW'";C$
270 CLS : PRINT : PRINT "A- TO EXIT TO MAIN PROGRAM.";TAB(38)"B- TO QUIT CHANGES
    AND EXECUTE PROGRAM."
280 PRINT "C- RISE É";M;"©";TAB(38)"D- RETURN É";L;"©"
290 PRINT "    1        PARABOLIC MOTION                11"
300 PRINT "    2        SIMPLE HARMONIC MOTION          12"
310 PRINT "    3        CYCLOIDAL MOTION                13"
320 PRINT "    4        3-4 POLYNOMIAL                  14"
330 PRINT "    5        3-4-5 POLYNOMIAL                15"
340 PRINT "    6        4-5-6-7 POLYNOMIAL              16"
350 PRINT "    7        MODIFIED TRAPEZOIDAL ACCEL.     17"
360 PRINT "    8        MODIFIED SINUSOIDAL ACCEL.      18"
370 PRINT "E- STROKE h (in.) É";H;"©"
380 PRINT "F- ECCENTRICITY E (+ or -) É";E;"©"
390 PRINT "G- DISTANCE A (in.) É";A;"©"
400 PRINT "H- TOTAL ANGLE FOR RISE (DEGREE) É";B1;"©"
410 PRINT "I- TOTAL ANGLE FOR RETURN (DEGREES) É";B2;"©"
420 PRINT "J- CAM ROTATION (1:CW;2:CCW) É";C$;"©"
430 PRINT : INPUT "WHICH VALUE WOULD YOU LIKE TO CHANGE (A..J)";CV$
440 IF CV$="A" OR CV$="a" GOTO 1920
450 IF CV$="B" OR CV$="b" GOTO 580
470 PRINT : INPUT "WHAT IS THE NEW VALUE ";NV
480 IF CV$="C" OR CV$="c" THEN M=NV : GOTO 270
490 IF CV$="D" OR CV$="d" THEN L=NV : GOTO 270
500 IF CV$="E" OR CV$="e" THEN H=NV : GOTO 270
510 IF CV$="F" OR CV$="f" THEN E=NV : GOTO 270
520 IF CV$="G" OR CV$="g" THEN A=NV : GOTO 270
530 IF CV$="H" OR CV$="h" THEN B1=NV : GOTO 270
540 IF CV$="I" OR CV$="i" THEN B2=NV : GOTO 270
550 IF CV$="J" AND NV=1 OR CV$="j" AND NV=1 THEN C$="CW" : GOTO 270
560 IF CV$="J" AND NV=2 OR CV$="j" AND NV=2 THEN C$="CCW" : GOTO 270
570 GOTO 270
580 IF C$="CW"OR C$="cw" THEN I=-1:ELSE I=1
590 N=60
600 T1=B1/360
610 T2=B2/360
620 PRINT"H=";H;"E=";E;"A=";A;"B1=";B1;"B2=";B2
630 LPRINT"H=";H;"E=";E;"A=";A;"B1=";B1;"B2=";B2
640 LPRINT
650 PRINT
660 IF M=1 THEN LPRINT"PARABOLIC MOTION; RISE"
670 IF M=1 THEN PRINT"PARABOLIC MOTION; RISE":GOTO 820
680 IF M=2 THEN LPRINT"SIMPLE HARMONIC MOTION; RISE"
690 IF M=2 THEN PRINT"SIMPLE HARMONIC MOTION; RISE":GOTO 920
```

```
700 IF M=3 THEN LPRINT"CYCLOIDAL MOTION; RISE"
710 IF M=3 THEN PRINT"CYCLOIDAL MOTION; RISE":GOTO 980
720 IF M=4 THEN LPRINT"3-4 POLYNOMIAL; RISE"
730 IF M=4 THEN PRINT"3-4 POLYNOMIAL; RISE":GOTO 1040
740 IF M=5 THEN LPRINT"3-4-5 POLYNOMIAL; RISE"
750 IF M=5 THEN PRINT"3-4-5 POLYNOMIAL; RISE":GOTO 1100
760 IF M=6 THEN LPRINT"4-5-6-7 POLYNOMIAL; RISE"
770 IF M=6 THEN PRINT"4-5-6-7 POLYNOMIAL; RISE":GOTO 1160
780 IF M=7 THEN LPRINT"MODIFIED TRAPEZOIDAL ACCEL.; RISE"
790 IF M=7 THEN PRINT "MODIFIED TRAPEZOIDAL ACCEL.; RISE" : GOTO 1930
800 IF M=8 THEN LPRINT"MODIFIED SINUSOIDAL ACCEL.; RISE"
810 IF M=8 THEN PRINT"MODIFIED SINUSOIDAL ACCEL.; RISE":GOTO 2250
820 FOR TA=0 TO B1/2 'PARABOLIC MOTION
830 Y=2*H*(TA/B1)^2
840 VB=4*H/T1*TA/B1
850 GOSUB 1220
860 NEXT TA
870 IF TA>B1/2 THEN TA=B1/2
880 Y=2*H*(TA/B1)^2
890 VB=4*H/T1*TA/B1
900 GOSUB 1220
910 GOTO 1270
920 FOR TA=0 TO B1/2 'SIMPLE HARMONIC MOTION
930 Y=H/2*(1-COS(3.14159*TA/B1))
940 VB=H/2/T1*3.14159*SIN(3.14159*TA/B1)
950 GOSUB 1220
960 NEXT TA
970 GOTO 1270
980 FOR TA=0 TO B1/2 'CYCLOIDAL MOTION
990 Y=H*(TA/B1-1/2/3.14159*SIN(6.28319*TA/B1))
1000 VB=H/T1*(1-COS(6.28319*TA/B1))
1010 GOSUB 1220
1020 NEXT TA
1030 GOTO 1270
1040 FOR TA=0 TO B1/2 '3-4 POLYNOMIAL
1050 Y=H*(8*(TA/B1)^3-8*(TA/B1)^4)
1060 VB=H/T1*(24*(TA/B1)^2-32*(TA/B1)^3
1070 GOSUB 1220
1080 NEXT TA
1090 GOTO 1270
1100 FOR TA=0 TO B1/2 '3-4-5 POLYNOMIAL
1110 Y=H*(10*(TA/B1)^3-15*(TA/B1)^4+6*(TA/B1)^5)
1120 VB=H/T1*(30*(TA/B1)^2-60*(TA/B1)^3+30*(TA/B1)^4)
1130 GOSUB 1220
1140 NEXT TA
1150 GOTO 1270
1160 FOR TA=0 TO B1/2 '4-5-6-7 POLYNOMIAL
1170 Y=H*(35*(TA/B1)^4-84*(TA/B1)^5+70*(TA/B1)^6-20*(TA/B1)^7)
1180 VB=H/T1*(140*(TA/B1)^3-420*(TA/B1)^4+420*(TA/B1)^5-140*(TA/B1)^6)
1190 GOSUB 1220
```

```
1200 NEXT TA
1210 GOTO 1270
1220 VC=I*SQR(E^2+(A+Y)^2)*6.28319
1230 T3=1.5708-ATN(E/(A+Y))
1240 TE=ATN((VB-VC*COS(T3))/VC/SIN(T3))
1250 IF ABS(TE)>ABS(T5) THEN T5=TE:TB=TA
1260 RETURN
1270 LPRINT"MAX. PRESSURE ANGLE BY RISE=";T5*57.3;"AT CAM ANGLE THETA=";TB
1280 PRINT"MAX. PRESSURE ANGLE BY RISE=";T5*57.3;"AT CAM SHAFT ANGLE THETA=";TB
1290 TB=0:T5=0
1300 IF L=11 THEN LPRINT"PARABOLIC MOTION; RETURN"
1310 IF L=11 THEN PRINT"PARABOLIC MOTION; RETURN":GOTO 1460
1320 IF L=12 THEN LPRINT"SIMPLE HARMONIC MOTION; RETURN"
1330 IF L=12 THEN PRINT"SIMPLE HARMONIC MOTION; RETURN":GOTO 1520
1340 IF L=13 THEN LPRINT"CYCLOIDAL MOTION; RETURN"
1350 IF L=13 THEN PRINT"CYCLOIDAL MOTION; RETURN":GOTO 1580
1360 IF L=14 THEN LPRINT"3-4 POLYNOMIAL; RETURN"
1370 IF L=14 THEN PRINT"3-4 POLYNOMIAL; RETURN":GOTO 1640
1380 IF L=15 THEN LPRINT"3-4-5 POLYNOMIAL; RETURN"
1390 IF L=15 THEN PRINT"3-4-5 POLYNOMIAL; RETURN":GOTO 1700
1400 IF L=16 THEN LPRINT"4-5-6-7 POLYNOMIAL; RETURN"
1410 IF L=16 THEN PRINT"4-5-6-7 POLYNOMIAL; RETURN":GOTO 1760
1420 IF L=17 THEN LPRINT"MODIFIED TRAPEZOIDAL ACCEL.; RETURN"
1430 IF L=17 THEN PRINT"MODIFIED TRAPEZOIDAL ACCEL.; RETURN":GOTO 2090
1440 IF L=18 THEN LPRINT"MODIFIED SINUSOIDAL ACCEL.; RETURN"
1450 IF L=18 THEN PRINT"MODIFIED SINUSOIDAL ACCEL.; RETURN":GOTO 2360
1460 FOR TA=B2/2 TO B2
1470 Y=2*H*((B2-TA)/B2)^2
1480 VB=-4*H/T2*(B2-TA)/B2
1490 GOSUB 1220
1500 NEXT TA
1510 GOTO 1810
1520 FOR TA=B2/2 TO B2
1530 Y=H/2*(1+COS(3.14159*TA/B2))
1540 VB=-H/2/T2*3.14159*SIN(3.14159*TA/B2)
1550 GOSUB 1220
1560 NEXT TA
1570 GOTO 1810
1580 FOR TA=B2/2 TO B2
1590 Y=H*(1-TA/B2+1/6.28319*SIN(6.28319*TA/B2))
1600 VB=H/T2*(COS(6.28319*TA/B2)-1)
1610 GOSUB 1220
1620 NEXT TA
1630 GOTO 1810
1640 FOR TA=B2/2 TO B2
1650 Y=H*(8*TA/B2-24*(TA/B2)^2+24*(TA/B2)^3-8*(TA/B2)^4)
1660 VB=H/T2*(8-48*TA/B2+72*(TA/B2)^2-32*(TA/B2)^3)
1670 GOSUB 1220
1680 NEXT TA
1690 GOTO 1810
```

```
1700  FOR TA=B2/2 TO B2
1710  Y=H*(1-10*(TA/B2)^3+15*(TA/B2)^4-6*(TA/B2)^5)
1720  VB=-H/T2*(30*(TA/B2)^2-60*(TA/B2)^3+30*(TA/B2)^4)
1730  GOSUB 1220
1740  NEXT TA
1750  GOTO 1810
1760  FOR TA=B2/2 TO B2
1770  Y=H*(1-35*(TA/B2)^4+84*(TA/B2)^5-70*(TA/B2)^6+20*(TA/B2)^7)
1780  VB=-H/T2*(140*(TA/B2)^3-420*(TA/B2)^4+420*(TA/B2)^5-140*(TA/B2)^6)
1790  GOSUB 1220
1800  NEXT TA
1810  LPRINT"MAX. PRESSURE ANGLE BY RETURN=";T5*57.3;"AT CAM ANGLE THETA=";TB
1820  PRINT"MAX. PRESSURE ANGLE BY RETURN=";T5*57.3;"AT CAM SHAFT ANGLE THETA=";T
B
1830  LPRINT
1840  LPRINT"STROKE H=";H
1850  LPRINT"EXCENTRICITY E=";E
1860  LPRINT"DISTANCE A=";A
1870  LPRINT"TOTAL ANGLE FOR RISE=";B1
1880  LPRINT"TOTAL ANGLE FOR RETURN=";B2
1890  IF C$="CW" THEN LPRINT"CAM ROTATES CW"
1900  IF C$="CCW" THEN LPRINT"CAM ROTATES CCW"
1910  GOTO 270
1920  CLS : RUN "MAIN" : END
1930  FOR TA=0 TO B1/8
1940  Y=H/5.14159*(2*TA/B1-1/6.28319*SIN(12.5664*TA/B1))
1950  VB=H/T1/5.14159*(2-2*COS(12.5664*TA/B1))
1960  GOSUB 1220
1970  NEXT TA
1980  FOR TA=B1/8 TO 3*B1/8
1990  Y=H/5.14159*(12.5664*(TA/B1)^2+(2-3.14159)*TA/B1+.03719)
2000  VB=H/T1/5.14159*(25.1327*TA/B1-1.14159)
2010  GOSUB 1220
2020  NEXT TA
2030  FOR TA=3*B1/8 TO B1/2
2040  Y=H/5.14159*(8.2832*TA/B1-1/6.28319*SIN(12.5664*(TA/B1-.25))-1.5708)
2050  VB=H/T1/5.14159*(8.2832-2*COS(12.5664*(TA/B1-.25)))
2060  GOSUB 1220
2070  NEXT TA
2080  GOTO 1270
2090  FOR TA=B2/2 TO 5*B2/8
2100  Y=H/5.14159*(6.7124-8.2832*TA/B2-1/6.28319*SIN(12.5664*(.75-TA/B2)))
2110  VB=H/T2/5.14159*(-8.2832+2*COS(12.5664*(.75-TA/B2)))
2120  GOSUB 1220
2130  NEXT TA
2140  FOR TA=5*B2/8 TO 7*B2/8
2150  Y=H/5.14159*(11.462-23.9911*TA/B2+12.5664*(TA/B2)^2)
2160  VB=H/T2/5.14159*(-23.9911+25.1327*TA/B2)
2170  GOSUB 1220
2180  NEXT TA
```

```
2190 FOR TA=7*B2/8 TO B2
2200 Y=H/5.14159*(2-2*TA/B2-1/6.28319*SIN(12.5664*(1-TA/B2)))
2210 VB=H/T2/5.14159*(-2+2*COS(12.5664*(1-TA/B2)))
2220 GOSUB 1220
2230 NEXT TA
2240 GOTO 1810
2250 FOR TA=0 TO B1/8
2260 Y=H/7.14159*(3.14159*TA/B1-.25*SIN(12.5664*TA/B1))
2270 VB=.4399*H/T1*(1-COS(12.5664*TA/B1))
2280 GOSUB 1220
2290 NEXT TA
2300 FOR TA=B1/8 TO B1/2
2310 Y=H/7.14159*(2+3.14159*TA/B1-9/4*SIN(3.14159/3+4/3*3.14159*TA/B1))
2320 VB=.4399*H/T1*(1-3*COS(3.14159/3+4/3*3.14159*TA/B1))
2330 GOSUB 1220
2340 NEXT TA
2350 GOTO 1270
2360 FOR TA=B2/2 TO 7*B2/8
2370 Y=H/7.14159*(5.14159-3.14159*TA/B2+9/4*SIN(3.14159/3+4/3*3.14159*TA/B2))
2380 VB=.4399*H/T2*(-1+3*COS(3.14159/3+4/3*3.14159*TA/B2))
2390 GOSUB 1220
2400 NEXT TA
2410 FOR TA=7*B2/8 TO B2
2420 Y=H/7.14159*(3.14159-3.14159*TA/B2+.25*SIN(12.5664*TA/B2))
2430 VB=.4399*H/T2*(-1+COS(12.5664*TA/B2))
2440 GOSUB 1220
2450 NEXT TA
2460 GOTO 1810




10 'THIS PROGRAM IS FILED UNDER "TRANS2"
20 CLS
30 PRINT"SYNTHESIS OF CAMS FOR PRESCRIBED MAX PRESSURE ANGLES"
40 PRINT"TRANSLATING ROLLER FOLLOWER"
50 PRINT
70 PRINT"TRANSLATING ROLLER FOLLOWER"
80 PRINT
90 PRINT"RISE                                    RETURN"
100 PRINT" 1   PARABOLIC MOTION                    11"
110 PRINT" 2   SIMPLE HARMONIC MOTION              12"
120 PRINT" 3   CYCLOIDAL MOTION                    13"
130 PRINT" 4   3-4 POLYNOMIAL                      14"
140 PRINT" 5   3-4-5 POLYNOMIAL                    15"
150 PRINT" 6   4-5-6-7 POLYNOMIAL                  16"
160 PRINT" 7   MODIFIED TRAPEZOIDAL ACCEL.         17"
170 PRINT" 8   MODIFIED SINUSOIDAL ACCEL.          18"
180 PRINT" 9                                       19"
190 PRINT
```

```
200 PRINT
210 INPUT"TIME--DISPLACEMENT CURVE FOR RISE (1-9)";M
220 INPUT"TIME--DISPLACEMENT CURVE FOR RETURN (11-19)";L
230 INPUT"STROKE H, IN.";H
240 INPUT"TOTAL ANGLE FOR RISE, DEGREES";B1
250 INPUT"TOTAL ANGLE FOR RETURN, DEGREES";B2
260 INPUT"MAX PRESSURE ANGLE FOR RISE, DEGREES";A1
270 INPUT"MAX PRESSURE ANGLE FOR RETURN, DEGREES";A2
280 INPUT"IF CAM ROTATION IS CLOCKWISE WRITE CW, ELSE WRITE CCW";C$
290 CLS : PRINT : PRINT "A- TO EXIT TO MAIN PROGRAM.";TAB(38)"B- TO QUIT CHANGES
    AND EXECUTE PROGRAM."
300 PRINT "C- RISE É";M;"©";TAB(38)"D- RETURN É";L;"©
310 PRINT "    1      PARABOLIC MOTION              11"
320 PRINT "    2      SIMPLE HARMOINC MOTION        12"
330 PRINT "    3      CYCLOIDAL MOTION              13"
340 PRINT "    4      3-4 POLYNOMIAL                14"
350 PRINT "    5      3-4-5 POLYNOMIAL              15"
360 PRINT "    6      4-5-6-7 POLYNOMIAL            16"
370 PRINT "    7      MODIFIED TRAPEZOIDAL ACCEL.   17"
380 PRINT "    8      MODIFIED SINUSOIDAL ACCEL.    18"
390 PRINT "E- STROKE h (in.) É";H;"©"
400 PRINT "F- TOTAL ANGLE FOR RISE (DEGREES) É";B1;"©"
410 PRINT "G- TOTAL ANGLE FOR RETURN (DEGREES) É";B2;"©"
420 PRINT "H- MAXIMUM PRESSURE ANGLE FOR RISE (DEGREES) É";A1;"©"
430 PRINT "I- MAXIMUM PRESSURE ANGLE FOR RETURN (DEGREES) É";A2;"©"
440 PRINT "J- CAM ROTATION (1:CW ; 2:CCW) É";C$;"©"
450 PRINT : INPUT "WHICH VALUE WOULD YOU LIKE TO CHANGE (A..J) ";CV$
460 IF CV$="A" OR CV$="a" GOTO 2650
470 IF CV$="B" OR CV$="b" GOTO 600
490 PRINT : INPUT "WHAT IS THE NEW VALUE ";NV
500 IF CV$="C" OR CV$="c" THEN M=NV : GOTO 290
510 IF CV$="D" OR CV$="d" THEN L=NV : GOTO 290
520 IF CV$="E" OR CV$="e" THEN H=NV : GOTO 290
530 IF CV$="F" OR CV$="f" THEN B1=NV : GOTO 290
540 IF CV$="G" OR CV$="g" THEN B2=NV : GOTO 290
550 IF CV$="H" OR CV$="h" THEN A1=NV : GOTO 290
560 IF CV$="I" OR CV$="i" THEN A2=NV : GOTO 290
570 IF CV$="J" AND NV=1 OR CV$="j" AND NV=1 THEN C$="CW" : GOTO 290
580 IF CV$="J" AND NV=2 OR CV$="j" AND NV=2 THEN C$="CCW" : GOTO 290
590 GOTO 290
600 IF M=1 THEN LPRINT"PARABOLIC MOTION: RISE"
610 IF M=1 THEN PRINT"PARABOLIC MOTION: RISE":GOTO 770
620 IF M=2 THEN LPRINT"SIMPLE HARMONIC MOTION; RISE"
630 IF M=2 THEN PRINT"SIMPLE HARMONIC MOTION; RISE":GOTO 810
640 IF M=3 THEN LPRINT"CYCLOIDAL MOTION; RISE"
650 IF M=3 THEN PRINT"CYCLOIDAL MOTION; RISE":GOTO 850
660 IF M=4 THEN LPRINT"3-4 POLYNOMIAL; RISE"
670 IF M=4 THEN PRINT"3-4 POLYNOMIAL; RISE": GOTO 890
680 IF M=5 THEN LPRINT"3-4-5 POLYNOMIAL; RISE"
690 IF M=5 THEN PRINT"3-4-5 POLYNOMIAL; RISE":GOTO 930
```

```
700 IF M=6 THEN LPRINT"4-5-6-7 POLYNOMIAL; RISE"
710 IF M=6 THEN PRINT"4-5-6-7 POLYNOMIAL; RISE":GOTO 970
720 IF M=7 THEN LPRINT"MODIFIED TRAPEZOIDAL ACCEL.; RISE"
730 IF M=7 THEN PRINT"MODIFIED TRAPEZOIDAL ACCEL.; RISE":GOTO 1010
740 IF M=8 THEN LPRINT"MODIFIED SINUSOIDAL ACCEL.; RISE"
750 IF M=8 THEN PRINT"MODIFIED SINUSOIDAL ACCEL.; RISE":GOTO 1070
770 GOSUB 2680
780 X0=V3/V2*Q*H*90/B1
790 Y0=-2*H*(TA/B1)^2    'PARABOLIC MOTION
800 GOTO 1110
810 GOSUB 2840
820 X0=V3/V2*Q*H*90/B1
830 Y0=-H/2*(1-COS(3.14159*TA/B1)) 'SIMPLE HARMONIC MOTION
840 GOTO 1110
850 GOSUB 3000
860 X0=V3/V2*Q*H*90/B1
870 Y0=-H*(TA/B1-SIN(6.28319*TA/B1)/6.28319)
880 GOTO 1110
890 GOSUB 3160
900 X0=V3/V2*Q*H*90/B1
910 Y0=-H*(8*(TA/B1)^3-8*(TA/B1)^4)
920 GOTO 1110
930 GOSUB 3320
940 X0=V3/V2*Q*H*90/B1
950 Y0=-H*(10*(TA/B1)^3-15*(TA/B1)^4+6*(TA/B1)^5)
960 GOTO 1110
970 GOSUB 3490
980 X0=V3/V2*Q*H*90/B1
990 Y0=-H*(35*(.5)^4-84*(.5)^5+70*(.5)^6-20*(.5)^7)
1000 GOTO 1110
1010 GOSUB 3630
1020 X0=V3/V2*Q*H*90/B1
1030 IF TA>=3*B1/8 THEN Y0=-H/5.14159*(8.2832*TA/B1-1/6.28319*SIN(12.5664*(TA/B1
-.25))-1.57):GOTO 1110
1040 IF TA>=B1/8 THEN Y0=-H/5.14159*(12.5664*(TA/B1)^2-1.14159*TA/B1+.03719):GOT
O 1110
1050 Y0=-H/5.14159*(2*TA/B1-1/6.28319*SIN(12.5664*TA/B1))
1060 GOTO 1110
1070 GOSUB 3930
1080 X0=V3/V2*Q*H*90/B1
1090 IF TA>=B1/8 THEN Y0=-H/7.14159*(2+3.14159*TA/B1-9/4*SIN(3.14159/3+4/3*3.141
59*TA/B1)):GOTO 1110
1100 Y0=-H/7.14159*(3.14159*TA/B1-.25*SIN(12.5664*TA/B1)):GOTO 1110
1110 IF L=11 THEN LPRINT"PARABOLIC MOTION;RETURN"
1120 IF L=11 THEN PRINT"PARABOLIC MOTION;RETURN":GOTO 1290
1130 IF L=12 THEN LPRINT"SIMPLE HARMONIC MOTION;RETURN"
1140 IF L=12 THEN PRINT"SIMPLE HARMONIC MOTION;RETURN":GOTO 1420
1150 IF L=13 THEN LPRINT"CYCLOIDAL MOTION;RETURN"
1160 IF L=13 THEN PRINT"CYCLOIDAL MOTION;RETURN":GOTO 1550
1170 IF L=14 THEN LPRINT"3-4 POLYNOMIAL;RETURN"
```

```
1180 IF L=14 THEN PRINT"3-4 POLYNOMIAL;RETURN":GOTO 1690
1190 IF L=15 THEN LPRINT"3-4-5 POLYNOMIAL;RETURN"
1200 IF L=15 THEN PRINT"3-4-5 POLYNOMIAL;RETURN":GOTO 1820
1210 IF L=16 THEN LPRINT"4-5-6-7 POLYNOMIAL;RETURN"
1220 IF L=16 THEN PRINT"4-5-6-7 POLYNOMIAL;RETURN":GOTO 1950
1230 IF L=17 THEN LPRINT"MODIFIED TRAPEZOIDAL ACCEL.;RETURN"
1240 IF L=17 THEN PRINT"MODIFIED TRAPEZOIDAL ACCEL.;RETURN":GOTO 2080
1250 IF L=18 THEN LPRINT"MODIFIED SINUSOIDAL ACCEL.; RETURN"
1260 IF L=18 THEN PRINT"MODIFIED SINUSOIDAL ACCEL.;RETURN":GOTO 2300
1270 IF L=19 THEN LPRINT
1280 IF L=19 THEN PRINT
1290 A3=A1 'STORE THE VALUE OF A1
1300 A1=A2 'CHANGE A1 TO THE VALUE OF A2
1310 B3=B1 'STORE THE VALUE OF B1
1320 B1=B2 'CHANGE B1 TO THE VALUE OF B2
1330 GOSUB 2680    'PARABOLIC MOTION
1340 X5=V3/V2*Q*H*90/B1
1350 Y5=-2*H*(TA/B1)^2
1360 A2=A1 'RESTORE THE VALUE OF A2
1370 A1=A3 'RESTORE THE VALUE OF A1
1380 B2=B1 'RESTORE THE VALUE OF B2
1390 B1=B3 'RESTORE THE VALUE OF B1
1400 GOSUB 2380
1410 GOTO 2460
1420 A3=A1
1430 A1=A2
1440 B3=B2
1450 B1=B2
1460 GOSUB 2840    'SIMPLE HARMONIC MOTION
1470 X5=V3/V2*Q*H*90/B1
1480 Y5=-H/2*(1-COS(3.14159*TA/B1))
1490 A2=A1
1500 A1=A3
1510 B2=B1
1520 B1=B3
1530 GOSUB 2380
1540 GOTO 2460
1550 A3=A1
1560 A1=A2
1570 B3=B1
1580 B3=B2
1590 B1=B2
1600 GOSUB 3000    'CYCLOIDAL MOTION
1610 X5=V3/V2*Q*H*90/B1
1620 Y5=-H*(TA/B1-SIN(6.28319*TA/B1)/6.28319)
1630 A2=A1
1640 A1=A3
1650 B2=B1
1660 B1=B3
1670 GOSUB 2380
```

```
1680 GOTO 2460
1690 A3=A1
1700 A1=A2
1710 B3=B1
1720 B1=B2
1730 GOSUB 3160   '3-4 POLYNOMIAL
1740 X5=V3/V2*Q*H*90/B1
1750 Y5=-H*(8*(TA/B1)^3-8*(TA/B1)^4)
1760 A2=A1
1770 A1=A3
1780 B2=B1
1790 B1=B3
1800 GOSUB 2380
1810 GOTO 2460
1820 A3=A1
1830 A1=A2
1840 B3=B1
1850 B1=B2
1860 GOSUB 3320   '3-4-5 POLYNOMIAL
1870 X5=V3/V2*Q*H*90/B1
1880 Y5=-H*(10*(TA/B1)^3-15*(TA/B1)^4+6*(TA/B1)^5)
1890 A2=A1
1900 A1=A3
1910 B2=B1
1920 B1=B3
1930 GOSUB 2380
1940 GOSUB 2460
1950 A3=A1
1960 A1=A2
1970 B3=B1
1980 B1=B2
1990 GOSUB 3490   '4-5-6-7 POLYNOMIAL
2000 X5=V3/V2*Q*H*90/B1
2010 Y5=-H*(35*(.5)^4-84*(.5)^5+70*(.5)^6-20*(.5)^7)
2020 A2=A1
2030 A1=A3
2040 B2=B1
2050 B1=B3
2060 GOSUB 2380
2070 GOTO 2460
2080 A3=A1
2090 A1=A2
2100 B3=B1
2110 B1=B2
2120 GOSUB 3630   'MODIFIED TRAPEZOIDAL ACCELERATION
2130 X5=V3/V2*Q*H*90/B1
2140 IF TA>=3*B1/8 THEN Y5=-H/5.14159*(8.2832*TA/B1-1/6.28319*SIN(12.5664*(TA/B1
-.25))-1.57):GOTO 2170
2150 IF TA>=B1/8 THEN Y5=-H/5.14159*(12.5664*(TA/B1)^2-1.14159*TA/B1+.03719):GOT
O 2170
```

```
2160 Y5=-H/5.14159*(2*TA/B1-1/6.28319*SIN(12.5664*TA/B1))
2170 A2=A1
2180 A1=A3
2190 B2=B1
2200 B1=B3
2210 GOSUB 2380
2220 GOTO 2460
2230 A3=A1
2240 A1=A2
2250 B3=B1
2260 B1=B2
2270 GOSUB 3930
2280 X5=V3/V2*Q*H*90/B1
2290 IF TA>=B1/8 THEN Y5=-H/7.14159*(2+3.14159*TA/B1-9/4*SIN(3.14159/3+4/3*3.141
59*TA/B1)):GOTO 2310
2300 Y5=-H/7.14159*(3.14159*TA/B1-.25*SIN(12.5664*TA/B1))
2310 A2=A1
2320 A1=A3
2330 B2=B1
2340 B1=B3
2350 GOSUB 2380
2360 GOTO 2460
2370 'INTERSECTION BETWEEN TWO LINES FOLLOWS
2380 D=TAN((90-A2)*.01745)-TAN((90+A1)*.01745)
2390 X=TAN((90-A2)*.01745)*(-X5)-Y5-TAN((90+A1)*.01745)*X0+Y0
2400 X=X/D
2410 Y=TAN((90+A1)*.01745)*(TAN((90-A2)*.01745)*(-X5)-Y5)
2420 Y=Y-TAN((90-A2)*.01745)*(TAN((90+A1)*.01745)*X0-Y0)
2430 Y=Y/D
2440 RETURN
2450 GOTO 50   'DUMMY RETURN-START
2460 IF C$="CW" THEN X=-X
2470 PRINT "A=";Y;"E=";X
2480 LPRINT "A=";Y;"E=";X
2490 IF C$="CW" THEN PRINT"CAM ROTATES CW"
2500 IF C$="CW" THEN LPRINT"CAM ROTATES CW"
2510 IF C$="CCW" THEN PRINT"CAM ROTATES CCW"
2520 IF C$="CCW" THEN LPRINT"CAM ROTATES CCW"
2530 PRINT"ANGLE FOR RISE=";B1
2540 PRINT"ANGLE FOR RETURN=";B2
2550 PRINT"MAX PRESSURE ANGLE BY RISE=";A1
2560 PRINT"MAX PRESSURE ANGLE BY RETURN=";A2
2570 LPRINT"ANGLE FOR RISE=";B1
2580 LPRINT"ANGLE FOR RETURN=";B2
2590 LPRINT"MAX PRESSURE ANGLE BY RISE=";A1
2600 LPRINT"MAX PRESSURE ANGLE BY RETURN=";A2
2610 LPRINT
2620 LPRINT
2630 LPRINT
2640 GOTO 290
```

```
2650 CLS : RUN "MAIN" : END
2660 GOTO 50 'DUMMY RETURN-START
2670 'PARABOLIC MOTION:
2680 Q=1.273  'CHARACTERISTIC CONSTANT
2690 V2=2*H  'UNITARY MAX VELOCITY AT MIDPOINT
2700 Y2=-H/2
2710 Z2=-H/2-TAN((90+Al)*.01745)*Q*H*90/Bl 'INTERSECTION WITH Y-AXIS
2720 FOR TA=Bl/2-.5 TO 0 STEP -.5
2730 V3=4*H*TA/Bl 'UNITARY VELOCITY (BY START: .5 DEG. BEFORE MIDPOINT)
2740 Y3=-2*H*(TA/Bl)^2
2750 X3=V3/V2*Q*H*90/Bl
2760 Z3=Y3-TAN((90+Al)*.01745)*V3/V2*Q*H*90/Bl  'INTERSECTION WITH Y-AXIS
2770 IF Z3>Z2 THEN Z2=Z3:ELSE 2790
2780 NEXT TA
2790 TA=TA+.5
2800 V3=4*H*TA/Bl
2810 RETURN
2820 GOTO 50  'DUMMY RETURN-START
2830 'SIMPLE HARMONIC MOTION
2840 Q=1
2850 V2=1.571*H
2860 Y2=-H/2
2870 Z2=-H/2-TAN((90+Al)*.01745)*Q*H*90/Bl
2880 FOR TA=Bl/2-.5 TO 0 STEP -.5
2890 V3=1.571*H*SIN(3.14159*TA/Bl)
2900 Y3=-H/2*(1-COS(3.14159*TA/Bl))
2910 X3=V3/V2*Q*H*90/Bl
2920 Z3=Y3-TAN((90+Al)*.01745)*V3/V2*Q*H*90/Bl
2930 IF Z3>Z2 THEN Z2=Z3:ELSE 2950
2940 NEXT TA
2950 TA=TA+.5
2960 V3=1.571*H*SIN(3.14159*TA/Bl)
2970 RETURN
2980 GOTO 50 'DUMMY RETURN-START
2990 'CYCLOIDAL MOTION
3000 Q=1.273
3010 V2=2*H
3020 Y2=-H/2
3030 Z2=-H/2-TAN((90+Al)*.01745)*Q*H*90/Bl
3040 FOR TA=Bl/2-.5 TO 0 STEP -.5
3050 V3=H*(1-COS(6.28319*TA/Bl))
3060 Y3=-H*(TA/Bl-SIN(6.28319*TA/Bl)/6.28319)
3070 X3=V3/V2*Q*H*90/Bl
3080 Z3=Y3-TAN((90+Al)*.01745)*V3/V2*Q*H*90/Bl
3090 IF Z3>Z2 THEN Z2=Z3:ELSE 3110
3100 NEXT TA
3110 TA=TA+.5
3120 V3=H*(1-COS(6.28319*TA/Bl))
3130 RETURN
```

```
3140 GOTO 50 'DUMMY RETURN-START
3150 '3-4 POLYNOMIAL
3160 Q=1.273
3170 V2=2*H
3180 Y2=-H/2
3190 Z2=-H/2-TAN((90+A1)*.01745)*Q*H*90/Bl
3200 FOR TA=Bl/2-.5 TO 0 STEP -.5
3210 V3=H*(24*(TA/Bl)^2-32*(TA/Bl)^3)
3220 Y3=-H*(8*(TA/Bl)^3-8*(TA/Bl)^4)
3230 X3=V3/V2*Q*H*90/Bl
3240 Z3=Y3-TAN((90+A1)*.01745)*V3/V2*Q*H*90/Bl
3250 IF Z3>Z2 THEN Z2=Z3:ELSE 3270
3260 NEXT TA
3270 TA=TA+.5
3280 V3=H*(24*(TA/Bl)^2-32*(TA/Bl)^3)
3290 RETURN
3300 GOTO 50 'DUMMY RETURN-START
3310 '3-4-5 POLYNOMIAL
3320 Q=1.193
3330 V2=1.875*H
3340 X2=Q*H*90/Bl
3350 Y2=-H/2
3360 Z2=-H/2-TAN((90+A1)*.01745)*Q*H*90/Bl
3370 FOR TA=Bl/2-.5 TO 0 STEP -.5
3380 V3=H*(30*(TA/Bl)^2-60*(TA/Bl)^3+30*(TA/Bl)^4)
3390 Y3=-H*(10*(TA/Bl)^3-15*(TA/Bl)^4+6*(TA/Bl)^5)
3400 X3=V3/V2*Q*H*90/Bl
3410 Z3=Y3-TAN((90+A1)*.01745)*V3/V2*Q*H*90/Bl
3420 IF Z3>Z2 THEN Z2=Z3:ELSE 3440
3430 NEXT TA
3440 TA=TA+.5
3450 V3=H*(30*(TA/Bl)^2-60*(TA/Bl)^3+30*(TA/Bl)^4)
3460 RETURN
3470 GOTO 50 'DUMMY RETURN-START
3480 '4-5-6-7 POLYNOMIAL
3490 Q=1.392
3500 V2=2.875*H
3510 Y2=-H/2
3520 Z2=-H/2-TAN((90+A1)*.01745)*Q*H*90/Bl
3530 FOR TA=Bl/2-.5   TO 0 STEP -.5
3540 V3=H*(140*(TA/Bl)^3-420*(TA/Bl)^4+420*(TA/Bl)^5-140*(TA/Bl)^6)
3550 Y3=-H*(35*(TA/Bl)^4-84*(TA/Bl)^5+70*(TA/Bl)^6-20*(TA/Bl)^7)
3560 X3=V3/V2*Q*H*90/Bl
3570 Z3=Y3-TAN((90+A1)*.01745)*V3/V2*Q*H*90/Bl
3580 IF Z3>Z2 THEN Z2=Z3:ELSE 3600
3590 NEXT TA
3600 TA=TA+.5
3610 V3=H*(140*(TA/Bl)^3-420*(TA/Bl)^4+420*(TA/Bl)^5-140*(TA/Bl)^6)
3620 RETURN
3630 Q=1.273
```

```
3640 V2=2*H
3650 Y2=-H/2
3660 Z2=-H/2-TAN((90+A1)*.01745)*Q*H*90/Bl
3670 FOR TA=Bl/2-.5 TO 3*Bl/8 STEP -.5
3680 V3=H/5.14159*(8.2832-2*COS(12.5664*(TA/Bl-.25)))
3690 Y3=-H/5.14159*(8.2832*TA/Bl-1/6.28319*SIN(12.5664*(TA/Bl-.25))-1.57)
3700 X3=V3/V2*Q*H*90/Bl
3710 Z3=Y3-TAN((90+A1)*.01745)*V3/V2*Q*H*90/Bl
3720 IF Z3>Z2 THEN Z2=Z3:ELSE 3880
3730 NEXT TA
3740 FOR TA=3*Bl/8-.5 TO Bl/8 STEP -.5
3750 V3=H/5.14159*(25.1327*TA/Bl-1.14159)
3760 Y3=-H/5.14159*(12.5664*(TA/Bl)^2-1.14159*TA/Bl+.03719)
3770 X3=V3/V2*Q*H*90/Bl
3780 Z3=Y3-TAN((90+A1)*.01745)*V3/V2*Q*H*90/Bl
3790 IF Z3>Z2 THEN Z2=Z3:ELSE 3880
3800 NEXT TA
3810 FOR TA=Bl/8-.5 TO 0 STEP -.5
3820 V3=H/5.14159*(2-2*COS(12.5664*TA/Bl))
3830 Y3=-H/5.14159*(2*TA/Bl-1/6.28319*SIN(12.5664*TA/Bl))
3840 X3=V3/V2*Q*H*90/Bl
3850 Z3=Y3-TAN((90+A1)*.01745)*V3/V2*Q*H*90/Bl
3860 IF Z3>Z2 THEN Z2=Z3:ELSE 3880
3870 NEXT TA
3880 TA=TA+.5
3890 IF TA>=3*Bl/8 THEN V3=H/5.14159*(8.2832-2*COS(12.5664*(TA/Bl-.25))):GOTO 39
20
3900 IF TA>=Bl/8 THEN V3=H/5.14159*(25.1327*TA/Bl-1.14159):GOTO 3920
3910 V3=H/5.14159*(2-2*COS(12.5664*TA/Bl))
3920 RETURN
3930 Q=1.12
3940 V2=1.76*H
3950 Y2=-H/2
3960 Z2=-H/2-TAN((90+A1)*.01745)*Q*H*90/Bl
3970 FOR TA=Bl/2-.5 TO Bl/8 STEP -.5
3980 V3=.4399*H*(1-3*COS(3.14159/3+4/3*3.14159*TA/Bl))
3990 Y3=-H/7.14159*(2+3.14159*TA/Bl-9/4*SIN(3.14159/3+4/3*.14159*TA/Bl))
4000 X3=V3/V2*Q*H*90/Bl
4010 Z3=Y3-TAN((90+A1)*.01745)*V3/V2*Q*H*90/Bl
4020 IF Z3>Z2 THEN Z2=Z3:ELSE 4110
4030 NEXT TA
4040 FOR TA=Bl/8-.5 TO 0 STEP -.5
4050 V3=.4399*H*(1-COS(12.5664*TA/Bl))
4060 Y3=-H/7.14159*(3.14159*TA/Bl-.25*SIN(12.5664*TA/Bl))
4070 X3=V3/V2*Q*H*90/Bl
4080 Z3=Y3-TAN((90+A1)*.01745)*V3/V2*Q*H*90/Bl
4090 IF Z3>Z2 THEN Z2=Z3:ELSE 4110
4100 NEXT TA
4110 TA=TA+.5
4120 IF TA>=Bl/8 THEN V3=.4399*H*(1-3*COS(3.14159/3+4/3*3.14159*TA/Bl)):GOTO 414
0
4130 V3=.4399*H*(1-COS(12.5664*TA/Bl))
```

```
4140 RETURN
10 'THIS PROGRAM IS FILED UNDER "TRANS3"
20 'FOR GIVEN TIME-DISPLACEMENT DIAGRAMS AND
30 'CAM PROPORTIONS FOR SWINGING ROLLER
40 'FOLLOWER THIS PROGRAM WILL CALCULATE THE
50 'RADIUS OF CURVATURE OF THE RELATIVE PATH
60 'OF THE ROLLER FOLLOWER AND THE COMPRESSIVE
70 'STRESS FOR EACH 1 DEGREE CAM ROTATION
80 'DURING RISE AND RETURN
90 'AT THE END OF RISE IT PRINTS OUT THE MAX.
100 'VALUES OF THE COMPRESSIVE STRESS DURING
110 'ACCELERATION AND DECELERATION
120 'IT THEN PRINTS OUT THE CORRESPONDING VALUES
130 'FOR RETURN.
140 CLS
150 PRINT"RISE                                      RETURN"
160 PRINT" 1   PARABOLIC MOTION                      11"
170 PRINT" 2   SIMPLE HARMONIC MOTION                12"
180 PRINT" 3   CYCLOIDAL MOTION                      13"
190 PRINT" 4   3-4 POLYNOMIAL                        14"
200 PRINT" 5   3-4-5 POLYNOMIAL                      15"
210 PRINT" 6   4-5-6-7 POLYNOMIAL                    16"
220 PRINT" 7 MODIFIED TRAPEZOIDAL ACCEL.            17"
230 PRINT" 8 MODIFIED SINUSOIDAL ACCEL.             18"
240 PRINT" 9                                        19"
250 PRINT
260 PRINT
270 INPUT"TIME-DISPLACEMENT CURVE FOR RISE (1-9)";M
280 INPUT"TIME-DISPLACEMENT CURVE FOR RETURN(11-19)";L
290 INPUT"STROKE H, IN.";H
300 INPUT"EXCENTRICITY E(+ OR -)";E
310 INPUT"DISTANCE A, IN.";A
320 INPUT"TOTAL ANGLE FOR RISE, DEGREES";B1
330 INPUT"TOTAL ANGLE FOR RETURN, DEGREES";B2
340 INPUT"ROLLER RADIUS Rf, IN.";RF
350 INPUT"ROLLER WIDTH B, IN.";B
360 INPUT"CAM SPED, RPM";N
370 INPUT"IF CAM ROTATION IS CLOCKWISE WRITE CW, ELSE WRITE CCW";C$
380 INPUT"TOTAL MASS TO BE MOVED";W
390 INPUT"CAM MATERIAL(STEEL=1,CAST IRON=2)";MAT
400 CLS : PRINT : PRINT "A- TO EXIT TO MAIN PROGRAM.";TAB(38)"B- TO QUIT CHANGES
    AND EXECUTE PROGRAM."
410 PRINT "C- RISE É";M;"©";TAB(38)"D- RETURN É";L;"©"
420 PRINT "    1       PARABOLIC MOTION                 11"
430 PRINT "    2       SIMPLE HARMONIC MOTION           12"
440 PRINT "    3       CYCLOIDAL MOTION                 13"
450 PRINT "    4       3-4 POLYNOMIAL                   14"
460 PRINT "    5       3-4-5 POLYNOMIAL                 15"
470 PRINT "    6       4-5-6-7 POLYNOMIAL               16"
480 PRINT "    7       MODIFIED TRAPEZOIDAL ACCEL.      17"
490 PRINT "    8       MODIFIED SINUSOIDAL ACCEL.       18"
```

```
500 PRINT "E- STROKE h (in.) É";H;"©";TAB(38)"F- ECCENTRICITY E (+ or -) É";E;"©
"
510 PRINT "G- DISTANCE A (in.) É";A;"©"
520 PRINT "H- TOTAL ANGLE FOR RISE (DEGREES) É";B1;"©"
530 PRINT "I- TOTAL ANGLE FOR RETURN (DEGREES) É";B2;"©"
540 PRINT "J- ROLLER RADIUS (in.) É";RF;"©";TAB(38)"K- ROLLER WIDTH (in.) É";B;"
©"
550 PRINT "L- CAM SPEED (rpm) É";N;"©"
560 PRINT "M- CAM ROTATION (1:CW ; 2:CCW) É";C$;"©"
570 PRINT "N- TOTAL MASS TO BE MOVED (lb.) É";W;"©"
580 PRINT "O- CAM MATERIAL (1:STEEL ; 2:CAST IRON) É";MAT;"©"
590 PRINT : INPUT "WHICH VALUE WOULD YOU LIKE TO CHANGE (A..O) ";CV$
600 IF CV$="A" OR CV$="a" GOTO 6020
610 IF CV$="B" OR CV$="b" GOTO 790
630 PRINT : INPUT "WHAT IS THE NEW VALUE ";NV
640 IF CV$="C" OR CV$="c" THEN M=NV : GOTO 400
650 IF CV$="D" OR CV$="d" THEN L=NV : GOTO 400
660 IF CV$="E" OR CV$="e" THEN H=NV : GOTO 400
670 IF CV$="F" OR CV$="f" THEN E=NV : GOTO 400
680 IF CV$="G" OR CV$="g" THEN A=NV : GOTO 400
690 IF CV$="H" OR CV$="h" THEN B1=NV : GOTO 400
700 IF CV$="I" OR CV$="i" THEN B2=NV : GOTO 400
710 IF CV$="J" OR CV$="j" THEN RF=NV : GOTO 400
720 IF CV$="K" OR CV$="k" THEN B=NV : GOTO 400
730 IF CV$="L" OR CV$="l" THEN N=NV : GOTO 400
740 IF CV$="M" AND NV=1 OR CV$="m" AND NV=1 THEN C$="CW" : GOTO 400
750 IF CV$="M" AND NV=2 OR CV$="m" AND NV=2 THEN C$="CCW" : GOTO 400
760 IF CV$="N" OR CV$="n" THEN W=NV : GOTO 400
770 IF CV$="O" OR CV$="o" THEN MAT=NV : GOTO 400
780 GOTO 400
790 OM=N*3.14159/30
800 IF C$="CW" THEN OM=-OM
810 T1=B1/N/6
820 T2=B2/N/6
830 LPRINT"H=";H;"E=";E;"A=";A;"B1=";B1;"B2=";B2;"RF=";RF;"B=";B;"CAM SPEED,RPM=
";N
840 IF M=1 THEN LPRINT"PARABOLIC MOTION; RISE":GOTO 1010
850 IF M=2 THEN LPRINT"SIMPLE HARMONIC MOTION; RISE":GOTO 1450
860 IF M=3 THEN LPRINT"CYCLOIDAL MOTION; RISE":GOTO 1890
870 IF M=4 THEN LPRINT"3-4 POLYNOMIAL;RISE":GOTO 2330
880 IF M=5 THEN LPRINT"3-4-5 POLYNOMIAL; RISE":GOTO 2760
890 IF M=6 THEN LPRINT"4-5-6-7 POLYNOMIAL; RISE":GOTO 3200
900 IF M=7 THEN LPRINT"MODIFIED TRAPEZOIDAL ACCEL.; RISE":GOTO 3640
910 IF M=8 THEN LPRINT"MODIFIED SINUSOIDAL ACCEL.; RISE":GOTO 5080
930 IF L=11 THEN LPRINT"PARABOLIC MOTION; RETURN":GOTO 1220
940 IF L=12 THEN LPRINT"SIMPLE HARMONIC MOTION";RETURN:GOTO 1660
950 IF L=13 THEN LPRINT"CYCLOIDAL MOTION; RETURN":GOTO 2100
960 IF L=14 THEN LPRINT"3-4 POLYNOMIAL; RETURN":GOTO 2530
970 IF L=15 THEN LPRINT"3-4-5 POLYNOMIAL; RETURN":GOTO 2970
980 IF L=16 THEN LPRINT"4-5-6-7 POLYNOMIAL; RETURN":GOTO 3410
```

```
990 IF L=17 THEN LPRINT"MODIFIED TRAPEZOIDAL ACCEL.; RETURN":GOTO 4210
1000 IF L=18 THEN LPRINT"MODIFIED SINUSOIDAL ACCEL.; RETURN":GOTO 5470
1010 LPRINT"THETA                    RHO        COMPRESSIVE STRESS"
1020 FOR TA=0 TO B1/2
1030 Y=2*H*(TA/B1)^2
1040 VB=4*H/T1*(TA/B1) 'PARABOLIC MOTION; 1ST HALF RISE"
1050 AB=4*H/(T1)^2
1060 GOSUB 4800
1070 GOSUB 4890
1080 LPRINT TA,RO,SC
1090 NEXT TA
1100 LPRINT
1110 FOR TA=B1/2 TO B1
1120 Y=H*(1-2*((B1-TA)/B1)^2)
1130 VB=4*H/T1*(B1-TA)/B1   'PARABOLIC MOTION; 2ND HALF RISE
1140 AB=-4*H/T1^2
1150 GOSUB 4800
1160 GOSUB 4990
1170 LPRINT TA,RO,SC
1180 NEXT TA
1190 LPRINT
1200 LPRINT"MAX. COMPRESSIVE STRESS DURING 1ST HALF RISE =";S1;" PSI";" AT THETA
    =";TB
1210 LPRINT"MAX. COMPRESSIVE STRESS DURING 2ND HALF RISE =";S2;" PSI";" AT THETA
    =";TC
1220 S1=0:S2=0
1230 LPRINT
1240 FOR TA=0 TO B2/2
1250 Y=H*(1-2*(TA/B2)^2)
1260 VB=-4*H/T2*(TA/B2) 'PARABOLIC MOTION; 1ST HALF RETURN
1270 AB=-4*H/T2^2
1280 GOSUB 4800
1290 GOSUB 4990
1300 LPRINT TA,RO,SC
1310 NEXT TA
1320 LPRINT
1330 FOR TA=B2/2 TO B2
1340 Y=2*H*((B2-TA)/B2)^2
1350 VB=-4*H/T2*(B2-TA)/B2 'PARABOLIC MOTION; 2ND HALF RETURN
1360 AB=4*H/T2^2
1370 GOSUB 4800
1380 GOSUB 4890
1390 LPRINT TA,RO,SC
1400 NEXT TA
1410 LPRINT
1420 LPRINT"MAX. COMPRESSIVE STRESS DURING 1ST HALF RETURN =";S2;" PSI";" AT THE
TA =";TB
1430 LPRINT"MAX. COMPRESSIVE STRESS DURING 2ND HALF RETURN =";S1;" PSI";" AT THE
TA =";TC
1440 GOTO 5880
```

```
1450 LPRINT"THETA                    RHO          COMPRESSIVE STRESS"
1460 FOR TA=0 TO B1/2
1470 Y=H/2*(1-COS(3.14159*TA/B1))
1480 VB=H/2*3.14159/T1*SIN(3.14159*TA/B1) 'SIMPLE HARMONIC MOTION; RISE
1490 AB=H/2*(3.14159/T1)^2*COS(3.14159*TA/B1)
1500 GOSUB 4800
1510 GOSUB 4890
1520 LPRINT TA,RO,SC
1530 NEXT TA
1540 LPRINT
1550 FOR TA=B1/2 TO B1
1560 Y=H/2*(1-COS(3.14159*TA/B1))
1570 VB=H/2*3.14159/T1*SIN(3.14159*TA/B1)
1580 AB=H/2*(3.14159/T1)^2*COS(3.14159*TA/B1)
1590 GOSUB 4800
1600 GOSUB 4990
1610 LPRINT TA,RO,SC
1620 NEXT TA
1630 LPRINT
1640 LPRINT"MAX. COMPRESSIVE STRESS DURING 1ST HALF RISE =";S1;" PSI";" AT CAM A
LGLE THETA =";TB
1650 LPRINT"MAX. COMPRESSIVE STRESS DURING 2ND HALF RISE =";S2;" PSI";" AT CAM A
NGLE THETA =";TC
1660 S1=0:S2=0
1670 PRINT
1680 FOR TA=0 TO B2/2
1690 Y=H/2*(1+COS(3.14159*TA/B2))
1700 VB=-H/2*3.14159/T2*SIN(3.14159*TA/B2) 'SIMPLE HARMONIC MOTION; RETURN
1710 AB=-H/2*(3.14159/T2)^2*COS(3.14159*TA/B2)
1720 GOSUB 4800
1730 GOSUB 4990
1740 LPRINT TA,RO,SC
1750 NEXT TA
1760 LPRINT
1770 FOR TA=B2/2 TO B2
1780 Y=H/2*(1+COS(3.14159*TA/B2))
1790 VB=-H/2*3.14159/T2*SIN(3.14159*TA/B2)
1800 AB=-H/2*(3.14159/T2)^2*COS(3.14159*TA/B2)
1810 GOSUB 4800
1820 GOSUB 4890
1830 LPRINT TA,RO,SC
1840 NEXT TA
1850 LPRINT
1860 LPRINT"MAX. COMPRESSIVE STRESS DURING 1ST HALF RETURN =";S2;" PSI AT CAM AN
GLE THETA =";TB
1870 LPRINT"MAX. COMPRESSIVE STRESS DURING 2ND HALF RETURN =";S1;" PSI AT CAM AN
GLE THETA =";TC
1880 GOTO 5880
1890 LPRINT"THETA                    RHO          COMPRESSIVE STRESS"
1900 FOR TA=0 TO B1/2
```

```
1910 Y=H*(TA/B1-1/6.28319*SIN(6.28319*TA/B1))
1920 VB=H/T1*(1-COS(6.28319*TA/B1)) 'CYCLOIDAL MOTION; RISE
1930 AB=6.28319*H/(T1)^2*SIN(6.28319*TA/B1)
1940 GOSUB 4800
1950 GOSUB 4890
1960 LPRINT TA,RO,SC
1970 NEXT TA
1980 LPRINT
1990 FOR TA=B1/2 TO B1
2000 Y=H*(TA/B1-1/6.28319*SIN(6.28319*TA/B1))
2010 VB=H/T1*(1-COS(6.28319*TA/B1))
2020 AB=6.28319*H/(T1)^2*SIN(6.28319*TA/B1)
2030 GOSUB 4800
2040 GOSUB 4990
2050 LPRINT TA,RO,SC
2060 NEXT TA
2070 LPRINT
2080 LPRINT"MAX. COMPRESSIVE STRESS DURING 1ST HALF RISE =";S1;" PSI AT CAM ANGL
E THETA =";TB
2090 LPRINT"MAX. COMPRESSIVE STRESS DURING 2ND HALF RISE =";S2;" PSI AT CAM ANGL
E THETA =";TC
2100 S1=0:S2=0
2110 LPRINT
2120 FOR TA=0 TO B2/2
2130 Y=H*(1-TA/B2+1/6.28139*SIN(6.28319*TA/B2))
2140 VB=H/T2*(COS(6.28319*TA/B2)-1) 'CYCLOIDAL MOTION; RETURN
2150 AB=-6.28319*H/(T2)^2*SIN(6.28319*TA/B2)
2160 GOSUB 4800
2170 GOSUB 4990
2180 LPRINT TA,RO,SC
2190 NEXT TA
2200 LPRINT
2210 FOR TA=B2/2 TO B2
2220 Y=H*(1-TA/B2+1/6.28319*SIN(6.28319*TA/B2))
2230 VB=H/T2*(COS(6.28319*TA/B2)-1)
2240 AB=-6.28319*H/(T2)^2*SIN(6.28319*TA/B2)
2250 GOSUB 4800
2260 GOSUB 4890
2270 LPRINT TA,RO,SC
2280 NEXT TA
2290 LPRINT
2300 LPRINT"MAX. COMPRESSIVE STRESS DURING 1ST HALF RETURN =";S2;" PSI AT CAM AN
GLE THETA =";TB
2310 LPRINT"MAX. COMPRESSIVE STRESS DURING 2ND HALF RETURN =";S1;" PSI AT CAM AN
GLE THETA =";TC
2320 GOTO 5880
2330 LPRINT"THETA              RHO         COMPRESSIVE STRESS"
2340 FOR TA=0 TO B1/2
2350 Y=H*(8*(TA/B1)^3-8*(TA/B1)^4)
```

```
2360 VB=H/T1*(24*(TA/B1)^2-32*(TA/B1)^3) '3-4 POLYNOMIAL; 1ST HALF RISE
2370 AB=H/(T1)^2*(48*TA/B1-96*(TA/B1)^2)
2380 GOSUB 4800
2390 GOSUB 4890
2400 LPRINT TA,RO,SC
2410 NEXT TA
2420 LPRINT
2430 FOR TA=B1/2 TO B1
2440 Y=H*(1-8*TA/B1+24*(TA/B1)^2-24*(TA/B1)^3+8*(TA/B1)^4)
2450 VB=H/T1*(-8+48*TA/B1-72*(TA/B1)^2+32*(TA/B1)^3) Ñ3-4 POLYNOMIAL;2ND HALF RI
SE
2460 AB=H/(T1)^2*(48-144*TA/B1+96*(TA/B1)^2)
2470 GOSUB 4800
2480 GOSUB 4990
2490 LPRINT TA,RO,SC
2500 NEXT TA
2510 LPRINT"MAX. COMPRESSIVE STRESS DURING 1ST HALF RISE =";S1;" PSI AT CAM ANGL
E THETA =";TB
2520 LPRINT"MAX. COMPRESSIVE STRESS DURING 2ND HALF RISE =";S1;" PSI AT CAM ANGL
E THETA =";TC
2530 S1=0:S2=0
2540 LPRINT
2550 FOR TA=0 TO B2/2
2560 Y=H*(1-8*(TA/B2)^3+8*(TA/B2)^4)
2570 VB=H/T2*(-24*(TA/B2)^2+32*(TA/B2)^3) '3-4 POLYNOMIAL; 1ST HALF RETURN
2580 AB=H/(T2)^2*(-48*TA/B2+96*(TA/B2)^2)
2590 GOSUB 4800
2600 GOSUB 4990
2610 LPRINT TA,RO,SC
2620 NEXT TA
2630 LPRINT
2640 FOR TA=B2/2 TO B2
2650 Y=H*(8*TA/B2-24*(TA/B2)^2+24*(TA/B2)^3-8*(TA/B2)^4)
2660 VB=H/T2*(8-48*TA/B2+72*(TA/B2)^2-32*(TA/B2)^3) '3-4 POLYNOMIAL; 2ND HALF RE
TURN
2670 AB=H/(T2)^2*(-48+144*TA/B2-96*(TA/B2)^2)
2680 GOSUB 4800
2690 GOSUB 4890
2700 LPRINT TA,RO,SC
2710 NEXT TA
2720 LPRINT
2730 LPRINT"MAX. COMPRESSIVE STRESS DURING 1ST HALF RETURN =";S2;" PSI AT CAM AN
GLE THETA =";TB
2740 LPRINT"MAX. COMPRESSIVE STRESS DURING 2ND HALF RETURN =";S1;" PSI AT CAM AN
GLE THETA =";TC
2750 GOTO 5880
2760 LPRINT"THETA                    RHO          COMPRESSIVE STRESS"
2770 FOR TA=0 TO B1/2
2780 Y=H*(10*(TA/B1)^3-15*(TA/B1)^4+6*(TA/B1)^5)
2790 VB=H/T1*(30*(TA/B1)^2-60*(TA/B1)^3+30*(TA/B1)^4) '3-4-5 POLYNOMIAL; RISE
```

```
2800 AB=H/(T1)^2*(60*TA/B1-180*(TA/B1)^2+120*(TA/B1)^3)
2810 GOSUB 4800
2820 GOSUB 4890
2830 LPRINT TA,RO,SC
2840 NEXT TA
2850 LPRINT
2860 FOR TA=B1/2 TO B1
2870 Y=H*(10*(TA/B1)^3-15*(TA/B1)^4+6*(TA/B1)^5)
2880 VB=H/T1*(30*(TA/B1)^2-60*(TA/B1)^3+30*(TA/B1)^4)
2890 AB=H/(T1)^2*(60*TA/B1-180*(TA/B1)^2+120*(TA/B1)^3)
2900 GOSUB 4800
2910 GOSUB 4990
2920 LPRINT TA,RO,SC
2930 NEXT TA
2940 LPRINT
2950 LPRINT"MAX. COMPRESSIVE STRESS DURING 1ST HALF RISE =";S1;" PSI AT CAM ANGL
E THETA =";TB
2960 LPRINT"MAX. COMPRESSIVE STRESS DURING 2ND HALF RISE =";S2;" PSI AT CAM ANGL
E THETA =";TC
2970 S1=0:S2=0
2980 LPRINT
2990 FOR TA=0 TO B2/2
3000 Y=H*(1-10*(TA/B2)^3+15*(TA/B2)^4-6*(TA/B2)^5)
3010 VB=H/T2*(-30*(TA/B2)^2+60*(TA/B2)^3-30*(TA/B2)^4) '3-4-5 POLYNOMIAL; RETURN

3020 AB=H/(T2)^2*(-60*TA/B2+180*(TA/B2)^2-120*(TA/B2)^3)
3030 GOSUB 4800
3040 GOSUB 4990
3050 LPRINT TA,RO,SC
3060 NEXT TA
3070 LPRINT
3080 FOR TA=B2/2 TO B2
3090 Y=H*(1-10*(TA/B2)^3+15*(TA/B2)^4-6*(TA/B2)^5)
3100 VB=H/T2*(-30*(TA/B2)^2+60*(TA/B2)^3-30*(TA/B2)^4)
3110 AB=H/(T2)^2*(-60*TA/B2+180*(TA/B2)^2-120*(TA/B2)^3)
3120 GOSUB 4800
3130 GOSUB 4890
3140 LPRINT TA,RO,SC
3150 NEXT TA
3160 LPRINT
3170 LPRINT"MAX. COMPRESSIVE STRESS DURING 1ST HALF RETURN =";S2;" PSI AT CAM AN
GLE THETA =";TB
3180 LPRINT"MAX. COMPRESSIVE STRESS DURING 2ND HALF RETURN =";S1;" PSI AT CAM AN
GLE THETA =";TC
3190 GOTO 5880
3200 LPRINT"THETA                    RHO           COMPRESSIVE STRESS"
3210 FOR TA=0 TO B1/2
3220 Y=H*(35*(TA/B1)^4-84*(TA/B1)^5+70*(TA/B1)^6-20*(TA/B1)^7)
3230 VB=H/T1*(140*(TA/B1)^3-420*(TA/B1)^4+420*(TA/B1)^5-140*(TA/B1)^6) '4-5-6-7
```

```
POLYNOMIAL; RISE
3240 AB=H/(T1)^2*(420*(TA/B1)^2-1680*(TA/B1)^3+2100*(TA/B1)^4-840*(TA/B1)^5)
3250 GOSUB 4800
3260 GOSUB 4890
3270 LPRINT TA,RO,SC
3280 NEXT TA
3290 LPRINT
3300 FOR TA=B1/2 TO B1
3310 Y=H*(35*(TA/B1)^4-84*(TA/B1)^5+70*(TA/B1)^6-20*(TA/B1)^7)
3320 VB=H/T1*(140*(TA/B1)^3-420*(TA/B1)^4+420*(TA/B1)^5-140*(TA/B1)^6)
3330 AB=H/(T1)^2*(420*(TA/B1)^2-1680*(TA/B1)^3+2100*(TA/B1)^4-840*(TA/B1)^5)
3340 GOSUB 4800
3350 GOSUB 4990
3360 LPRINT TA,RO,SC
3370 NEXT TA
3380 LPRINT
3390 LPRINT"MAX. COMPRESSIVE STRESS DURING 1ST HALF RISE =";S1;" PSI AT CAM ANGL
E THETA =";TB
3400 LPRINT"MAX. COMPRESSIVE STRESS DURING 2ND HALF RISE =";S2;" PSI AT CAM ANGL
E THETA =";TC
3410 S1=0:S2=0
3420 LPRINT
3430 FOR TA=0 TO B2/2
3440 Y=H*(1-35*(TA/B2)^4+84*(TA/B2)^5-70*(TA/B2)^6+20*(TA/B2)^7)
3450 VB=H/T2*(-140*(TA/B2)^3+420*(TA/B2)^4-420*(TA/B2)^5+140*(TA/B2)^6)  '4-5-6-7
POLYNOMIAL; RETURN
3460 AB=H/(T2)^2*(-420*(TA/B2)^2+1680*(TA/B2)^3-2100*(TA/B2)^4+840*(TA/B2)^5))
3470 GOSUB 4800
3480 GOSUB 4990
3490 LPRINT TA,RO,SC
3500 NEXT TA
3510 LPRINT
3520 FOR TA=B2/2 TO B2
3530 Y=H*(1-35*(TA/B2)^4+84*(TA/B2)^5-70*(TA/B2)^6+20*(TA/B2)^7)
3540 VB=H/T2*(-140*(TA/B2)^3+420*(TA/B2)^4-420*(TA/B2)^5+140*(TA/B2)^6)
EDIT 910
3550 AB=H/(T2)^2*(-420*(TA/B2)^2+1680*(TA/B2)^3-2100*(TA/B2)^4+840*(TA/B2)^5)
3560 GOSUB 4800
3570 GOSUB 4890
3580 LPRINT TA,RO,SC
3590 NEXT TA
3600 LPRINT
3610 LPRINT"MAX. COMPRESSIVE STRESS DURING 1ST HALF RETURN =";S2;" PSI AT CAM AN
GLE THETA =";TB
3620 LPRINT"MAX. COMPRESSIVE STRESS DURING 2ND HALF RETURN =";S1;" PSI AT CAM AN
GLE THETA =";TC
3630 GOTO 5880
3640 LPRINT"THETA                    RHO           COMPRESSIVE STRESS"
3650 FOR TA=0 TO B1/8
3660 Y=H/5.14159*/92*TA/B1-1/6.28319*SIN(12.5664*TA/B1))
```

```
3670  VB=H/T1/5.14159*(2-2*COS(12.5664*TA/Bl))
3680  AB=4.88812*H/(T1)^2*SIN(12.5664*TA/Bl)
3690  GOSUB 4800
3700  GOSUB 4890
3710  LPRINT TA,RO,SC
3720  NEXT TA
3730  LPRINT
3740  FOR TA=Bl/8 TO 3*Bl/8
3750  Y=H/5.14159*(12.5664*(TA/Bl)^2-1.1416*TA/Bl+.037195)
3760  VB=H/T1/5.14159*(25.1327*TA/Bl-1.14159)
3770  AB=4.88812*H/(T1)^2
3780  GOSUB 4800
3790  GOSUB 4890
3800  LPRINT TA,RO,SC
3810  NEXT TA
3820  LPRINT
3830  FOR TA=3*Bl/8 TO Bl/2
3840  Y=H/5.14159*(8.2832*TA/Bl-1/6.28319*SIN(12.5664*(TA/Bl-.25))-1.5708)
3850  VB=H/T1/5.14159*(8.2832-2*COS(12.5664*(TA/Bl-.25)))
3860  AB=4.88812*H/(T1)^2*SIN(12.5664*(TA/Bl-.25))
3870  GOSUB 4800
3880  GOSUB 4890
3890  LPRINT TA,RO,SC
3900  NEXT TA
3910  LPRINT
3920  FOR TA=Bl/2 TO 5*Bl/8
3930  Y=H/5.14159*(-1.5708+8.2832*TA/Bl+1/6.28319*SIN(12.5664*(.75-TA/Bl)))
3940  VB=H/5.14159/T1*(8.2832-2*COS(12.5664*(.75-TA/Bl)))
3950  AB=-4.8882*H/(T1)^2*SIN(12.5664*(.75-TA/Bl))
3960  GOSUB 4800
3970  GOSUB 4990
3980  LPRINT TA,RO,SC
3990  NEXT TA
4000  LPRINT
4010  FOR TA=5*Bl/8 TO 7*Bl/8
4020  Y=H/5.14159*(-6.32038+23.991*TA/Bl-12.5664*(TA/Bl)^2)
4030  VB=H/5.14159/T1*(23.991-25.133*TA/Bl)
4040  AB=4.88812*H/(T1)^2
4050  GOSUB 4800
4060  GOSUB 4990
4070  LPRINT TA,RO,SC
4080  NEXT TA
4090  LPRINT
4100  FOR TA=7*Bl/8 TO Bl
4110  Y=H/5.14159*(3.14159+2*TA/Bl+1/6.28319*SIN(12.5664*(1-TA/Bl)))
4120  VB=H/5.14159/T1*(2-2*COS(12.5664*(1-TA/Bl)))
4130  AB=-4.88812*H/(T1)^2*SIN(12.5664*(1-TA/Bl))
4140  GOSUB 4800
4150  GOSUB 4990
4160  LPRINT TA,RO,SC
```

```
4170 NEXT TA
4180 LPRINT
4190 LPRINT"MAX. COMPRESSIVE STRESS DURING 1ST HALF RISE =";S1;" PSI AT CAM ANGL
E THETA =";TB
4200 LPRINT"MAX. COMPRESSIVE STRESS DURING 2ND HALF RISE =";S2;" PSI AT CAM ANGL
E THETA =";TC:GOTO 930
4210 S1=0:S2=0
4220 LPRINT
4230 FOR TA=0 TO B2/8
4240 Y=H/5.14159*(5.14159-2*TA/B2+1/6.28319*SIN(12.5664*TA/B2))
4250 VB=H/5.14159/T2*(-2+2*COS(12.5664*TA/B2))
4260 AB=-4.88812*H/(T2)^2*SIN(12.5664*TA/B2)
4270 GOSUB 4800
4280 GOSUB 4990
4290 LPRINT TA,RO,SC
4300 NEXT TA
4310 LPRINT
4320 FOR TA=B2/8 TO 3*B2/8
4330 Y=H/5.14159*(5.14159-12.5564*(TA/B2)^2+1.14159*TA/B2-.037195)
4340 VB=H/T2/5.14159*(-25.1327*TA/B2+1.14159)
4350 AB=-4.88812*H/(T2)^2
4360 GOSUB 4800
4370 GOSUB 4990
4380 LPRINT TA,RO,SC
4390 NEXT TA
4400 LPRINT
4410 FOR TA=3*B2/8 TO B2/2
4420 Y=H/5.14159*(6.7124-8.2832*TA/B2+1/6.28319*SIN(12.5664*(TA/B2-.25)))
4430 VB=H/T2/5.14159*(-8.28319+2*COS(12.5664*(TA/B2-.25)))
4440 AB=-4.88812*H/(T2)^2*SIN(12.5664*(TA/B2-.25))
4450 GOSUB 4800
4460 GOSUB 4990
4470 LPRINT TA,RO,SC
4480 NEXT TA
4490 LPRINT
4500 FOR TA=B2/2 TO 5*B2/8
4510 Y=H/5.14159*(6.7124-8.2832*TA/B2+1/6.28319*SIN(12.5664*(.75-TA/B2)))
4520 VB=H/T2/5.14159*(-8.28319+2*COS(12.5664*(.75-TA/B2)))
4530 AB=4.88812*H/(T2)^2*SIN(12.5664*(.75-TA/B2))
4540 GOSUB 4800
4550 GOSUB 4890
4560 LPRINT TA,RO,SC
4570 NEXT TA
4580 LPRINT
4590 FOR TA=5*B2/8 TO 7*B2/8
4600 Y=H/5.14159*(11.462-23.9912*TA/B2+12.5664*(TA/B2)^2)
4610 VB=H/T2/5.14159*(-23.9912+25.1327*TA/B2)
4620 AB=4.88812*H/(T2)^2
4630 GOSUB 4800
4640 GOSUB 4890
```

```
4650 LPRINT TA,RO,SC
4660 NEXT TA
4670 LPRINT
4680 FOR TA=7*B2/8 TO B2
4690 Y=H/5.14159*(2-2*TA/B2-1/6.28319*SIN(12.5664*(1-TA/B2)))
4700 VB=H/T2/5.14159*(-2+2*COS(12.5664*(1-TA/B2)))
4710 AB=4.88812*H/(T2)^2*SIN(12.5664*(1-TA/B2))
4720 GOSUB 4800
4730 GOSUB 4890
4740 LPRINT TA,RO,SC
4750 NEXT TA
4760 LPRINT
4770 LPRINT"MAX. COMPRESSIVE STRESS DURING 1ST HALF RETURN =";S2;" PSI AT CAM AN
GLE THETA =";TB
4780 LPRINT"MAX. COMPRESSIVE STRESS DURING 2ND HALF RETURN =";S1;" PSI AT CAM AN
GLE THETA =";TC
4790 GOTO 5880
4800 R=SQR(E^2+(A+Y)^2)
4810 VC=R*OM
4820 T3=1.5708-ATN(E/(A+Y))
4830 TE=ATN((VB-VC*COS(T3))/VC/SIN(T3))
4840 V3=VC*SIN(T3)/COS(TE)
4850 A3=AB*SIN(1.571-T4)+R*OM^2*SIN(T3-TE)-2*V3*OM
4860 RO=V3^2/A3
4870 F1=ABS(W/386*AB/COS(TE))
4880 RETURN
4890 IF RO>0 THEN RK=RO+RF:ELSE GOTO 4920
4900 IF MAT=1 THEN SC=2290*SQR(ABS(F1/B*(1/RK-1/RF))):GOTO 4970
4910 IF MAT=2 THEN SC=1850*SQR(ABS(F1/B*(1/RK-1/RF))):GOTO 4970
4920 IF RO<0 THEN RK=RO+RF
4930 IF RK>=0 THEN GOTO 6030
4940 RK=ABS(RK)
4950 IF MAT=1 THEN SC=2290*SQR(F1/B*(1/RK+1/RF))
4960 IF MAT=2 THEN SC=1850*SQR(F1/B*(1/RK+1/RF))
4970 IF SC>S1 THEN S1=SC:TB=TA
4980 RETURN
4990 IF RO>0 THEN RK=RO-RF:ELSE GOTO 5030
5000 IF RK<=0 THEN GOTO 6030
5010 IF MAT=1 THEN SC=2290*SQR(F1/B*(1/RK+1/RF)):GOTO 5060
5020 IF MAT=2 THEN SC=1850*SQR(F1/B*(1/RK+1/RF)):GOTO 5060
5030 IF RO<0 THEN RK=ABS(RO-RF)
5040 IF MAT=1 THEN SC=2290*SQR(ABS(F1/B*(1/RK-1/RF)))
5050 IF MAT=2 THEN SC=1850*SQR(ABS(FI/B*(1/RK-1/RF)))
5060 IF SC>S2 THEN S2=SC:TC=TA
5070 RETURN
5080 LPRINT"THETA                    RHO          COMPRESSIVE STRESS"
5090 FOR TA=0 TO B1/8
5100 Y=H/7.14159*(3.14159*TA/B1-.25*SIN(12.5664*TA/B1))
5110 VB=.4399*H/T1*(1-COS(12.5664*TA/B1))
5120 AB=5.528*H/(T1)^2*SIN(12.5664*TA/B1)
```

```
5130 GOSUB 4800
5140 GOSUB 4890
5150 LPRINT TA,RO,SC
5160 NEXT TA
5170 LPRINT
5180 FOR TA=B1/8 TO B1/2
5190 Y=H/7.14159*(2+3.14159*TA/B1-9/4*SIN(3.14159/3+4/3*3.14159*TA/B1))
5200 VB=.4399*H/T1*(1-3*COS(3.14159/3+4/3*3.14159*TA/B1))
5210 AB=5.528*H/(T1)^2*SIN(3.14159/3+4/3*3.14159*TA/B1)
5220 GOSUB 4800
5230 GOSUB 4890
5240 LPRINT TA,RO,SC
5250 NEXT TA
5260 LPRINT
5270 FOR TA=B1/2 TO 7*B1/8
5280 Y=H/7.14159*(2+3.14159*TA/B1-9/4*SIN(3.14159/3+4/3*3.14159*TA/B1))
5290 VB=.4399*H/T1*(1-3*COS(3.14159/3+4/3*3.14159*TA/B1))
5300 AB=5.528*H/(T1)^2*SIN(3.14159/3+4/3*3.14159*TA/B1)
5310 GOSUB 4800
5320 GOSUB 4990
5330 LPRINT TA,RO,SC
5340 NEXT TA
5350 LPRINT
5360 FOR TA=7*B1/8 TO B1
5370 Y=H/7.14159*(4+3.14159*TA/B1-.25*SIN(12.5664*TA/B1))
5380 VB=.4399*H/T1*(1-COS(12.5664*TA/B1))
5390 AB=5.528*H/(T1)^2*SIN(12.5664*TA/B1)
5400 GOSUB 4800
5410 GOSUB 4990
5420 LPRINT TA,RO,SC
5430 NEXT TA
5440 LPRINT
5450 LPRINT"MAX. COMPRESSIVE STRESS DURING 1ST HALF RISE =";S1;" PSI AT CAM ANGL
E THETA =";TB
5460 LPRINT"MAX. COMPRESSIVE STRESS DURING 2ND HALF RISE =";S2;" PSI AT CAM ANGL
E THETA =";TC:GOTO 930
5470 S1=0:S2=0
5480 LPRINT
5490 FOR TA=0 TO B2/8
5500 Y=H/7.14159*(7.14159-3.14159*TA/B2+.25*SIN(12.5664*TA/B2))
5510 VB=.4399*H/T2*(-1+COS(12.5664*TA/B2))
5520 AB=-5.528*H/(T2)^2*SIN(12.5664*TA/B2)
5530 GOSUB 4800
5540 GOSUB 4990
5550 LPRINT TA,RO,SC
5560 NEXT TA
5570 LPRINT
5580 FOR TA=B2/8 TO B2/2
5590 Y=H/7.14159*(5.14159-3.14159*TA/B2+9/4*SIN(3.14159/3+4/3.14159*TA/B2))
5600 VB=.4399*H/T2*(-1+3*COS(3.14159/3+4/3*3.14159*TA/B2))
```

```
5610 AB=-5.528*H/(T2)^2*SIN(3.14159/3+4/3*3.14159*TA/B2)
5620 GOSUB 4800
5630 GOSUB 4990
5640 LPRINT TA,RO,SC
5650 NEXT TA
5660 LPRINT
5670 FOR TA=B2/2 TO 7*B2/8
5680 Y=H/7.14159*(5.14159-3.14159*TA/B2+9/4*SIN(3.14159/3+4/3*3.14159*TA/B2))
5690 VB=.4399*H/T2*(-1+3*COS(3.14159/3+4/3*3.14159*TA/B2))
5700 AB=-5.528*H/(T2)^2*SIN(3.14159/3+4/3*3.14159*TA/B2)
5710 GOSUB 4800
5720 GOSUB 4890
5730 LPRINT TA,RO,SC
5740 NEXT TA
5750 LPRINT
5760 FOR TA=7*B2/8 TO B2
5770 Y=H/7.14159*(3.14159-3.14159*TA/B2+.25*SIN(12.5664*TA/B2))
5780 VB=.4399*H/T2*(-1+COS(12.5664*TA/B2))
5790 AB=-5.528*H/(T2)^2*SIN(12.5664*TA/B2)
5800 GOSUB 4800
5810 GOSUB 4890
5820 LPRINT TA,RO,SC
5830 NEXT TA
5840 LPRINT
5850 LPRINT"MAX. COMPRESSIVE STRESS DURING 1ST HALF RETURN =";S2;" PSI AT CAM AN
GLE THETA =";TB
5860 LPRINT"MAX. COMPRESSIVE STRESS DURING 2ND HALF RETURN =";S1;" PSI AT CAM AN
GLE THETA =";TC
5870 GOTO 5880
5880 LPRINT"STROKE H=";H;" IN."
5890 LPRINT"EXCENTRICITY E=";E;" IN."
5900 LPRINT"DISTANCE A=";A;" IN."
5910 LPRINT"TOTAL ANGLE FOR RISE =";B1;" DEGREES"
5920 LPRINT"TOTAL ANGLE FOR RETURN =";B2;" DEGREES"
5930 LPRINT"ROLLER RADIUS Rf =";RF;" IN."
5940 LPRINT"ROLLER WIDTH B =";B;" IN."
5950 LPRINT"CAM SPEED, N =";N;" RPM"
5960 IF C$="CCW" THEN LPRINT"CAM ROTATES CCW"
5970 IF C$="CW" THEN LPRINT"CAM ROTATES CW"
5980 IF MAT=1 THEN LPRINT"CAM MATERIAL IS STEEL"
5990 IF MAT=2 THEN LPRINT"CAM MATERIAL IS CAST IRON"
6000 LPRINT"WEIGHT OF MASS TO BE MOVED =";W;" LB"
6010 GOTO 400
6020 CLS : RUN "MAIN" : END
6030 LPRINT"UNDERCUTTING OCCURS FOR CAM ANGLE THETA =";TA;"THEREFORE OTHER"
6040 LPRINT"CAM PROPORTIONS MUST BE CHOSEN"
```

```
10 'THIS PROGRAM IS FILED UNDER "SWING4"
20 CLS
30 'SWINGING ROLLER FOLLOWER'
40 'FINDING MAX. PRESSURE ANGLES BY RISE AND RETURN FOR GIVEN
50 'TIME-DISPLACEMENT DIAGRAMS AND CAM PROPORTIONS
60 PRINT"FINDING MAX. PRESSURE ANGLES BY RISE AND RETURN FOR GIVEN"
70 PRINT"TIME-DISPLACEMENT DIAGRAMS AND CAM PROPORTIONS"
80 PRINT"RISE                                RETURN"
90 PRINT" 1  PARABOLIC MOTION                  11"
100 PRINT" 2  SIMPLE HARMONIC MOTION            12"
110 PRINT" 3  CYCLOIDAL MOTION                  13"
120 PRINT" 4  3-4 POLYNOMIAL                    14"
130 PRINT" 5  3-4-5 POLYNOMIAL                  15"
140 PRINT" 6  4-5-6-7 POLYNOMIAL                16"
150 PRINT" 7  MODIFIED TRAPEZOIDAL ACCEL.       17"
160 PRINT" 8  MODIFIED SINUSOIDAL ACCEL.        18"
170 PRINT
180 PRINT
190 INPUT"TIME-DISPLACEMENT CURVE FOR RISE (1-9)";M
200 INPUT"TIME-DISPLACEMENT CURVE FOR RETURN(11-19)";L
210 INPUT"DISTANCE MO=";MO
220 INPUT"LENGTH OF FOLLOWER ARM LF=";LF
230 INPUT"MINIMUM RADIUS OF CAM(TO CENTER OF ROLLER) RM=";RM
240 INPUT"TOTAL ANGLE OF OSCILLATING FOLLOWER MOVEMENT FO=";FO
250 INPUT"TOTAL ANGLE FOR RISE, DEGREES"; B1
260 INPUT"TOTAL ANGLE FOR RETURN, DEGREES";B2
270 INPUT"IF CAM ROTATION IS CW WRITE 'CW':ELSE WRITE 'CCW'";C$
280 CLS : PRINT : PRINT "A- TO EXIT TO MAIN PROGRAM.";TAB(38)"B- TO QUIT CHANGES
  AND EXECUTE PROGRAM."
290 PRINT "C- RISE É";M;"©";TAB(38)"D- RETURN É";L;"©"
300 PRINT "      1       PARABOLIC MOTION               11"
310 PRINT "      2       SIMPLE HARMONIC MOTION          12"
320 PRINT "      3       CYCLOIDAL MOTION                13"
330 PRINT "      4       3-4 POLYNOMIAL                  14"
340 PRINT "      5       3-4-5 POLYNOMIAL                15"
350 PRINT "      6       4-5-6-7 POLYNOMIAL              16"
360 PRINT "      7       MODIFIED TRAPEZOIDAL ACCEL.     17"
370 PRINT "      8       MODIFIED SINUSOIDAL ACCEL.      18"
380 PRINT "E- DISTANCE MO É";MO;"©";TAB(38)"F- LENGTH OF FOLLOWER ARM É";LF;"©"
390 PRINT "G- MIN. RADIUS OF CAM TO CENTER OF ROLLER.) É";RM;"©"
400 PRINT "H- TOTAL ANGLE OF OSCILLATING FOLLOWER MOVEMENT É";FO;"©"
410 PRINT "I- TOTAL ANGLE FOR RISE (DEGREES) É";B1;"©"
420 PRINT "J- TOTAL ANGLE FOR RETURN (DEGREES) É";B2;"©"
430 PRINT "K- CAM ROTATION (1:CW;2:CCW) É";C$;"©"
440 PRINT : INPUT "WHICH VALUE WOULD YOU LIKE TO CHANGE (A..K)";CV$
450 IF CV$="A" OR CV$="a" GOTO 2080
460 IF CV$="B" OR CV$="b" GOTO 600
480 PRINT : INPUT "WHAT IS THE NEW VALUE ";NV
490 IF CV$="C" OR CV$="c" THEN M=NV : GOTO 280
500 IF CV$="D" OR CV$="d" THEN L=NV : GOTO 280
510 IF CV$="E" OR CV$="e" THEN MO=NV : GOTO 280
```

```
520 IF CV$="F" OR CV$="f" THEN LF=NV : GOTO 280
530 IF CV$="G" OR CV$="g" THEN RM=NV : GOTO 280
540 IF CV$="H" OR CV$="h" THEN F0=NV : GOTO 280
550 IF CV$="I" OR CV$="i" THEN B1=NV : GOTO 280
560 IF CV$="J" OR CV$="j" THEN B2=NV : GOTO 280
570 IF CV$="K" AND NV=1 OR CV$="k" AND NV=1 THEN C$="CW" : GOTO 280
580 IF CV$="K" AND NV=2 OR CV$="k" AND NV=2 THEN C$="CCW" : GOTO 280
590 GOTO 280
600 IF C$="CW" THEN I=-1:ELSE I=1
610 N=60
620 OM=N*3.14159/30
630 T1=B1/360
640 T2=B2/360
650 H=LF*F0*.01745
660 PRINT"MO=";MO;"LF=";LF;"RM=";RM;"F0=";F0;"B1=";B1;"B2=";B2
670 LPRINT"MO=";MO;"LF=";LF;"RM=";RM;"F0=";F0;"B1=";B1;"B2=";B2
680 LPRINT
690 PRINT
700 IF M=1 THEN LPRINT"PARABOLIC MOTION; RISE"
710 IF M=1 THEN PRINT"PARABOLIC MOTION; RISE":GOTO 860
720 IF M=2 THEN LPRINT"SIMPLE HARMONIC MOTION; RISE"
730 IF M=2 THEN PRINT"SIMPLE HARMONIC MOTION; RISE":GOTO 960
740 IF M=3 THEN LPRINT"CYCLOIDAL MOTION; RISE"
750 IF M=3 THEN PRINT"CYCLOIDAL MOTION; RISE":GOTO 1020
760 IF M=4 THEN LPRINT"3-4 POLYNOMIAL; RISE"
770 IF M=4 THEN PRINT"3-4 POLYNOMIAL; RISE":GOTO 1080
780 IF M=5 THEN LPRINT"3-4-5 POLYNOMIAL; RISE"
790 IF M=5 THEN PRINT"3-4-5 POLYNOMIAL; RISE":GOTO 1140
800 IF M=6 THEN LPRINT"4-5-6-7 POLYNOMIAL; RISE"
810 IF M=6 THEN PRINT"4-5-6-7 POLYNOMIAL; RISE":GOTO 1200
820 IF M=7 THEN LPRINT"MODIFIED TRAPEZOIDAL ACCEL.; RISE"
830 IF M=7 THEN PRINT"MODIFIED TRAPEZOIDAL ACCEL.; RISE":GOTO 2090
840 IF M=8 THEN LPRINT"MODIFIED SINUSOIDAL ACCEL.; RISE"
850 IF M=8 THEN PRINT"MODIFIED SINUSOIDAL ACCEL.; RISE":GOTO 2410
860 FOR TA=0 TO B1/2 'PARABOLIC MOTION
870 Y=2*H*(TA/B1)^2
880 VB=4*H/T1*TA/B1
890 GOSUB 1260
900 NEXT TA
910 IF TA>B1/2 THEN TA=B1/2
920 Y=2*H*(TA/B1)^2
930 VB=4*H/T1*TA/B1
940 GOSUB 1260
950 GOTO 1420
960 FOR TA=0 TO B1/2 'SIMPLE HARMONIC MOTION
970 Y=H/2*(1-COS(3.14159*TA/B1))
980 VB=H/2/T1*3.14159*SIN(3.14159*TA/B1)
990 GOSUB 1260
1000 NEXT TA
1010 GOTO 1420
```

```
1020 FOR TA=0 TO B1/2 'CYCLOIDAL MOTION
1030 Y=H*(TA/B1-1/2/3.14159*SIN(6.28319*TA/B1))
1040 VB=H/T1*(1-COS(6.28319*TA/B1))
1050 GOSUB 1260
1060 NEXT TA
1070 GOTO 1420
1080 FOR TA=0 TO B1/2     '3-4 POLYNOMIAL
1090 Y=H*(8*(TA/B1)^3-8*(TA/B1)^4)
1100 VB=H/T1*(24*(TA/B1)^2-32*(TA/B1)^3)
1110 GOSUB 1260
1120 NEXT TA
1130 GOTO 1420
1140 FOR TA=0 TO B1/2 '3-4-5 POLYNOMIAL
1150 Y=H*(10*(TA/B1)^3-15*(TA/B1)^4+6*(TA/B1)^5)
1160 VB=H/T1*(30*(TA/B1)^2-60*(TA/B1)^3+30*(TA/B1)^4)
1170 GOSUB 1260
1180 NEXT TA
1190 GOTO 1420
1200 FOR TA=0 TO B1/2 '4-5-6-7 POLYNOMIAL
1210 Y=H*(35*(TA/B1)^4-84*(TA/B1)^5+70*(TA/B1)^6-20*(TA/B1)^7)
1220 VB=H/T1*(140*(TA/B1)^3-420*(TA/B1)^4+420*(TA/B1)^5-140*(TA/B1)^6)
1230 GOSUB 1260
1240 NEXT TA
1250 GOTO 1420
1260 FI=Y/H*F0*.01745
1270 S0=(MO^2+LF^2-RM^2)/(2*MO*LF)
1280 S0=-ATN(S0/SQR(-S0*S0+1))+1.5708
1290 D1=3.14159-S0-FI
1300 OC=SQR(MO^2+LF^2-2*MO*LF*COS(S0+FI))
1310 D2=(OC^2+MO^2-LF^2)/(2*OC*MO)
1320 D2=-ATN(D2/SQR(-D2*D2+1))+1.5708
1330 VC=I*OC*OM
1340 D3=-ATN((VB*COS(D1)+VC*COS(D2))/(VB*SIN(D1)+VC*SIN(D2)))
1350 BC=(VB*SIN(D1)+VC*SIN(D2))/COS(D3)
1360 IF VB=0 THEN VB=.1
1370 MY=(BC^2+VB^2-VC^2)/(2*BC*VB)
1380 MY=-ATN(MY/SQR(-MY*MY+1))+1.5708
1390 PA=1.5708-MY
1400 IF ABS(PA)>ABS(PB) THEN PB=PA:TB=TA
1410 RETURN
1420 LPRINT"MAX. PRESSURE ANGLE BY RISE=";PB*57.3;"AT CAM ANGLE THETA=";TB
1430 PRINT"MAX. PRESSURE ANGLE BY RISE=";PB*57.3;"AT CAM SHAFT ANGLE THETA=";TB
1440 TB=0:PB=0
1450 IF I=11 THEN LPRINT"PARABOLIC MOTION; RETURN"
1460 IF I=11 THEN PRINT"PARABOLIC MOTION; RETURN":GOTO 1610
1470 IF I=12 THEN LPRINT"SIMPLE HARMONIC MOTION; RETURN"
1480 IF I=12 THEN PRINT"SIMPLE HARMONIC MOTION; RETURN":GOTO 1670
1490 IF I=13 THEN LPRINT"CYCLOIDAL MOTION; RETURN"
1500 IF I=13 THEN PRINT"CYCLOIDAL MOTION; RETURN":GOTO 1730
1510 IF I=14 THEN LPRINT"3-4 POLYNOMIAL; RETURN"
```

```
1520 IF L=14 THEN PRINT"3-4 POLYNOMIAL; RETURN":GOTO 1790
1530 IF L=15 THEN LPRINT"3-4-5 POLYNOMIAL; RETURN"
1540 IF L=15 THEN PRINT"3-4-5 POLYNOMIAL; RETURN":GOTO 1850
1550 IF L=16 THEN LPRINT"4-5-6-7 POLYNOMIAL; RETURN"
1560 IF L=16 THEN PRINT"4-5-6-7 POLYNOMIAL; RETURN":GOTO 1910
1570 IF L=17 THEN LPRINT"MODIFIED TRAPEZOIDAL ACCEL.; RETURN"
1580 IF L=17 THEN PRINT"MODIFIED TRAPEZOIDAL ACCEL.; RETURN":GOTO 2250
1590 IF L=18 THEN LPRINT"MODIFIED SINUSOIDAL ACCEL.; RETURN"
1600 IF L=18 THEN PRINT"MODIFIED SINUSOIDAL ACCEL.; RETURN":GOTO 2520
1610 FOR TA=B2/2 TO B2
1620 Y=2*H*((B2-TA)/B2)^2
1630 VB=-4*H/T2*(B2-TA)/B2
1640 GOSUB 1260
1650 NEXT TA
1660 GOTO 1960
1670 FOR TA=B2/2 TO B2
1680 Y=H/2*(1+COS(3.14159*TA/B2))
1690 VB=-H/2/T2*3.14159*SIN(3.14159*TA/B2)
1700 GOSUB 1260
1710 NEXT TA
1720 GOTO 1960
1730 FOR TA=B2/2 TO B2
1740 Y=H*(1-TA/B2+1/6.28319*SIN(6.28319*TA/B2))
1750 VB=H/T2*(COS(6.28319*TA/B2)-1)
1760 GOSUB 1260
1770 NEXT TA
1780 GOTO 1960
1790 FOR TA=B2/2 TO B2
1800 Y=H*(8*TA/B2-24*(TA/B2)^2+24*(TA/B2)^3-8*(TA/B2)^4)
1810 VB=H/T2*(8-48*TA/B2+72*(TA/B2)^2-32*(TA/B2)^3)
1820 GOSUB 1260
1830 NEXT TA
1840 GOTO 1960
1850 FOR TA=B2/2 TO B2
1860 Y=H*(1-10*(TA/B2)^3+15*(TA/B2)^4-6*(TA/B2)^5)
1870 VB=-H/T2*(30*(TA/B2)^2-60*(TA/B2)^3+30*(TA/B2)^4)
1880 GOSUB 1260
1890 NEXT TA
1900 GOTO 1960
1910 FOR TA=B2/2 TO B2
1920 Y=H*(1-35*(TA/B2)^4+84*(TA/B2)^5-70*(TA/B2)^6+20*(TA/B2)^7)
1930 VB=-H/T2*(140*(TA/B2)^3-420*(TA/B2)^4+420*(TA/B2)^5-140*(TA/B2)^6)
1940 GOSUB 1260
1950 NEXT TA
1960 LPRINT"MAX. PRESSURE ANGLE BY RETURN=";PB*57.3;"AT CAM ANGLE THETA=";TB
1970 PRINT"MAX. PRESSURE ANGLE BY RETURN=";PB*57.3;"AT CAM SHAFT ANGLE THETA=";T
B
1980 LPRINT
1990 LPRINT"LENGTH OF FOLLOWER ARM LF=";LF
2000 LPRINT"DISTANCE MO=";MO
```

```
2010 LPRINT"MINIMUM RADIUS OF CAM (TO ROLLER CENTER) RM=";RM
2020 LPRINT"TOTAL ANGLE OF SWINGING ROLLER FOLLOWER FO=";FO
2030 LPRINT"TOTAL ANGLE FOR RISE=";B1
2040 LPRINT"TOTAL ANGLE FOR RETURN=";B2
2050 IF C$="CW" THEN LPRINT"CAM ROTATES CW"
2060 IF C$="CCW" THEN LPRINT"CAM ROTATES CCW"
2070 GOTO 280
2080 CLS : RUN "MAIN" : END
2090 FOR TA=0 B1/8
2100 Y=H/5.14159*(2*TA/B1/6.28319*SIN(12.5664*TA/B1))
2110 VB=H/T1/5.14159*(2-2*COS(12.5664*TA/B1))
2120 GOSUB 1260
2130 NEXT TA
2140 FOR TA=B1/8 TO 3*B1/8
2150 Y=H/5.14159*(12.5664*(TA/B1)^2+(2-3.14159)*TA/B1+.03719)
2160 VB=H/T1/5.14159*(25.1327*TA/B1-1.14159)
2170 GOSUB 1260
2180 NEXT TA
2190 FOR TA=3*B1/8 TO B1/2
2200 Y=H/5.14159*(8.2832*TA/B1-1/6.28319*SIN(12.5664*(TA/B1-.25))-1.5708)
2210 VB=H/T1/5.14159*(8.2832-2*COS(12.5664*(TA/B1-.25)))
2220 GOSUB 1260
2230 NEXT TA
2240 GOTO 1420
2250 FOR TA=B2/2 TO 5*B2/8
2260 Y=H/5.14159*(6.7124-8.2832*TA/B2-1/6.28319*SIN(12.5664*(.75-TA/B2)))
2270 VB=H/T2/5.14159*(-8.2832+2*COS(12.5664*(.75-TA/B2)))
2280 GOSUB 1260
2290 NEXT TA
2300 FOR TA=5*B2/8 TO 7*B2/8
2310 Y=H/5.14159*(11.462-23.9911*TA/B2+12.5664*(TA/B2)^2)
2320 VB=H/T2/5.14159*(-23.9911+25.1327*TA/B2)
2330 GOSUB 1260
2340 NEXT TA
2350 FOR TA=7*B2/8 TO B2
2360 Y=H/5.14159*(2-2*TA/B2-1/6.28319*SIN(12.5664*(1-TA/B2)))
2370 VB=H/T2/5.14159*(-2+2*COS(12.5664*(1-TA/B2)))
2380 GOSUB 1260
2390 NEXT TA
2400 GOTO 1960
2410 FOR TA=0 TO B1/8
2420 Y=H/7.14159*(3.14159*TA/B1-.25*SIN(12.5664*TA/B1))
2430 VB=.4399*H/T1*(1-COS(12.5664*TA/B1))
2440 GOSUB 1260
2450 NEXT TA
2460 FOR TA=B1/8 TO B1/2
2470 Y=H/7.14159*(2+3.14159*TA/B1-9/4*SIN(3.14159/3+4/3*3.14159*TA/B1))
2480 VB=.4399*H/T1*(1-3*COS(3.14159/3+4/3*3.14159*TA/B1))
2490 GOSUB 1260
2500 NEXT TA
```

```
2510 GOTO 1420
2520 FOR TA=B2/2 TO 7*B2/8
2530 Y=H/7.14159*(5.14159-3.14159*TA/B2+9/4*SIN(3.14159/3+4/3*3.14159*TA/B2))
2540 VB=.4399*H/T2*(-1+3*COS(3.14159/3+4/3*3.14159*TA/B2))
2550 GOSUB 1260
2560 NEXT TA
2570 FOR TA=7*B2/8 TO B2
2580 Y=H/7.14159*(3.14159-3.14159*TA/B2+.25*SIN(12.5664*TA/B2))
2590 VB=.4399*H/T2*(-1+COS(12.5664*TA/B2))
2600 GOSUB 1260
2610 NEXT TA
2620 GOTO 1960
2630 FOR TA=0 TO 13 STEP 2
2640 PRINT TA
2650 NEXT TA
2660 PRINT TA
2670 GOTO 280
```

```
10 'THIS PROGRAM IS FILED UNDER "SWING5"
20 CLS
30 'SYNTHESIS OF CAMS FOR PRESCRIBED MAX. PRESSURE ANGLES
40 'SWINGING ROLLER FOLLOWER
50 PRINT"SYNTHESIS OF CAMS FOR PRESCRIBED MAX. PRESSURE ANGLES"
60 PRINT"SWINGING ROLLER FOLLOWER"
70 PRINT"RISE                                        RETURN"
80 PRINT" 1   PARABOLIC MOTION                        11"
90 PRINT" 2   SIMPLE HARMONIC MOTION                  12"
100 PRINT" 3   CYCLOIDAL MOTION                       13"
110 PRINT" 4   3-4 POLYNOMIAL                         14"
120 PRINT" 5   3-4-5 POLYNOMIAL                       15"
130 PRINT" 6   4-5-6-7 POLYNOMIAL                     16"
140 PRINT" 7   MODIFIED TRAPEZOIDAL ACCEL.            17"
150 PRINT" 8   MODIFIED SINUSOIDAL ACCEL.             18"
160 PRINT
170 PRINT
180 INPUT"TIME-DISPLACEMENT CURVE FOR RISE (1-9)";M
190 INPUT"TIME-DISPLACEMENT CURVE FOR RETURN(11-19)";L
200 INPUT"LENGTH OF FOLLOWER ARM LF=";LF
210 INPUT"TOTAL ANGLE OF OSCILLATING FOLLOWER MOVEMENT FO=";FO
220 INPUT"TOTAL ANGLE FOR RISE, DEGREES";B1
230 INPUT"TOTAL ANGLE FOR RETURN, DEGREES";B2
240 INPUT"MAX. PRESSURE ANGLE FOR RISE";A1
250 INPUT"MAX. PRESSURE ANGLE FOR RETURN";A2
260 INPUT"IF CAM ROTATION IS CW WRITE 'CW':ELSE WRITE 'CCW'";C$
270 CLS : PRINT : PRINT "A- TO EXIT TO MAIN PROGRAM.";TAB(38)"B- TO QUIT CHANGES
    AND EXECUTE PROGRAM."
280 PRINT "C- RISE É";M;"©";TAB(38)"D- RETURN É";L;"©"
```

```
290 PRINT "   1      PARABOLIC MOTION                  11"
300 PRINT "   2      SIMPLE HARMONIC MOTION            12"
310 PRINT "   3      CYCLOIDAL MOTION                  13"
320 PRINT "   4      3-4 POLYNOMIAL                    14"
330 PRINT "   5      3-4-5 POLYNOMIAL                  15"
340 PRINT "   6      4-5-6-7 POLYNOMIAL                16"
350 PRINT "   7      MODIFIED TRAPEZOIDAL ACCEL.       17"
360 PRINT "   8      MODIFIED SINUSOIDAL ACCEL.        18"
370 PRINT "E- LENGTH OF FOLLOWER ARM É";LF;"©"
380 PRINT "F- TOTAL ANGLE OF OSCILLATING FOLLOWER MOVEMENT É";F0;"©"
390 PRINT "G- TOTAL ANGLE FOR RISE (DEGREES) É";Bl;"©"
400 PRINT "H- TOTAL ANGLE FOR RETURN (DEGREES) É";B2;"©"
410 PRINT "I- MAXIMUM PRESSURE ANGLE FOR RISE É";Al;"©"
420 PRINT "J- MAXIMUM PRESSURE ANGLE FOR RETURN É";A2;"©"
430 PRINT "K- CAM ROTATION (1:CW;2:CCW) É";C$;"©"
440 PRINT : INPUT "WHICH VALUE WOULD YOU LIKE TO CHANGE (A..K)";CV$
450 IF CV$="A" OR CV$="a" GOTO 1180
460 IF CV$="B" OR CV$="b" GOTO 600
480 PRINT : INPUT "WHAT IS THE NEW VALUE ";NV
490 IF CV$="C" OR CV$="c" THEN M=NV : GOTO 270
500 IF CV$="D" OR CV$="d" THEN L=NV : GOTO 270
510 IF CV$="E" OR CV$="e" THEN LF=NV : GOTO 270
520 IF CV$="F" OR CV$="f" THEN F0=NV : GOTO 270
530 IF CV$="G" OR CV$="g" THEN Bl=NV : GOTO 270
540 IF CV$="H" OR CV$="h" THEN B2=NV : GOTO 270
550 IF CV$="I" OR CV$="i" THEN Al=NV : GOTO 270
560 IF CV$="J" OR CV$="j" THEN A2=NV : GOTO 270
570 IF CV$="K" AND NV=1 OR CV$="k" AND NV=1 THEN C$="CW" : GOTO 270
580 IF CV$="K" AND NV=2 OR CV$="k" AND NV=2 THEN C$="CCW" : GOTO 270
590 GOTO 270
600 IF C$="CW" THEN I=-1:ELSE I=1
650 F0=F0*.01745
660 H=LF*F0
670 PRINT"LF=";LF;"F0=";F0/.01745;"Bl=";Bl;"B2=";B2;"Al=";Al;"A2=";A2
680 Al=Al*.01745
690 A2=A2*.01745
695 TM=0
700 FOR TA=0 TO Bl/2 STEP Bl/10
710 IF M=1 THEN GOSUB 1190 : GOTO 790
720 IF M=2 THEN GOSUB 1320:GOTO 790
730 IF M=3 THEN GOSUB 1450:GOTO 790
740 IF M=4 THEN GOSUB 1580:GOTO 790
750 IF M=5 THEN GOSUB 1710:GOTO 790
760 IF M=6 THEN GOSUB 1840:GOTO 790
770 IF M=7 THEN GOSUB 1970:GOTO 790
780 IF M=8 THEN GOSUB 2400
790 FOR TB=0 TO B2/2 STEP B2/10
800 IF L=11  THEN GOSUB 1250:GOTO 880
810 IF L=12 THEN GOSUB 1380:GOTO 880
820 IF L=13 THEN GOSUB 1510:GOTO 880
```

```
830 IF L=14 THEN GOSUB 1640:GOTO 880
840 IF L=15 THEN GOSUB 1770:GOTO 880
850 IF L=16 THEN GOSUB 1900:GOTO 880
860 IF L=17 THEN GOSUB 2170:GOTO 880
870 IF L=18 THEN GOSUB 2530
880 RM=SQR((X3-LF*COS(F0/2))^2+(Y3-LF*SIN(F0/2))^2)
885 IF RM>TM THEN TM=RM:X5=X3:Y5=Y3
890 NEXT TB
900 NEXT TA
910 MO=SQR(X5^2+Y5^2)
930 IF M=1 THEN LPRINT"PARABOLIC MOTION; RISE"
940 IF M=2 THEN LPRINT"SIMPLE HARMONIC MOTION; RISE"
950 IF M=3 THEN LPRINT"CYCLOIDAL MOTION; RISE"
960 IF M=4 THEN LPRINT"3-4 POLYNOMIAL; RISE"
970 IF M=5 THEN LPRINT"3-4-5 POLYNOMIAL; RISE"
980 IF M=6 THEN LPRINT"4-5-6-7 POLYNOMIAL; RISE"
990 IF M=7 THEN LPRINT"MODIFIED TRAPEZOIDAL ACCEL.; RISE"
1000 IF M=8 THEN LPRINT"MODIFIED SINUSOIDAL ACCEL.; RISE"
1010 IF L=11 THEN LPRINT"PARABOLIC MOTION; RETURN"
1020 IF L=12 THEN LPRINT"SIMPLE HARMONIC MOTION; RETURN"
1030 IF L=13 THEN LPRINT"CYCLOIDAL MOTION; RETURN"
1040 IF L=14 THEN LPRINT"3-4 POLYNOMIAL; RETURN"
1050 IF L=15 THEN LPRINT"3-4-5 POLYNOMIAL; RETURN"
1060 IF L=16 THEN LPRINT"4-5-6-7 POLYNOMIAL; RETURN"
1070 IF L=17 THEN LPRINT"MODIFIED TRAPEZOIDAL ACCEL.; RETURN"
1080 IF L=18 THEN LPRINT"MODIFIED SINUSOIDAL ACCEL.; RETURN"
1090 LPRINT"LENGTH OF FOLLOWER ARM Lf=";LF
1100 LPRINT"TOTAL ANGLE OF SWINGING ROLLER FOLLOWER MOEMENT  F=";F0/.01745
1110 LPRINT"TOTAL ANGLE FOR RISE BETA1=";B1
1120 LPRINT"TOTAL ANGLE FOR RETURN BETA2 =";B2
1130 LPRINT"MAX. PRESSURE ANGLE FOR RISE =";A1/.01745
1140 LPRINT"MAX. PRESSURE ANGLE FOR RETURN =";A2/.01745
1150 IF C$="CCW" THEN LPRINT"CAM ROTATES CCW":ELSE LPRINT"CAM ROTATES CW"
1160 LPRINT"DISTANCE MO=";MO
1170 LPRINT"MINIMUM RADIUS OF CAM (TO CENTER OF ROLLER) Rmin =";TM
1175 GOTO 270
1180 CLS : RUN "MAIN" : END
1190 V2=2*H                      'PARABOLIC MOTION
1200 V3=4*H*TA/B1
1210 O1=I*V3/V2*1.273*H*90/B1
1220 Y8=2*H*(TA/B1)^2
1230 F8=F0*Y8/H
1240 RETURN
1250 V2=2*H
1260 V4=4*H*TB/B2
1270 Q1=-I*V4/V2*1.273*H*90/B2
1280 Y9=2*H*(TB/B2)^2
1290 F9=F0*Y9/H
1300 GOSUB 2680
1310 RETURN
```

```
1320 V2=1.5708*H                    'SIMPLE HARMONIC MOTION
1330 V3=1.5708*H*SIN(3.14159*TA/B1)
1340 O1=I*V3/V2*H*90/B1
1350 Y8=H/2*(1-COS(3.14159*TA/B1))
1360 F8=F0*Y8/H
1370 RETURN
1380 V2=1.5708*H
1390 V4=1.5708*H*SIN(3.14159*TB/B2)
1400 Q1=-I*V4/V2*H*90/B2
1410 Y9=H/2*(1-COS(3.14159*TB/B2))
1420 F9=F0*Y9/H
1430 GOSUB 2680
1440 RETURN
1450 V2=2*H                    'CYCLOIDAL MOTIO
1460 V3=H*(1-COS(6.28319*TA/B1))
1470 O1=I*V3/V2*1.273*H*90/B1
1480 Y8=H*(TA/B1-1/6.28319*SIN(6.28319*TA/B1))
1490 F8=F0*Y8/H
1500 RETURN
1510 V2=2*H
1520 V4=H*(1-COS(6.28319*TB/B2))
1530 Q1=-I*V4/V2*1.273*H*90/B2
1540 Y9=H*(TB/B2-1/6.28319*SIN(6.28319*TB/B2))
1550 F9=F0*Y9/H
1560 GOSUB 2680
1570 RETURN
1580 V2=2*H                    '3-4 POLYNOMIAL
1590 V3=H*(24*(TA/B1)^2-32*(TA/B1)^3)
1600 O1=I*V3/V2*1.273*H*90/B1
1610 Y8=H*(8*(TA/B1)^3-8*(TA/B1)^4)
1620 F8=F0*Y8/H
1630 RETURN
1640 V2=2*H
1650 V4=H*(24*(TB/B2)^2-32*(TB/B2)^3)
1660 Q1=-I*V4/V2*1.273*H*90/B2
1670 Y9=H*(8*(TB/B2)^3-8*(TB/B2)^4)
1680 F9=F0*Y9/H
1690 GOSUB 2680
1700 RETURN
1710 V2=1.875*H                    '3-4-5- POLYNOMIAL
1720 V3=H*(30*(TA/B1)^2-60(TA/B1)^3+30*(TA/B1)^4)
1730 O1=I*V3/V2*1.193*H*90/B1
1740 Y8=H*(10*(TA/B1)^3-15*(TA/B1)^4+6*(TA/B1)^5)
1750 F8=F0*Y8/H
1760 RETURN
1770 V2=1.875*H
1780 V4=H*(30*(TB/B2)^2-60(TB/B2)^3+30*(TB/B2)^4)
1790 Q1=-I*V4/V2*1.193*H*90/B2
1800 Y9=H*(10*(TB/B2)^3-15*(TB/B2)^4+6*(TB/B2)^5)
1810 F9=F0*Y9/H
```

```
1820 GOSUB 2680
1830 RETURN
1840 V2=2.1875*H                 '4-5-6-7 POLYNOMIAL
1850 V3=H*(140*(TA/Bl)^3-420*(TA/Bl)^4+420*(TA/Bl)^5-140*(TA/Bl)^6)
1860 O1=I*V3/V2*1.392*H*90/Bl
1870 Y8=H*(35*(TA/Bl)^4-84*(TA/Bl)^5+70*(TA/Bl)^6-20*(TA/Bl)^7)
1880 F8=F0*Y8/H
1890 RETURN
1900 V2=2.1875*H
1910 V4=H*(140*(TB/B2)^3-420*(TB/B2)^4+420*(TB/B2)^5-140*(TB/B2)^6)
1920 Q1=-I*V4/V2*1.392*H*90/B2
1930 Y9=H*(35*(TB/B2)^4-84*(TB/B2)^5+70*(TB/B2)^6-20*(TB/B2)^7)
1940 F9=F0*Y9/H
1950 GOSUB 2680
1960 RETURN
1970 IF TA<=Bl/8 THEN:ELSE GOTO 2040        'MODIFIED TRAPEZOIDAL ACCELERATION
1980 V2=2*H
1990 V3=H/5.14159*(2-2*COS(12.5664*TA/Bl))
2000 O1=I*V3/V2*1.273*H*90/Bl
2010 Y8=H/5.14159*(2*TA/Bl-1/6.28319*SIN(12.5664*TA/Bl))
2020 F8=F0*Y8/H
2030 RETURN
2040 IF TA<=3*Bl/8 THEN:ELSE GOTO 2110
2050 V2=2*H
2060 V3=H/5.14159*(25.133*TA/Bl-1.14159)
2070 O1=I*V3/V2*1.273*H*90/Bl
2080 Y8=H/5.14159*(12.5664*(TA/Bl)^2-1.14159*TA/Bl+.0372)
2090 F8=F0*Y8/H
2100 RETURN
2110 V2=2*H
2120 V3=H/5.14159*(8.2832-2*COS(12.5664*(TA/Bl-.25)))
2130 O1=I*V3/V2*1.273*H*90/Bl
2140 Y8=H/5.14159*(8.2832*TA/Bl-1/6.28319*SIN(12.5664*(TA/Bl-.25))-1.5708)
2150 F8=F0*Y8/H
2160 RETURN
2170 IF TB<=B2/8 THEN:ELSE 2250
2180 V2=2*H
2190 V4=H/5.14159*(2-2*COS(12.5664*TB/B2))
2200 Q1=-I*V4/V2*1.273*H*90/B2
2210 Y9=H/5.14159*(2*TB/Bl-1/6.28319*SIN(12.5664*TB/B2))
2220 F9=F0*Y9/H
2230 GOSUB 2680
2240 RETURN
2250 IF TB<=3*B2/8 THEN:ELSE 2330
2260 V2=2*H
2270 V4=H/5.14159*(25.133*TB/B2-1.14159)
2280 Q1=-I*V4/V2*1.273*H*90/B2
2290 Y9=H/5.14159*(12.5664*(TB/B2)^2-1.14159*TB/B2+.0372)
2300 F9=F0*Y9/H
2310 GOSUB 2680
```

```
2320 RETURN
2330 V2=2*H
2340 V4=H/5.14159*(8.2832-2*COS(12.5664*(TB/B2-.25)))
2350 Q1=-I*V4/V2*1.273*H*90/B2
2360 Y9=H/5.14159*(8.2832*TB/B2-1/6.28319*SIN(12.5664*(TB/B2-.25))-1.5708)
2370 F9=F0*Y9/H
2380 GOSUB 2680
2390 RETURN
2400 IF TA<=B1/8 THEN:ELSE 2470          'MODIFIED SINUSOIDAL ACCEL.
2410 V2=1.76*H
2420 V3=.44*H*(1-COS(12.5664*TA/B1))
2430 O1=I*V3/V2*1.12*H*90/B1
2440 Y8=H/7.14159*(3.14159*TA/B1-.25*SIN(12.5664*TA/B1))
2450 F8=F0*Y8/H
2460 RETURN
2470 V2=1.76*H
2480 V3=.44*H*(1-3*COS(3.14159/3+4/3*3.14159*TA/B1))
2490 O1=I*V3/V2*1.12*H*90/B1
2500 Y8=H/7.14159*(2+3.14159*TA/B1-9/4*SIN(3.14159/3+4/3*3.14159*TA/B1))
2510 F8=F0*Y8/H
2520 RETURN
2530 IF TB<=B2/8 THEN:ELSE 2610
2540 V2=1.76*H
2550 V4=.44*H*(1-COS(12.5664*TB/B2))
2560 Q1=-I*V4/V2*1.12*H*90/B2
2570 Y9=.44*H*(3.14159*TB/B2-.25*SIN(12.5664*TB/B2))
2580 F9=F0*Y9/H
2590 GOSUB 2680
2600 RETURN
2610 V2=1.76*H
2620 V4=.44*H*(1-3*COS(3.14159/3+4/3*3.14159*TB/B2))
2630 Q1=-I*V4/V2*1.12*H*90/B2
2640 Y9=H/7.14159*(2+3.14159*TB/B2-9/4*SIN(3.14159/3+4/3*3.14159*TB/B2))
2650 F9=F0*Y9/H
2660 GOSUB 2680
2670 RETURN
2680 X1=(LF+O1)*COS(F0/2-F8)
2690 Y1=(LF+O1)*SIN(F0/2-F8)
2700 X2=(LF+Q1)*COS(F0/2-F9)
2710 Y2=(LF+Q1)*SIN(F0/2-F9)
2720 D=TAN(F0/2-F9+I*A2)-TAN(F0/2-F8-I*A1)
2730 DY=TAN(F0/2-F8-I*A1)*(TAN(F0/2-F9+I*A2)*X2-Y2)
2740 DY=DY-TAN(F0/2-F9+I*A2)*(TAN(F0/2-F8-I*A1)*X1-Y1)
2750 DX=TAN(F0/2-F9+I*A2)*X2-Y2-TAN(F0/2-F8-I*A1)*X1+Y1
2760 Y3=DY/D
2770 X3=DX/D
2780 RETURN
```

```
10 'THIS PROGRAM IS FILED UNDER "SWING6"
20 'FOR GIVEN TIME-DISPLACEMENT DIAGRAMS AND
30 'CAM PROPORTIONS FOR SWINGING ROLLER
40 'FOLLOWER THIS PROGRAM WILL CALCULATE THE
50 'RADIUS OF CURVATURE OF THE RELATIVE PATH
60 'OF THE ROLLER FOLLOWER AND THE COMPRESSIVE
70 'STRESS FOR EACH 1 DEGREE CAM ROTATION
80 'DURING RISE AND RETURN
90 'AT THE END OF RISE IT PRINTS OUT THE MAX.
100 'VALUES OF THE COMPRESSIVE STRESS DURING
110 'ACCELERATION AND DECELERATION
120 'IT THEN PRINTS OUT THE CORRESPONDING VALUES
130 'FOR RETURN.
140 CLS
150 PRINT"RISE                                    RETURN"
160 PRINT" 1   PARABOLIC MOTION                    11"
170 PRINT" 2   SIMPLE HARMONIC MOTION              12"
180 PRINT" 3   CYCLOIDAL MOTION                    13"
190 PRINT" 4   3-4 POLYNOMIAL                      14"
200 PRINT" 5   3-4-5 POLYNOMIAL                    15"
210 PRINT" 6   4-5-6-7 POLYNOMIAL                  16"
220 PRINT" 7 MODIFIED TRAPEZOIDAL ACCEL.          17"
230 PRINT" 8 MODIFIED SINUSOIDAL ACCEL.           18"
240 PRINT" 9                                      19"
250 PRINT
260 PRINT
270 INPUT"TIME-DISPLACEMENT CURVE FOR RISE (1-9)";M
280 INPUT"TIME-DISPLACEMENT CURVE FOR RETURN(11-19)";L
290 INPUT"DISTANCE MO=";MO
300 INPUT"LENGTH OF FOLLOWER ARM LF=";LF
310 INPUT"MIN. RADIUS OF CAM (TO CENTER OF ROLLER) RM=";RM
320 INPUT"TOTAL ANGLE OF SWINGING ROLLER FOLLOWER MOVEMENT FO=";FO
330 INPUT"TOTAL ANGLE FOR RISE, DEGREES";B1
340 INPUT"TOTAL ANGLE FOR RETURN, DEGREES";B2
350 INPUT"ROLLER RADIUS Rf, IN.";RF
360 INPUT"ROLLER WIDTH B, IN.";B
370 INPUT"CAM SPED, RPM";N
380 INPUT"IF CAM ROTATION IS CLOCKWISE WRITE CW, ELSE WRITE CCW";C$
390 INPUT"TOTAL EQUIVALENT MASS TO BE MOVED";W
400 INPUT"CAM MATERIAL(STEEL=1,CAST IRON=2)";MAT
410 CLS:PRINT:PRINT"A- TO EXIT TO MAIN PROGRAM.";
420 PRINT TAB(38)"B- TO QUIT CHANGES AND EXECUTE PROGRAM."
430 PRINT"C- RISE É";M;"©";TAB(38)"D- RETURN É";L;"©"
440 PRINT"     1        PARABOLIC MOTION            11"
450 PRINT"     2        SIMPLE HARMONIC MOTION      12"
460 PRINT"     3        CYCLOIDAL MOTION            13"
470 PRINT"     4        3-4 POLYNOMIAL              14"
480 PRINT"     5        3-4-5 POLYNOMIAL            15"
490 PRINT"     6        4-5-6-7 POLYNOMIAL          16"
500 PRINT"     7        MODIFIED TRAPEZOIDAL ACCEL. 17"
```

```
510 PRINT"    8      MODIFIED SINUSOIDAL ACCEL.        18"
520 PRINT"E- DISTANCE MO É";MO;"©";TAB(38)"F- LENGTH OF FOLLOWER ARM É";LF;"©"
530 PRINT"G- MIN. RADIUS OF CAM (TO CENTER OF ROLLER.) É";RM;"©"
540 PRINT"H- TOTAL ANGLE OF SWINGING ROLLER FOLLOWER MOVEMENT É";FO;"©"
550 PRINT"I- TOTAL ANGLE FOR RISE (DEGREES.) É";B1;"©"
560 PRINT"J- TOTAL ANGLE FOR RETURN (DEGREES.) É";B2;"©"
570 PRINT"K- ROLLER RADIUS (IN.) É";RF;"©";TAB(38)"L- ROLLER WIDTH (IN.) É";B;"©"

580 PRINT"M- CAM SPEED (RPM) É";N;"©" : PRINT"N- CAM ROTATION (1:CW;2:CCW) É";C$
;"©"
590 PRINT"O- TOTAL EQUIVALENT MASS TO BE MOVED É";W;"©"
600 PRINT"P- CAM MATERIAL (1:STEEL;2:CAST IRON) É";MAT;"©"
610 PRINT:INPUT"WHICH VALUE WOULD YOU LIKE TO CHANGE (A..P)";CV$
620 IF CV$="A" OR CV$="a" GOTO 6180
630 IF CV$="B" OR CV$="b" GOTO 810
640 PRINT:INPUT"WHAT IS THE NEW VALUE? ";NV
650 IF CV$="C" OR CV$="c" THEN M=NV : GOTO 410
660 IF CV$="D" OR CV$="d" THEN L=NV : GOTO 410
670 IF CV$="E" OR CV$="e" THEN MO=NV : GOTO 410
680 IF CV$="F" OR CV$="f" THEN LF=NV : GOTO 410
690 IF CV$="G" OR CV$="g" THEN RM=NV : GOTO 410
700 IF CV$="H" OR CV$="h" THEN FO=NV : GOTO 410
710 IF CV$="I" OR CV$="i" THEN B1=NV : GOTO 410
720 IF CV$="J" OR CV$="j" THEN B2=NV : GOTO 410
730 IF CV$="K" OR CV$="k" THEN RF=NV : GOTO 410
740 IF CV$="L" OR CV$="l" THEN B=NV : GOTO 410
750 IF CV$="M" OR CV$="m" THEN N=NV : GOTO 410
760 IF CV$="N" AND NV=1 OR CV$="n" AND NV=1 THEN C$="CW" : GOTO 410
770 IF CV$="N" AND NV=2 OR CV$="n" AND NV=2 THEN C$="CCW" : GOTO 410
780 IF CV$="O" OR CV$="o" THEN W=NV : GOTO 410
790 IF CV$="P" THEN MAT=NV
800 GOTO 410
810 CLS : H=LF*FO*.01745
820 OM=N*3.14159/30
830 IF C$="CW" THEN OM=-OM
840 T1=B1/N/6
850 T2=B2/N/6
860 LPRINT"MO=";MO;"LF=";LF;"RM=";RM;"FO=";FO;"B1=";B1;"B2=";B2;"RF=";RF;"B=";B;
"CAM SPEED,RPM=";N
870 IF M=1 THEN LPRINT"PARABOLIC MOTION; RISE":GOTO 1040
880 IF M=2 THEN LPRINT"SIMPLE HARMONIC MOTION; RISE":GOTO 1480
890 IF M=3 THEN LPRINT"CYCLOIDAL MOTION; RISE:GOTO 1000
900 IF M=4 THEN LPRINT"3-4 POLYNOMIAL;RISE":GOTO 2360
910 IF M=5 THEN LPRINT"3-4-5 POLYNOMIAL; RISE":GOTO 2790
920 IF M=6 THEN LPRINT"4-5-6-7 POLYNOMIAL; RISE":GOTO 3230
930 IF M=7 THEN LPRINT"MODIFIED TRAPEZOIDAL ACCEL.; RISE":GOTO 3670
940 IF M=8 THEN LPRINT"MODIFIED SINUSOIDAL ACCEL.; RISE":GOTO 5230
950 IF M=9
960 IF L=11 THEN LPRINT"PARABOLIC MOTION; RETURN":GOTO 1250
970 IF L=12 THEN LPRINT"SIMPLE HARMONIC MOTION";RETURN:GOTO 1690
```

```
980 IF L=13 THEN LPRINT"CYCLOIDAL MOTION; RETURN":GOTO 2130
990 IF L=14 THEN LPRINT"3-4 POLYNOMIAL; RETURN":GOTO 2560
1000 IF L=15 THEN LPRINT"3-4-5 POLYNOMIAL; RETURN":GOTO 3000
1010 IF L=16 THEN LPRINT"4-5-6-7 POLYNOMIAL; RETURN":GOTO 3440
1020 IF L=17 THEN LPRINT"MODIFIED TRAPEZOIDAL ACCEL.; RETURN":GOTO 4240
1030 IF L=18 THEN LPRINT"MODIFIED SINUSOIDAL ACCEL.; RETURN":GOTO 5620
1040 LPRINT"THETA                    RHO         COMPRESSIVE STRESS"
1050 FOR TA=0 TO B1/2
1060 Y=2*H*(TA/B1)^2
1070 VB=4*H/T1*(TA/B1) 'PARABOLIC MOTION; 1ST HALF RISE"
1080 AB=4*H/(T1)^2
1090 GOSUB 4830
1100 GOSUB 5040
1110 LPRINT TA,RO,SC
1120 NEXT TA
1130 LPRINT
1140 FOR TA=B1/2 TO B1
1150 Y=H*(1-2*((B1-TA)/B1)^2)
1160 VB=4*H/T1*(B1-TA)/B1  'PARABOLIC MOTION; 2ND HALF RISE
1170 AB=-4*H/T1^2
1180 GOSUB 4830
1190 GOSUB 5140
1200 LPRINT TA,RO,SC
1210 NEXT TA
1220 LPRINT
1230 LPRINT"MAX. COMPRESSIVE STRESS DURING 1ST HALF RISE =";S1;" PSI";" AT THETA
 =";TB
1240 LPRINT"MAX. COMPRESSIVE STRESS DURING 2ND HALF RISE =";S2;" PSI";" AT THETA
 =";TC
1250 S1=0:S2=0
1260 LPRINT
1270 FOR TA=0 TO B2/2
1280 Y=H*(1-2*(TA/B2)^2)
1290 VB=-4*H/T2*(TA/B2) 'PARABOLIC MOTION; 1ST HALF RETURN
1300 AB=-4*H/T2^2
1310 GOSUB 4830
1320 GOSUB 5140
1330 LPRINT TA,RO,SC
1340 NEXT TA
1350 LPRINT
1360 FOR TA=B2/2 TO B2
1370 Y=2*H*((B2-TA)/B2)^2
1380 VB=-4*H/T2*(B2-TA)/B2 'PARABOLIC MOTION; 2ND HALF RETURN
1390 AB=4*H/T2^2
1400 GOSUB 4830
1410 GOSUB 5040
1420 LPRINT TA,RO,SC
1430 NEXT TA
1440 LPRINT
1450 LPRINT"MAX. COMPRESSIVE STRESS DURING 1ST HALF RETURN =";S2;" PSI";" AT THE
```

```
TA =";TB
1460 LPRINT"MAX. COMPRESSIVE STRESS DURNIG 2ND HALF RETURN =";S1;" PSI";" AT THE
TA =";TC
1470 GOTO 6030
1480 LPRINT"THETA                RHO        COMPRESSIVE STRESS"
1490 FOR TA=0 TO B1/2
1500 Y=H/2*(1-COS(3.14159*TA/B1))
1510 VB=H/2*3.14159/T1*SIN(3.14159*TA/B1) 'SIMPLE HARMONIC MOTION; RISE
1520 AB=H/2*(3.14159/T1)^2*COS(3.14159*TA/B1)
1530 GOSUB 4830
1540 GOSUB 5040
1550 LPRINT TA,RO,SC
1560 NEXT TA
1570 LPRINT
1580 FOR TA=B1/2 TO B1
1590 Y=H/2*(1-COS(3.14159*TA/B1))
1600 VB=H/2*3.14159/T1*SIN(3.14159*TA/B1)
1610 AB=H/2*(3.14159/T1)^2*COS(3.14159*TA/B1)
1620 GOSUB 4830
1630 GOSUB 5140
1640 LPRINT TA,RO,SC
1650 NEXT TA
1660 LPRINT
1670 LPRINT"MAX. COMPRESSIVE STRESS DURING 1ST HALF RISE =";S1;" PSI";" AT CAM A
LGLE THETA =";TB
LGLE THETA =";TB
1680 LPRINT"MAX. COMPRESSIVE STRESS DURING 2ND HALF RISE =";S2;" PSI";" AT CAM A
NGLE THETA =";TC
1690 S1=0:S2=0
1700 LPRINT
1710 FOR TA=0 TO B2/2
1720 Y=H/2*(1+COS(3.14159*TA/B2))
1730 VB=-H/2*3.14159/T2*SIN(3.14159*TA/B2) 'SIMPLE HARMONIC MOTION; RETURN
1740 AB=-H/2*(3.14159/T2)^2*COS(3.14159*TA/B2)
1750 GOSUB 4830
1760 GOSUB 5140
1770 LPRINT TA,RO,SC
1780 NEXT TA
1790 LPRINT
1800 FOR TA=B2/2 TO B2
1810 Y=H/2*(1+COS(3.14159*TA/B2))
1820 VB=-H/2*3.14159/T2*SIN(3.14159*TA/B2)
1830 AB=-H/2*(3.14159/T2)^2*COS(3.14159*TA/B2)
1840 GOSUB 4830
1850 GOSUB 5040
1860 LPRINT TA,RO,SC
1870 NEXT TA
1880 LPRINT
1890 LPRINT"MAX. COMPRESSIVE STRESS DURING 1ST HALF RETURN =";S2;" PSI AT CAM AN
GLE THETA =";TB
```

```
1900 LPRINT"MAX. COMPRESSIVE STRESS DURING 2ND HALF RETURN =";S1;" PSI AT CAM AN
GLE THETA =";TC
1910 GOTO 6030
1920 LPRINT"THETA                  RHO           COMPRESSIVE STRESS"
1930 FOR TA=0 TO B1/2
1940 Y=H*(TA/B1-1/6.28319*SIN(6.28319*TA/B1))
1950 VB=H/T1*(1-COS(6.28319*TA/B1)) 'CYCLOIDAL MOTION; RISE
1960 AB=6.28319*H/(T1)^2*SIN(6.28319*TA/B1)
1970 GOSUB 4830
1980 GOSUB 5040
1990 LPRINT TA,RO,SC
2000 NEXT TA
2010 LPRINT
2020 FOR TA=B1/2 TO B1
2030 Y=H*(TA/B1-1/6.28319*SIN(6.28319*TA/B1))
2040 VB=H/T1*(1-COS(6.28319*TA/B1))
2050 AB=6.28319*H/(T1)^2*SIN(6.28319*TA/B1)
2060 GOSUB 4830
2070 GOSUB 5140
2080 LPRINT TA,RO,SC
2090 NEXT TA
2100 LPRINT
2110 LPRINT"MAX. COMPRESSIVE STRESS DURING 1ST HALF RISE =";S1;" PSI AT CAM ANGL
E THETA =";TB
2120 LPRINT"MAX. COMPRESSIVE STRESS DURING 2ND HALF RISE =";S2;" PSI AT CAM ANGL
E THETA =";TC
2130 S1=0:S2=0
2140 LPRINT
2150 FOR TA=0 TO B2/2
2160 Y=H*(1-TA/B2+1/6.28139*SIN(6.28319*TA/B2))
2170 VB=H/T2*(COS(6.28319*TA/B2)-1) 'CYCLOIDAL MOTION; RETURN
2180 AB=-6.28319*H/(T2)^2*SIN(6.28319*TA/B2)
2190 GOSUB 4830
2200 GOSUB 5140
2210 LPRINT TA,RO,SC
2220 NEXT TA
2230 LPRINT
2240 FOR TA=B2/2 TO B2
2250 Y=H*(1-TA/B2+1/6.28319*SIN(6.28319*TA/B2))
2260 VB=H/T2*(COS(6.28319*TA/B2)-1)
2270 AB=-6.28319*H/(T2)^2*SIN(6.28319*TA/B2)
2280 GOSUB 4830
2290 GOSUB 5040
2300 LPRINT TA,RO,SC
2310 NEXT TA
2320 LPRINT
2330 LPRINT"MAX. COMPRESSIVE STRESS DURING 1ST HALF RETURN =";S2;" PSI AT CAM AN
GLE THETA =";TB
2340 LPRINT"MAX. COMPRESSIVE STRESS DURING 2ND HALF RETURN =";S1;" PSI AT CAM AN
GLE THETA =";TC
```

```
2350 GOTO 6030
2360 LPRINT"THETA                 RHO        COMPRESSIVE STRESS"
2370 FOR TA=0 TO B1/2
2380 Y=H*(8*(TA/B1)^3-8*(TA/B1)^4)
2390 VB=H/T1*(24*(TA/B1)^2-32*(TA/B1)^3) '3-4 POLYNOMIAL; 1ST HALF RISE
2400 AB=H/(T1)^2*(48*TA/B1-96*(TA/B1)^2)
2410 GOSUB 4830
2420 GOSUB 5040
2430 LPRINT TA,RO,SC
2440 NEXT TA
2450 LPRINT
2460 FOR TA=B1/2 TO B1
2470 Y=H*(1-8*TA/B1+24*(TA/B1)^2-24*(TA/B1)^3+8*(TA/B1)^4)
2480 VB=H/T1*(-8+48*TA/B1-72*(TA/B1)^2+32*(TA/B1)^3) Ñ3-4 POLYNOMIAL;2ND HALF RI
SE
2490 AB=H/(T1)^2*(48-144*TA/B1+96*(TA/B1)^2)
2500 GOSUB 4830
2510 GOSUB 5140
2520 LPRINT TA,RO,SC
2530 NEXT TA
2540 LPRINT"MAX. COMPRESSIVE STRESS DURING 1ST HALF RISE =";S1;" PSI AT CAM ANG
E THETA =";TB
2550 LPRINT"MAX. COMPRESSIVE STRESS DURING 2ND HALF RISE =";S1;" PSI AT CAM ANG
E THETA =";TC
2560 S1=0:S2=0
2570 LPRINT
2580 FOR TA=0 TO B2/2
2590 Y=H*(1-8*(TA/B2)^3+8*(TA/B2)^4)
2600 VB=H/T2*(-24*(TA/B2)^2+32*(TA/B2)^3) '3-4 POLYNOMIAL; 1ST HALF RETURN
2610 AB=H/(T2)^2*(-48*TA/B2+96*(TA/B2)^2)
2620 GOSUB 4830
2630 GOSUB 5140
2640 LPRINT TA,RO,SC
2650 NEXT TA
2660 LPRINT
2670 FOR TA=B2/2 TO B2
2680 Y=H*(8*TA/B2-24*(TA/B2)^2+24*(TA/B2)^3-8*(TA/B2)^4)
2690 VB=H/T2*(8-48*TA/B2+72*(TA/B2)^2-32*(TA/B2)^3) '3-4 POLYNOMIAL; 2ND HALF R
TURN
2700 AB=H/(T2)^2*(-48+144*TA/B2-96*(TA/B2)^2)
2710 GOSUB 4830
2720 GOSUB 5040
2730 LPRINT TA,RO,SC
2740 NEXT TA
2750 LPRINT
2760 LPRINT"MAX. COMPRESSIVE STRESS DURING 1ST HALF RETURN =";S2;" PSI AT CAM A
GLE THETA =";TB
2770 LPRINT"MAX. COMPRESSIBLE STRESS DURING 2ND HALF RETURN =";S1;" PSI AT CAM
NGLE THETA =";TC
2780 GOTO 6030
```

```
2790 LPRINT"THETA                    RHO          COMPRESSIVE STRESS"
2800 FOR TA=0 TO B1/2
2810 Y=H*(10*(TA/B1)^3-15*(TA/B1)^4+6*(TA/B1)^5)
2820 VB=H/T1*(30*(TA/B1)^2-60*(TA/B1)^3+30*(TA/B1)^4) '3-4-5 POLYNOMIAL; RISE
2830 AB=H/(T1)^2*(60*TA/B1-180*(TA/B1)^2+120*(TA/B1)^3)
2840 GOSUB 4830
2850 GOSUB 5040
2860 LPRINT TA,RO,SC
2870 NEXT TA
2880 LPRINT
2890 FOR TA=B1/2 TO B1
2900 Y=H*(10*(TA/B1)^3-15*(TA/B1)^4+6*(TA/B1)^5)
2910 VB=H/T1*(30*(TA/B1)^2-60*(TA/B1)^3+30*(TA/B1)^4)
2920 AB=H/(T1)^2*(60*TA/B1-180*(TA/B1)^2+120*(TA/B1)^3)
2930 GOSUB 4830
2940 GOSUB 5140
2950 LPRINT TA,RO,SC
2960 NEXT TA
2970 LPRINT
2980 LPRINT"MAX. COMPRESSIVE STRESS DURING 1ST HALF RISE =";S1;" PSI AT CAM ANGL
E THETA =";TB
2990 LPRINT"MAX. COMPRESSIVE STRESS DURING 2ND HALF RISE =";S2;" PSI AT CAM ANGL
E THETA =";TC
3000 S1=0:S2=0
3010 LPRINT
3020 FOR TA=0 TO B2/2
3030 Y=H*(1-10*(TA/B2)^3+15*(TA/B2)^4-6*(TA/B2)^5)
3040 VB=H/T2*(-30*(TA/B2)^2+60*(TA/B2)^3-30*(TA/B2)^4) '3-4-5 POLYNOMIAL; RETURN

3050 AB=H/(T2)^2*(-60*TA/B2+180*(TA/B2)^2-120*(TA/B2)^3)
3060 GOSUB 4830
3070 GOSUB 5140
3080 LPRINT TA,RO,SC
3090 NEXT TA
3100 LPRINT
3110 FOR TA=B2/2 TO B2
3120 Y=H*(1-10*(TA/B2)^3+15*(TA/B2)^4-6*(TA/B2)^5)
3130 VB=H/T2*(-30*(TA/B2)^2+60*(TA/B2)^3-30*(TA/B2)^4)
3140 AB=H/(T2)^2*(-60*TA/B2+180*(TA/B2)^2-120*(TA/B2)^3)
3150 GOSUB 4830
3160 GOSUB 5040
3170 LPRINT TA,RO,SC
3180 NEXT TA
3190 LPRINT
3200 LPRINT"MAX. COMPRESSIVE STRESS DURING 1ST HALF RETURN =";S2;" PSI AT CAM AN
GLE THETA =";TB
3210 LPRINT"MAX. COMPRESSIVE STRESS DURING 2ND HALF RETURN =";S1;" PSI AT CAM AN
GLE THETA =";TC
3220 GOTO 6030
3230 LPRINT"THETA                    RHO          COMPRESSIVE STRESS"
```

```
3240 FOR TA=0 TO Bl/2
3250 Y=H*(35*(TA/Bl)^4-84*(TA/Bl)^5+70*(TA/Bl)^6-20*(TA/Bl)^7)
3260 VB=H/Tl*(140*(TA/Bl)^3-420*(TA/Bl)^4+420*(TA/Bl)^5-140*(TA/Bl)^6) '4-5-6-7
POLYNOMIAL; RISE
3270 AB=H/(Tl)^2*(420*(TA/Bl)^2-1680*(TA/Bl)^3+2100*(TA/Bl)^4-840*(TA/Bl)^5)
3280 GOSUB 4830
3290 GOSUB 5040
3300 LPRINT TA,RO,SC
3310 NEXT TA
3320 LPRINT
3330 FOR TA=Bl/2 TO Bl
3340 Y=H*(35*(TA/Bl)^4-84*(TA/Bl)^5+70*(TA/Bl)^6-20*(TA/Bl)^7)
3350 VB=H/Tl*(140*(TA/Bl)^3-420*(TA/Bl)^4+420*(TA/Bl)^5-140*(TA/Bl)^6)
3360 AB=H/(Tl)^2*(420*(TA/Bl)^2-1680*(TA/Bl)^3+2100*(TA/Bl)^4-840*(TA/Bl)^5)
3370 GOSUB 4830
3380 GOSUB 5140
3390 LPRINT TA,RO,SC
3400 NEXT TA
3410 LPRINT
3420 LPRINT"MAX. COMPRESSIVE STRESS DURING 1ST HALF RISE =";S1;" PSI AT CAM ANGL
E THETA =";TB
3430 LPRINT"MAX. COMPRESSIVE STRESS DURING 2ND HALF RISE =";S2;" PSI AT CAM ANGL
E THETA =";TC
3440 S1=0:S2=0
3450 LPRINT
3460 FOR TA=0 TO B2/2
3470 Y=H*(1-35*(TA/B2)^4+84*(TA/B2)^5-70*(TA/B2)^6+20*(TA/B2)^7)
3480 VB=H/T2*(-140*(TA/B2)^3+420*(TA/B2)^4-420*(TA/B2)^5+140*(TA/B2)^6) '4-5-6-7
POLYNOMIAL; RETURN
3490 AB=H/(T2)^2*(-420*(TA/B2)^2+1680*(TA/B2)^3-2100*(TA/B2)^4+840*(TA/B2)^5))
3500 GOSUB 4830
3510 GOSUB 5140
3520 LPRINT TA,RO,SC
3530 NEXT TA
3540 LPRINT
3550 FOR TA=B2/2 TO B2
3560 Y=H*(1-35*(TA/B2)^4+84*(TA/B2)^5-70*(TA/B2)^6+20*(TA/B2)^7)
3570 VB=H/T2*(-140*(TA/B2)^3+420*(TA/B2)^4-420*(TA/B2)^5+140*(TA/B2)^6)
EDIT 940
3580 AB=H/(T2)^2*(-420*(TA/B2)^2+1680*(TA/B2)^3-2100*(TA/B2)^4+840*(TA/B2)^5)
3590 GOSUB 4830
3600 GOSUB 5040
3610 LPRINT TA,RO,SC
3620 NEXT TA
3630 LPRINT
3640 LPRINT"MAX. COMPRESSIVE STRESS DURING 1ST HALF RETURN =";S2;" PSI AT CAM AN
GLE THETA =";TB
3650 LPRINT"MAX. COMPRESSIVE STRESS DURING 2ND HALF RETURN =";S1;" PSI AT CAM AN
GLE THETA =";TC
3660 GOTO 6030
```

```
3670 LPRINT"THETA                RHO        COMPRESSIVE STRESS"
3680 FOR TA=0 TO Bl/8
3690 Y=H/5.14159*/92*TA/Bl-1/6.28319*SIN(12.5664*TA/Bl))
3700 VB=H/Tl/5.14159*(2-2*COS(12.5664*TA/Bl))
3710 AB=4.88812*H/(Tl)^2*SIN(12.5664*TA/Bl)
3720 GOSUB 4830
3730 GOSUB 5040
3740 LPRINT TA,RO,SC
3750 NEXT TA
3760 LPRINT
3770 FOR TA=Bl/8 TO 3*Bl/8
3780 Y=H/5.14159*(12.5664*(TA/Bl)^2-1.1416*TA/Bl+.037195)
3790 VB=H/Tl/5.14159*(25.1327*TA/Bl-1.14159)
3800 AB=4.88812*H/(Tl)^2
3810 GOSUB 4830
3820 GOSUB 5040
3830 LPRINT TA,RO,SC
3840 NEXT TA
3850 LPRINT
3860 FOR TA=3*Bl/8 TO Bl/2
3870 Y=H/5.14159*(8.2832*TA/Bl-1/6.28319*SIN(12.5664*(TA/Bl-.25))-1.5708)
3880 VB=H/Tl/5.14159*(8.2832-2*COS(12.5664*(TA/Bl-.25)))
3890 AB=4.88812*H/(Tl)^2*SIN(12.5664*(TA/Bl-.25))
3900 GOSUB 4830
3910 GOSUB 5040
3920 LPRINT TA,RO,SC
3930 NEXT TA
3940 LPRINT
3950 FOR TA=Bl/2 TO 5*Bl/8
3960 Y=H/5.14159*(-1.5708+8.2832*TA/Bl+1/6.28319*SIN(12.5664*(.75-TA/Bl)))
3970 VB=H/5.14159/Tl*(8.2832-2*COS(12.5664**(.75-TA/Bl)))
3980 AB=-4.8882*H/(Tl)^2*SIN(12.5664*(.75-TA/Bl))
3990 GOSUB 4830
4000 GOSUB 5140
4010 LPRINT TA,RO,SC
4020 NEXT TA
4030 LPRINT
4040 FOR TA=5*Bl/8 TO 7*Bl/8
4050 Y=H/5.14159*(-6.32038+23.991*TA/Bl-12.5664*(TA/Bl)^2)
4060 VB=H/5.14159/Tl*(23.991-25.133*TA/Bl)
4070 AB=4.88812*H/(Tl)^2
4080 GOSUB 4830
4090 GOSUB 5140
4100 LPRINT TA,RO,SC
4110 NEXT TA
4120 LPRINT
4130 FOR TA=7*Bl/8 TO Bl
4140 Y=H/5.14159*(3.14159+2*TA/Bl+1/6.28319*SIN(12.5664*(1-TA/Bl)))
4150 VB=H/5.14159/Tl*(2-2*COS(12.5664*(1-TA/Bl)))
4160 AB=-4.88812*H/(Tl)^2*SIN(12.5664*(1-TA/Bl))
```

```
4170 GOSUB 4830
4180 GOSUB 5140
4190 LPRINT TA,RO,SC
4200 NEXT TA
4210 LPRINT
4220 LPRINT"MAX. COMPRESSIVE STRESS DURING 1ST HALF RISE =";S1;" PSI AT CAM ANGL
E THETA =";TB
4230 LPRINT"MAX. COMPRESSIVE STRESS DURING 2ND HALF RISE =";S2;" PSI AT CAM ANGL
E THETA =";TC:GOTO 960
4240 S1=0:S2=0
4250 LPRINT
4260 FOR TA=0 TO B2/8
4270 Y=H/5.14159*(5.14159-2*TA/B2+1/6.28319*SIN(12.5664*TA/B2))
4280 VB=H/5.14159/T2*(-2+2*COS(12.5664*TA/B2))
4290 AB=-4.88812*H/(T2)^2*SIN(12.5664*TA/B2)
4300 GOSUB 4830
4310 GOSUB 5140
4320 LPRINT TA,RO,SC
4330 NEXT TA
4340 LPRINT
4350 FOR TA=B2/8 TO 3*B2/8
4360 Y=H/5.14159*(5.14159-12.5564*(TA/B2)^2+1.14159*TA/B2-.037195)
4370 VB=H/T2/5.14159*(-25.1327*TA/B2+1.14159)
4380 AB=-4.88812*H/(T2)^2
4390 GOSUB 4830
4400 GOSUB 5140
4410 LPRINT TA,RO,SC
4420 NEXT TA
4430 LPRINT
4440 FOR TA=3*B2/8 TO B2/2
4450 Y=H/5.14159*(6.7124-8.2832*TA/B2+1/6.28319*SIN(12.5664*(TA/B2-.25)))
4460 VB=H/T2/5.14159*(-8.28319+2*COS(12.5664*(TA/B2-.25)))
4470 AB=-4.88812*H/(T2)^2*SIN(12.5664*(TA/B2-.25))
4480 GOSUB 4830
4490 GOSUB 5140
4500 LPRINT TA,RO,SC
4510 NEXT TA
4520 LPRINT
4530 FOR TA=B2/2 TO 5*B2/8
4540 Y=H/5.14159*(6.7124-8.2832*TA/B2+1/6.28319*SIN(12.5664*(.75-TA/B2)))
4550 VB=H/T2/5.14159*(-8.28319+2*COS(12.5664*(.75-TA/B2)))
4560 AB=4.88812*H/(T2)^2*SIN(12.5664*(.75-TA/B2))
4570 GOSUB 4830
4580 GOSUB 5040
4590 LPRINT TA,RO,SC
4600 NEXT TA
4610 LPRINT
4620 FOR TA=5*B2/8 TO 7*B2/8
4630 Y=H/5.14159*(11.462-23.9912*TA/B2+12.5664*(TA/B2)^2)
4640 VB=H/T2/5.14159*(-23.9912+25.1327*TA/B2)
```

```
4650 AB=4.88812*H/(T2)^2
4660 GOSUB 4830
4670 GOSUB 5040
4680 LPRINT TA,RO,SC
4690 NEXT TA
4700 LPRINT
4710 FOR TA=7*B2/8 TO B2
4720 Y=H/5.14159*(2-2*TA/B2-1/6.28319*SIN(12.5664*(1-TA/B2)))
4730 VB=H/T2/5.14159*(-2+2*COS(12.5664*(1-TA/B2)))
4740 AB=4.88812*H/(T2)^2*SIN(12.5664*(1-TA/B2))
4750 GOSUB 4830
4760 GOSUB 5040
4770 LPRINT TA,RO,SC
4780 NEXT TA
4790 LPRINT
4800 LPRINT"MAX. COMPRESSIVE STRESS DURING 1ST HALF RETURN =";S2;" PSI AT CAM AN
GLE THETA =";TB
4810 LPRINT"MAX. COMPRESSIVE STRESS DURING 2ND HALF RETURN =";S1;" PSI AT CAM AN
GLE THETA =";TC
4820 GOTO 6030
4830 FI=Y/H*F0*.01745
4840 S0=(MO^2+LF^2-RM^2)/(2*MO*LF)
4850 S0=-ATN(S0/SQR(-S0*S0+1))+1.5708
4860 D1=3.14159-S0-FI
4870 OC=SQR(MO^2+LF^2-2*MO*LF*COS(S0+FI))
4880 D2=(OC^2+MO^2-LF^2)/(2*OC*MO)
4890 D2=-ATN(D2/SQR(-D2*D2+1))+1.5708
4900 VC=OC*OM
4910 D3=-ATN((VB*COS(D1)+VC*COS(D2))/(VB*SIN(D1)+VC*SIN(D2)))
4920 BC=(VB*SIN(D1)+VC*SIN(D2))/COS(D3)
4930 IF VB=0 THEN VB=.1
4940 MY=(BC^2+VB^2-VC^2)/(2*BC*VB)
4950 MY=-ATN(MY/SQR(-MY*MY+1))+1.5708
4960 PA=1.5708-MY
4970 V3=(VB*SIN(D1)+VC*SIN(D2))/COS(D3)
4980 BN=VB^2/LF
4990 CN=OC*OM^2
5000 BC=BN*SIN(D3-D1)-AB*COS(D3-D1)+CN*SIN(D2-D3)-2*V3*OM
5010 RO=V3^2/BC
5020 F1=ABS(W/386*AB/COS(PA))
5030 RETURN
5040 IF RO>0 THEN RK=RO+RF:ELSE GOTO 5070
5050 IF MAT=1 THEN SC=2290*SQR(ABS(F1/B*(1/RK-1/RF))):GOTO 5120
5060 IF MAT=2 THEN SC=1850*SQR(ABS(F1/B*(1/RK-1/RF))):GOTO 5120
5070 IF RO<0 THEN RK=RO+RF
5080 IF RK>=0 THEN GOTO 6190
5090 RK=ABS(RK)
5100 IF MAT=1 THEN SC=2290*SQR(F1/B*(1/RK+1/RF))
5110 IF MAT=2 THEN SC=1850*SQR(F1/B*(1/RK+1/RF))
5120 IF SC>S1 THEN S1=SC:TB=TA
```

```
5130 RETURN
5140 IF RO>0 THEN RK=RO-RF:ELSE GOTO 5180
5150 IF RK<=0 THEN GOTO 6190
5160 IF MAT=1 THEN SC=2290*SQR(F1/B*(1/RK+1/RF)):GOTO 5210
5170 IF MAT=2 THEN SC=1850*SQR(ABS(F1/B*(1/RK+1/RF)):GOTO 5210
5180 IF RO<0 THEN RK=ABS(RO-RF)
5190 IF MAT=1 THEN SC=2290*SQR(ABS(F1/B*(1/RK-1/RF)))
5200 IF MAT=2 THEN SC=1850*SQR(ABS(FI/B*(1/RK-1/RF)))
5210 IF SC>S2 THEN S2=SC:TC=TA
5220 RETURN
5230 LPRINT"THETA                    RHO          COMPRESSIVE STRESS"
5240 FOR TA=0 TO B1/8
5250 Y=H/7.14159*(3.14159*TA/B1-.25*SIN(12.5664*TA/B1))
5260 VB=.4399*H/T1*(1-COS(12.5664*TA/B1))
5270 AB=5.528*H/(T1)^2*SIN(12.5664*TA/B1)
5280 GOSUB 4830
5290 GOSUB 5040
5300 LPRINT TA,RO,SC
5310 NEXT TA
5320 LPRINT
5330 FOR TA=B1/8 TO B1/2
5340 Y=H/7.14159*(2+3.14159*TA/B1-9/4*SIN(3.14159/3+4/3*3.14159*TA/B1))
5350 VB=.4399*H/T1*(1-3*COS(3.14159/3+4/3*3.14159*TA/B1))
5360 AB=5.528*H/(T1)^2*SIN(3.14159/3+4/3*3.14159*TA/B1)
5370 GOSUB 4830
5380 GOSUB 5040
5390 LPRINT TA,RO,SC
5410 LPRINT
5420 FOR TA=B1/2 TO 7*B1/8
5430 Y=H/7.14159*(2+3.14159*TA/B1-9/4*SIN(3.14159/3+4/3*3.14159*TA/B1))
5440 VB=.4399*H/T1*(1-3*COS(3.14159/3+4/3*3.14159*TA/B1))
5450 AB=5.528*H/(T1)^2*SIN(3.14159/3+4/3*3.14159*TA/B1)
5460 GOSUB 4830
5470 GOSUB 5140
5480 LPRINT TA,RO,SC
5490 NEXT TA
5500 LPRINT
5510 FOR TA=7*B1/8 TO B1
5520 Y=H/7.14159*(4+3.14159*TA/B1-.25*SIN(12.5664*TA/B1))
5530 VB=.4399*H/T1*(1-COS(12.5664*TA/B1))
5540 AB=5.528*H/(T1)^2*SIN(12.5664*TA/B1)
5550 GOSUB 4830
5560 GOSUB 5140
5570 LPRINT TA,RO,SC
5580 NEXT TA
5590 LPRINT
5600 LPRINT"MAX. COMPRESSIVE STRESS DURING 1ST HALF RISE =";S1;" PSI AT CAM ANGL
E THETA =";TB
5610 LPRINT"MAX. COMPRESSIVE STRESS DURING 2ND HALF RISE =";S2;" PSI AT CAM ANGL
E THETA =";TC:GOTO 960
```

```
5620 S1=0:S2=0
5630 LPRINT
5640 FOR TA=0 TO B2/8
5650 Y=H/7.14159*(7.14159-3.14159*TA/B2+.25*SIN(12.5664*TA/B2))
5660 VB=.4399*H/T2*(-1+COS(12.5664*TA/B2))
5670 AB=-5.528*H/(T2)^2*SIN(12.5664*TA/B2)
5680 GOSUB 4830
5690 GOSUB 5140
5700 LPRINT TA,RO,SC
5710 NEXT TA
5720 LPRINT
5730 FOR TA=B2/8 TO B2/2
5740 Y=H/7.14159*(5.14159-3.14159*TA/B2+9/4*SIN(3.14159/3+4/3.14159*TA/B2))
5750 VB=.4399*H/T2*(-1+3*COS(3.14159/3+4/3*3.14159*TA/B2))
5760 AB=-5.528*H/(T2)^2*SIN(3.14159/3+4/3*3.14159*TA/B2)
5770 GOSUB 4830
5780 GOSUB 5140
5790 LPRINT TA,RO,SC
5800 NEXT TA
5810 LPRINT
5820 FOR TA=B2/2 TO 7*B2/8
5830 Y=H/7.14159*(5.14159-3.14159*TA/B2+9/4*SIN(3.14159/3+4/3*3.14159*TA/B2))
5840 VB=.4399*H/T2*(-1+3*COS(3.14159/3+4/3*3.14159*TA/B2))
5850 AB=-5.528*H/(T2)^2*SIN(3.14159/3+4/3*3.14159*TA/B2)
5860 GOSUB 4830
5870 GOSUB 5040
5880 LPRINT TA,RO,SC
5890 NEXT TA
5900 LPRINT
5910 FOR TA=7*B2/8 TO B2
5920 Y=H/7.14159*(3.14159-3.14159*TA/B2+.25*SIN(12.5664*TA/B2))
5930 VB=.4399*H/T2*(-1+COS(12.5664*TA/B2))
5940 AB=-5.528*H/(T2)^2*SIN(12.5664*TA/B2)
5950 GOSUB 4830
5960 GOSUB 5040
5970 LPRINT TA,RO,SC
5980 NEXT TA
5990 LPRINT
6000 LPRINT"MAX. COMPRESSIVE STRESS DURING 1ST HALF RETURN =";S2;" PSI AT CAM AN
GLE THETA =";TB
6010 LPRINT"MAX. COMPRESSIVE STRESS DURING 2ND HALF RETURN =";S1;" PSI AT CAM AN
GLE THETA =";TC
6020 GOTO 6030
6030 LPRINT"DISTANCE MO=";MO
6040 LPRINT"LENGTH OF FOLLOWER ARM LF=";LF
6050 LPRINT"MINIMUM RADIUS OF CAM (TO CENTER OF ROLLER) RM=";RM
6060 LPRINT"TOTAL ANGLE OF SWINGING ROLLER FOLLOWER MOVEMENT FO=";FO
6070 LPRINT"TOTAL ANGLE FOR RISE =";B1;" DEGREES"
6080 LPRINT"TOTAL ANGLE FOR RETURN =";B2;" DEGREES"
6090 LPRINT"ROLLER RADIUS Rf =";RF;" IN."
```

```
6100 LPRINT"ROLLER WIDTH B =";B;" IN."
6110 LPRINT"CAM SPEED, N =";N;" RPM"
6120 IF C$="CCW" THEN LPRINT"CAM ROTATES CCW"
6130 IF C$="CW" THEN LPRINT"CAM ROTATES CW"
6140 IF MAT=1 THEN LPRINT"CAM MATERIAL IS STEEL"
6150 IF MAT=2 THEN LPRINT"CAM MATERIAL IS CAST IRON"
6160 LPRINT"WEIGHT OF MASS TO BE MOVED =";W;" LB"
6170 GOTO 410
6180 RUN "MAIN" : END
6190 LPRINT"UNDERCUTTING OCCURS FOR CAM ANGLE THETA =";TA;"THEREFORE OTHER"
6200 LPRINT"CAM PROPORTIONS MUST BE CHOSEN" : GOTO 410
```

Bibliography

In order to facilitate the use of the bibliography, the references have been numbered and the numbers are listed under the following topics:

DISPLACEMENT CURVES

1, 11, 15, 18, 19, 20, 21, 28, 29, 31, 34, 36, 39, 40, 41, 42, 46, 60, 61, 62, 64, 67, 72, 73, 83, 87, 89, 90, 92, 96, 98, 107, 108, 109, 110, 111, 117, 120, 122, 126, 127, 129, 130, 133, 134, 136, 139, 142, 143, 145, 146, 149, 150, 156, 166, 170, 173, 176, 180, 181, 182, 183, 184, 187, 188, 189, 191, 196, 198, 203, 204, 207, 209, 213, 218, 222, 236, 238, 239, 241, 251, 269, 273, 276, 277, 278, 279, 280, 281, 282, 284, 290, 293, 296, 297, 298, 300, 304, 310, 311, 312, 314, 320, 322, 330, 331, 337, 340, 341, 356, 359, 360, 361, 362, 366, 367, 380, 384, 386, 387, 388, 393, 399, 400, 401, 403, 404, 406, 407, 408, 409, 412, 413, 416, 417, 418, 419, 424, 425, 427, 430, 431, 432, 439, 440, 441, 442, 443, 447, 448, 449, 451, 452, 453, 456, 457, 458, 459, 460, 463, 469, 474, 480, 481, 482, 485, 488, 490, 491, 492, 494, 497, 508, 520, 521, 522, 523, 526, 527, 528, 534, 537, 541, 543, 544, 547, 551, 554, 555, 557, 560, 561, 562, 569, 579, 580, 581, 586, 588, 591, 594, 597, 600, 607, 611, 612, 613, 615, 617, 618, 619, 620, 622, 625, 627, 629, 630, 631, 635, 636, 640, 641, 647, 650, 651, 652, 653, 654, 658, 668, 669, 684, 688, 694, 698, 699, 703, 706, 708, 709, 710, 711, 713, 714, 715, 723, 727, 730, 734, 736, 737, 743, 745, 769, 774, 775, 776, 784, 789, 790, 791, 792, 797, 798, 801, 819, 822, 823, 824, 835, 837, 845, 846, 847, 849, 850, 852, 853, 855, 856, 857, 858, 859, 860, 861, 863, 864, 865, 866, 870, 871, 880, 881, 882, 883, 888, 891, 894, 900, 901, 902, 904, 907, 909, 910, 914, 915, 916, 918, 922, 929, 930, 931, 932, 933, 934, 938, 939, 940, 943, 944, 946, 953, 954, 955, 957, 958, 970, 974, 978, 984, 986, 987, 989, 992, 993, 997, 998, 999, 1000, 1001, 1005, 1012, 1013, 1015, 1016, 1017, 1019, 1023, 1030, 1038, 1041, 1043, 1044, 1045, 1047, 1051, 1053, 1058, 1061, 1063, 1066, 1072,

1073, 1075, 1076, 1078, 1082, 1083, 1084, 1086, 1087, 1088, 1096, 1100,
1102, 1103, 1104, 1105, 1114, 1116, 1117, 1118, 1121, 1122, 1123, 1124,
1125, 1126, 1127, 1128, 1130, 1131, 1132, 1133, 1135, 1136, 1137, 1139,
1140, 1147, 1148, 1150, 1152, 1159, 1160, 1166, 1167, 1169, 1177, 1178,
1179, 1180, 1181, 1182, 1183, 1188, 1189, 1190, 1191, 1192, 1194, 1196,
1197, 1198, 1199, 1201, 1202, 1203, 1204, 1206, 1207, 1211, 1214, 1215,
1217, 1218, 1219, 1221, 1226, 1228, 1229, 1231, 1237, 1238, 1239, 1249,
1250, 1252, 1255, 1256, 1257, 1259, 1260, 1261, 1265, 1266, 1268, 1270,
1271, 1272, 1274, 1275, 1276, 1277, 1278, 1280, 1281, 1283, 1285, 1286,
1290, 1292, 1293, 1295, 1297, 1298, 1299, 1300, 1307, 1308, 1309, 1310,
1311, 1317, 1319, 1320, 1322, 1324, 1325, 1327, 1332, 1334, 1336, 1338,
1341, 1344, 1348, 1355, 1359, 1364, 1379, 1389, 1392, 1394, 1398, 1399,
1400, 1408, 1409, 1420, 1421, 1422, 1423, 1426, 1430, 1431, 1432, 1439,
1441, 1443, 1447, 1448, 1453, 1455, 1461, 1462, 1463, 1465, 1466, 1467,
1471, 1481, 1483, 1484, 1486, 1489, 1490, 1491, 1492, 1493, 1494, 1495,
1498, 1504, 1520, 1521, 1523, 1525, 1526, 1533, 1537, 1539, 1554, 1584,
1674, 1690, 1738, 1748, 1763, 1770.

CAM PROFILE

Circular Arc and Straight-Line Profile
Cutter Coordinates
Moment of Inertia
Balancing

3, 12, 15, 24, 38, 42, 56, 65, 77, 84, 99, 102, 116, 121, 125, 153, 161, 180,
196, 198, 235, 251, 322, 323, 340, 358, 377, 379, 388, 413, 439, 441, 447,
452, 461, 464, 465, 523, 528, 529, 541, 550, 579, 584, 590, 595, 596, 597,
619, 638, 643, 647, 695, 699, 704, 706, 711, 720, 723, 724, 725, 727, 734,
744, 775, 789, 794, 810, 816, 852, 862, 868, 871, 899, 901, 911, 915, 916,
917, 932, 938, 944, 946, 947, 960, 986, 994, 999, 1000, 1002, 1003, 1023,
1029, 1038, 1047, 1050, 1051, 1073, 1077, 1083, 1084, 1087, 1088, 1089,
1117, 1118, 1119, 1120, 1121, 1129, 1140, 1152, 1157, 1159, 1160, 1173,
1174, 1175, 1176, 1178, 1180, 1181, 1182, 1183, 1185, 1186, 1187, 1188,
1199, 1200, 1201, 1203, 1208, 1239, 1252, 1259, 1263, 1265, 1275, 1289,
1290, 1293, 1295, 1300, 1302, 1303, 1310, 1311, 1316, 1317, 1320, 1323,
1336, 1338, 1349, 1355, 1366, 1371, 1379, 1400, 1425, 1432, 1440, 1481,
1495, 1534, 1535, 1565, 1609, 1625, 1626, 1627, 1680, 1770.

CAM SIZE

Pressure Angle
Radius of Curvature

13, 14, 27, 37, 76, 77, 86, 105, 106, 112, 113, 116, 121, 131, 132, 154, 164, 210, 222, 258, 259, 270, 274, 282, 293, 303, 354, 355, 357, 358, 366, 388, 397, 398, 428, 438, 453, 489, 492, 498, 499, 500, 502, 503, 506, 509, 547, 557, 564, 565, 574, 575, 576, 577, 583, 611, 612, 630, 642, 661, 670, 671, 685, 689, 695, 696, 699, 700, 704, 705, 706, 724, 803, 804, 805, 807, 808, 821, 822, 836, 838, 839, 840, 842, 858, 859, 860, 864, 865, 886, 887, 904, 905, 911, 932, 985, 993, 1004, 1048, 1078, 1093, 1103, 1121, 1144, 1145, 1149, 1152, 1168, 1173, 1176, 1197, 1210, 1243, 1253, 1261, 1271, 1292, 1340, 1360, 1371, 1394, 1410, 1434, 1464, 1470, 1490, 1491, 1497, 1531, 1533, 1535, 1538, 1605, 1624, 1646, 1647.

FOLLOWER, SPRING, ROLLER DESIGN

Follower D
Spring Design

43, 58, 74, 75, 77, 135, 147, 339, 374, 388, 531, 613, 639, 648, 649, 699, 706, 777, 802, 890, 914, 928, 935, 995, 1040, 1056, 1057, 1171, 1247, 1267, 1279, 1293, 1307, 1308, 1437, 1706.

CAM DYNAMICS AND PERFORMANCE AT HIGH SPEED

Vibrations
Cam Forces
Shaft Torque
Balancing

12, 22, 25, 33, 45, 50, 54, 55, 57, 59, 61, 62, 69, 77, 78, 81, 88, 91, 100, 101, 122, 128, 138, 146, 147, 155, 163, 182, 197, 199, 200, 201, 202, 208, 242, 248, 290, 294, 299, 311, 318, 320, 332, 335, 336, 338, 339, 359, 363, 377, 378, 379, 390, 391, 395, 396, 404, 414, 415, 417, 455, 458, 462, 465, 467, 515, 531, 532, 543, 545, 551, 573, 590, 598, 599, 600, 604, 607, 615, 635, 641, 644, 699, 706, 707, 708, 717, 718, 719, 720, 721, 722, 724, 725, 726, 738, 739, 740, 741, 774, 778, 779, 780, 781, 782, 783, 785, 786, 787, 793,

806, 809, 814, 815, 818, 829, 830, 831, 832, 833, 835, 837, 848, 854, 862, 868, 876, 880, 891, 900, 901, 902, 928, 936, 939, 950, 951, 952, 954, 970, 977, 981, 984, 988, 995, 996, 1012, 1015, 1034, 1035, 1036, 1037, 1048, 1050, 1052, 1053, 1054, 1057, 1059, 1071, 1085, 1090, 1091, 1092, 1107, 1115, 1127, 1129, 1133, 1155, 1162, 1164, 1184, 1192, 1193, 1199, 1200, 1202, 1209, 1212, 1213, 1219, 1221, 1223, 1225, 1229, 1230, 1233, 1238, 1241, 1244, 1245, 1246, 1248, 1252, 1254, 1260, 1284, 1294, 1296, 1313, 1321, 1348, 1352, 1390, 1397, 1401, 1413, 1414, 1415, 1416, 1418, 1428, 1435, 1437, 1452, 1472, 1482, 1487, 1501, 1502, 1503, 1512, 1517, 1523, 1532, 1537, 1560.

MATERIALS AND WEAR

Materials
Surface Treatment
Wear
Lubrication
Surface Stress

6, 17, 44, 70, 79, 80, 141, 189, 205, 206, 215, 216, 217, 224, 252, 254, 257, 301, 309, 328, 329, 392, 394, 411, 416, 436, 454, 510, 533, 539, 583, 585, 592, 593, 594, 603, 608, 609, 610, 699, 706, 731, 735, 816, 834, 851, 878, 912, 913, 923, 925, 942, 956, 968, 1009, 1052, 1053, 1054, 1070, 1099, 1110, 1111, 1112, 1172, 1205, 1235, 1240, 1242, 1280, 1297, 1298, 1309, 1318, 1339, 1343, 1354, 1403, 1404, 1405, 1407, 1411, 1451, 1468, 1469, 1477, 1480, 1514, 1515, 1536, 1543, 1558, 1593, 1603, 1631, 1650, 1666, 1676, 1677, 1682, 1693, 1708, 1731, 1733, 1751, 1764, 1765, 1768, 1769, 1775, 1790, 1807, 1812, 1813, 1814.

CAM MANUFACTURE

Manufacture
Inspection

5, 7, 23, 30, 32, 35, 38, 39, 44, 47, 97, 104, 114, 115, 123, 137, 144, 148, 152, 157, 158, 159, 160, 162, 167, 168, 169, 172, 174, 176, 177, 178, 179, 185, 190, 192, 193, 194, 195, 219, 220, 221, 222, 223, 225, 227, 228, 229, 230, 231, 244, 245, 246, 247, 249, 250, 253, 255, 256, 257, 258, 259, 263, 264, 265, 280, 283, 287, 288, 289, 292, 293, 302, 315, 317, 319, 321, 324, 325,

326, 327, 333, 364, 365, 368, 370, 371, 375, 382, 383, 388, 389, 405, 410, 420, 421, 422, 423, 425, 426, 433, 434, 435, 536, 437, 439, 444, 447, 450, 460, 468, 511, 512, 514, 516, 517, 518, 519, 535, 538, 540, 542, 545, 552, 553, 558, 560, 561, 571, 578, 582, 587, 589, 605, 606, 614, 616, 622, 623, 626, 628, 634, 645, 646, 650, 651, 652, 654, 664, 665, 684, 699, 712, 728, 733, 735, 746, 749, 767, 772, 773, 795, 817, 820, 823, 824, 825, 841, 843, 847, 872, 874, 877, 884, 885, 889, 892, 893, 895, 897, 900, 902, 903, 906, 908, 919, 920, 921, 924, 926, 929, 930, 941, 948, 961, 963, 964, 965, 966, 969, 973, 980, 982, 983, 990, 1007, 1008, 1014, 1017, 1018, 1020, 1021, 1022, 1024, 1027, 1028, 1032, 1049, 1055, 1062, 1067, 1068, 1074, 1077, 1094, 1095, 1097, 1098, 1110, 1111, 1112, 1122, 1143, 1151, 1153, 1154, 1156, 1161, 1165, 1170, 1189, 1195, 1216, 1224, 1227, 1234, 1237, 1240, 1249, 1255, 1262, 1264, 1269, 1273, 1274, 1285, 1289, 1291, 1301, 1304, 1305, 1306, 1312, 1314, 1315, 1318, 1319, 1328, 1329, 1330, 1332, 1333, 1334, 1335, 1337, 1342, 1345, 1348, 1351, 1353, 1361, 1362, 1363, 1367, 1372, 1373, 1375, 1376, 1380, 1382, 1383, 1384, 1395, 1396, 1406, 1409, 1411, 1417, 1419, 1429, 1436, 1438, 1442, 1444, 1445, 1446, 1449, 1450, 1451, 1458, 1459, 1460, 1473, 1484, 1485, 1486, 1488, 1499, 1504, 1508, 1509, 1510, 1511, 1515, 1518, 1519, 1522, 1524, 1527, 1530, 1545, 1547, 1548, 1549, 1555, 1556, 1569, 1570, 1577, 1579, 1583, 1584, 1594, 1595, 1596, 1597, 1598, 1599, 1600, 1601, 1602, 1603, 1606, 1607, 1611, 1616, 1618, 1620, 1621, 1622, 1629, 1630, 1631, 1632, 1633, 1635, 1638, 1639, 1640, 1641, 1642, 1643, 1644, 1645, 1652, 1653, 1654, 1655, 1657, 1659, 1660, 1661, 1664, 1665, 1666, 1667, 1668, 1669, 1670, 1673, 1676, 1677, 1678, 1682, 1683, 1684, 1686, 1687, 1688, 1689, 1695, 1696, 1697, 1699, 1700, 1701, 1702, 1703, 1704, 1705, 1707, 1708, 1711, 1712, 1713, 1715, 1717, 1718, 1719, 1720, 1722, 1723, 1724, 1727, 1728, 1731, 1732, 1738, 1739, 1740, 1743, 1744, 1745, 1746, 1747, 1752, 1753, 1754, 1755, 1759, 1760, 1761, 1762, 1764, 1766, 1767, 1768, 1773, 1775, 1776, 1777, 1778, 1787, 1792, 1793, 1794, 1797, 1798, 1799, 1801, 1804, 1805, 1806, 1809, 1810, 1815, 1816.

APPLICATION OF CAMS

2, 6, 8, 9, 10, 16, 23, 48, 51, 52, 53, 60, 66, 68, 71, 81, 85, 94, 95, 99, 102, 103, 118, 140, 165, 171, 186, 212, 214, 232, 233, 234, 237, 240, 243, 260, 261, 262, 266, 267, 268, 275, 286, 291, 295, 308, 313, 316, 342, 343, 344, 345, 346, 347, 348, 349, 350, 351, 352, 353, 369, 372, 373, 376, 385, 401, 402, 429, 445, 446, 449, 451, 466, 471, 472, 473, 474, 475, 477, 478, 479, 486, 487, 493, 495, 496, 501, 504, 505, 507, 513, 525, 527, 530, 536, 549,

558, 559, 560, 561, 563, 565, 566, 567, 568, 570, 572, 588, 601, 602, 618, 620, 621, 623, 624, 625, 633, 646, 655, 656, 657, 659, 660, 662, 663, 666, 667, 672, 673, 674, 675, 676, 677, 678, 679, 680, 681, 682, 683, 687, 690, 691, 692, 693, 697, 699, 706, 715, 732, 743, 747, 748, 750, 751, 752, 753, 754, 755, 756, 757, 758, 759, 760, 761, 762, 763, 764, 765, 766, 768, 771, 788, 791, 792, 796, 797, 799, 800, 811, 812, 813, 828, 844, 866, 867, 869, 873, 877, 879, 883, 896, 934, 937, 943, 945, 949, 959, 962, 967, 970, 971, 972, 974, 976, 979, 987, 1006, 1010, 1011, 1016, 1019, 1025, 1026, 1031, 1033, 1039, 1042, 1045, 1058, 1061, 1064, 1065, 1069, 1081, 1082, 1101, 1106, 1109, 1113, 1125, 1128, 1131, 1132, 1134, 1137, 1138, 1139, 1141, 1142, 1158, 1165, 1166, 1169, 1174, 1175, 1176, 1177, 1179, 1180, 1182, 1183, 1184, 1194, 1198, 1206, 1220, 1232, 1236, 1251, 1256, 1268, 1270, 1282, 1287, 1316, 1324, 1326, 1331, 1336, 1341, 1346, 1347, 1350, 1356, 1357, 1358, 1365, 1368, 1369, 1370, 1374, 1377, 1378, 1381, 1385, 1386, 1387, 1388, 1393, 1402, 1412, 1429, 1433, 1443, 1454, 1456, 1473, 1474, 1475, 1476, 1494, 1496, 1497, 1500, 1505, 1506, 1507, 1513, 1515, 1516, 1520, 1526, 1529, 1540, 1541, 1542, 1543, 1544, 1546, 1550, 1551, 1552, 1553, 1557, 1559, 1560, 1561, 1562, 1563, 1564, 1566, 1567, 1568, 1571, 1572, 1573, 1574, 1575, 1576, 1578, 1580, 1581, 1582, 1585, 1586, 1587, 1588, 1589, 1590, 1591, 1592, 1604, 1609, 1610, 1612, 1613, 1614, 1615, 1617, 1619, 1623, 1628, 1633, 1634, 1635, 1636, 1637, 1648, 1649, 1651, 1656, 1658, 1662, 1663, 1670, 1671, 1672, 1675, 1679, 1681, 1685, 1692, 1698, 1700, 1703, 1704, 1706, 1709, 1710, 1713, 1714, 1716, 1721, 1725, 1729, 1730, 1734, 1735, 1736, 1737, 1741, 1742, 1749, 1750, 1756, 1757, 1758, 1765, 1769, 1772, 1773, 1774, 1779, 1780, 1781, 1782, 1783, 1784, 1785, 1786, 1788, 1789, 1791, 1798, 1802, 1803, 1808, 1811, 1817.

CAMS FOR IC ENGINES

18, 19, 20, 21, 25, 26, 28, 69, 96, 118, 137, 142, 143, 148, 149, 150, 151, 158, 175, 176, 189, 221, 226, 229, 230, 232, 235, 249, 263, 268, 271, 276, 305, 306, 309, 317, 319, 326, 327, 328, 329, 333, 334, 337, 338, 354, 364, 365, 380, 386, 392, 394, 405, 410, 419, 429, 450, 459, 510, 518, 519, 524, 539, 547, 548, 555, 556, 591, 592, 593, 634, 637, 640, 641, 644, 699, 727, 728, 729, 731, 785, 819, 845, 912, 913, 914, 922, 924, 926, 927, 940, 946, 947, 948, 970, 971, 991, 1014, 1043, 1044, 1063, 1067, 1070, 1074, 1075, 1077, 1078, 1079, 1080, 1086, 1087, 1092, 1094, 1099, 1105, 1106, 1108, 1111, 1112, 1114, 1126, 1191, 1192, 1196, 1237, 1239, 1257, 1258, 1261, 1262, 1263, 1266, 1281, 1284, 1288, 1318, 1379, 1390, 1391, 1413, 1418, 1424, 1425, 1428, 1437, 1451, 1456, 1463, 1473, 1477, 1478, 1479, 1495, 1528, 1544, 1546, 1548, 1549, 1556, 1566, 1570, 1594, 1595, 1596, 1597, 1598,

1599, 1600, 1601, 1602, 1603, 1608, 1615, 1635, 1650, 1655, 1656, 1657, 1661, 1667, 1676, 1677, 1681, 1688, 1691, 1695, 1699, 1708, 1709, 1712, 1718, 1725, 1726, 1729, 1730, 1731, 1733, 1736, 1737, 1739, 1745, 1746, 1749, 1751, 1752, 1753, 1754, 1764, 1768, 1771, 1776, 1778, 1781, 1787, 1789, 1790, 1792, 1795, 1796, 1800, 1805, 1808, 1809, 1812, 1814, 1816.

3-D CAM MECHANISMS

42, 110, 135, 195, 272, 274, 283, 285, 356, 399, 400, 503, 520, 546, 572, 631, 699, 770, 836, 849, 894, 905, 955, 975, 987, 1159, 1160, 1225, 1427, 1552, 1734, 1793.

ROLLING-CONTACT MECHANISMS

111, 211, 308, 470, 476, 483, 484, 686, 826, 827, 1123, 1427, 1457.

1. Ainsworth, W. E., "Calculations for Screw Machine Cams," *Am. Mach.* January 6, 1944, pp. 91—93.
2. Ainsworth, W. E., "Cam Actuated Fixture Sharpens Countersinks," *Am. Mach.* 89, August 30, 1945, p. 103.
3. Allais, D. C., "Cycloidal vs. Modified Trapezoidal Cams," *Mach. Des.* 35, January 31, 1963, pp. 92—96.
4. Allais, D. C., "Mirror—Image Cams," *Mach. Des.* 31, January 22, 1959, p. 136.
5. Allen, E. C., "Table for Use in Milling Cams for B. & S. Automatic Screw Machines," *Am. Mach.* 58, March 8, 1923, pp. 371—372.
6. Aller, F., and E. Weisenfels," Kurventrieb mit hydraulischer Kraftver-stärkung," *Ölhydraulik und Pneumatik* 18, 1974, No. 6, pp. 468—70 ("Cam Mechanism with Hydraulic Force Amplification").
7. Alstrom, A. I., "Tooling is the Key to Sintered—Iron Cams," *Am. Mach.* 97, November 23, 1953, p. 124.
8. Alt, H., "Getriebe der Papierverarbeitung," *Maschinenbau* 21, 1929 ("Cams and Linkages in Paper Packaging Machines").
9. Alt, H., "Getriebe bei Glasverarbeitungsmaschinen," *VDI—Z* 77, 1933, pp. 409—413. ("Cam and Linkage Mechanisms in Glass Manufacturing Ma-chinery").
10. Alt, H., "Abfüll – und Verpackmachinen," *VDI—Z* 81, 1937, p. 1111 ("Filling and Packaging Machinery").
11. Alt, H., "Zur Getriebetechnik der Verpackmaschinen," *VDI—Z* 75, 1931, pp. 245—254 ("On Mechanism Technology in Packaging Machines").
12. Alt, H., "Der Ruck," *Maschinenbau*, 1936, pp. 581—582, ("The Jerk").

13. Alt, H., "Der Übertragungswinkel und seine Bedeutung für das Konstru-
    ieren periodischer Getriebe," *Werkstatttechnik* 26, 1932, pp. 61—64
    ("The Transmission Angle and Its Importance in the Design of Linkages").

14. Alt, H., "Die Güte der Bewegungsübertragung bei periodischen Getrie-
    ben," *VDI–Z* 96, 1954, pp. 238—244.

15. Altschul, R., "Zur Massynthese und Beschleunigungsermittlung von Wälz-
    kurvengetrieben," *Maschinenbautechnik* 15, November 1, 1966, pp. 605—
    609 ("Dimensional Synthesis and Determination of Accelerations of
    Roller–Cam Mechanisms").

16. Amarnath, C., and B. K. Gupta, "Novel Cam–Linkage Mechanisms for
    Multiple Dwell Generations," Rees Jones: Cam and Cam Mechanisms, pp.
    123—127.

17. Ambros, G., "Verschleissuntersuchungen an Kurvenscheiben," *Maschinen-
    bautechnik* 6, 1957, pp. 575—576 ("Wear Studies on Disk Cams").

18. Ambs, O., "Bogennocken," *ATZ*, 1943, pp. 155—160 ("Arc Cams").

19. Ambs, O., "Entwurf und kinematische Untersuchung der Nocken mit
    Flachstössel," *ATZ*, 1940, pp. 476—480 ("Design and Kinematic Study of
    Cams with Flat Followers").

20. Ambs, O., "Sonderausführung des Tangentennockens mit Rolle und
    Stössel," *ATZ*, 1942, pp. 541—547 ("Special Design of the Tangential
    Cam with Roller and Slide Tappet").

21. Ambs, O., "Normalausführung des Tangentennockens mit Rolle und
    Stössel," *ATZ*, 1942, pp. 433—438 ("Standard Design of the Tangential
    Cam with Roller and Slide Tappet").

22. Anderson, D. G., "Cam Dynamics," *Product Eng.* 24, October 1953, pp.
    170—176.

23. Andrews, S. H., "Cam Standardization Ups Forming Speed 50 Percent,"
    Am. Mach. 94 October 2, 1950, p. 96.

24. Angeles, J., and O. Arteaga, "Optimal Synthesis of Cam Mechanisms via
    the Method of Newton–Raphson and Runge–Kutta," *IFToMM*, Vol. III,
    No. 1, 1977, pp. 1—12.

25. Angle, G. D., "Increase in Cam Tip Radius Effects Reduction in Inertia
    Forces," *Automotive Eng.* 48, April 12, 1923, pp. 824—827.

26. Angle, G. D., "Radial Engine Cam Design Simplified," *Automotive Ind.*,
    99, p. 15, 1948, pp. 28—31.

27. Antuma, H. J., "Pressure Angle and Undercutting with Cams Having a
    Sinusoidal Trend of the Acceleration," (in Dutch), *Ingenieur* 78, July,
    1966, pp. 175—185.

28. Appel, W. D., "Designing of Cams Is Much Simplified by Graphical
    Method," *Automotive Ind.* 52, March 19, 1925, pp. 544—546, Discussion,
    52, May 28, 1925, p. 948.

29. Applegate, R. B., "Good Cam Design," *SAE J.* February 1953, pp. 112—
    113.

30. Apter, R. H., and H. W. Smith, Jr., "Flame Hardening with City Gas; Full–
    Automatic Set–Ups," *Am. Mach.* 86, October 1, 1942, pp. 1088—1089.

31. Archer, N. B., "Layout of Cam Locks; Reference Book Sheet," *Am. Mach.* 90, April 25, 1946, p. 139.

32. Archer, R. H., "Solving a Problem with a Pot Type Hardening Furnace." *Am. Mach.* 69, July 26, 1928, pp. 149–150.

33. Ardayfio, D. D., "Kinetothermoelastodynamic (KITED) Synthesis of Cam Profiles," *Mechanism and Machine Theory* 15, 1980, No. 1, pp. 1–4.

34. Aref, M. H., and Badawy, E. M., "Application of the Concept of Equivalent Mechanisms in Cam Design," ASME Paper 64–Mech–21.

35. Arnold, J., H. Freitag, and W. Hirsack, "Durchgängige Prozessrationalisierung der Konstruktion, Technologie und Fertigung von Kurvenkörpern mit gesicherten Qualitätsmerkmalen," *Maschinenbautechnik* 28, 1979, No. 9, pp. 388–397 ("Process Simplification of the Design, Technology and Manufacture of Cams with Prescribed Quality").

36. Arnott, A. S., "How to Design Locking Cams," *Can. Machy.* 67, July 7, 1956, pp. 112–113; *Machine and Tool Blue Book* 51, October 1956, pp. 117–118, 120, 122 and 124. Schematic.

37. Astely, B. P., "Minimum Hub Diameters," *Am. Mach.* 92, December 30, 1948, p. 129.

38. Astrop, A. W., "Automatic High–Speed Inspection of Variable–Pitch Cams for Zoom Lenses," *Mach. and Prod. Eng.* June 1967, pp. 1360–1364.

39. Aurin, G., and Gartner, H., "Die Fertigung von Urnocken mit sprungfreiem Beschleunigungsverlauf bei Betrieb mit Rollenstössel" ("Designing Basic Cam Units with Non–Oscillating Acceleration in Roller–Type Drives"), *Kraftfahrzeugtechnik* 12, 1962, No. 4, pp. 142–146).

40. AWF–VDMA 641–643, "Konstruktion von Kurvenscheiben." Jahr, W., and Knechtel, P., 1932 ("Design of Cams").

41. AWF–VDMA 634–636, "Ebene Kurventriebe" ("Plane Cam Mechanisms").

42. AWF–VDMA 644–645, "Räumliche Kurventriebe" ("3–D Cam Mechanisms").

43. Babbit, A. B., "Location of Cam Followers," *Am. Mach.* 47, October 4, 1917, pp. 20, 504–506, 591–594.

44. Bachner, G. L., "How Service Conditions and Costs Affect Design of Sintered Cams," *Precision Metal Molding* 11, February 1953, pp. 32–34.

45. Badawy, E. M., and Maharen, N. A., Raafat. "Flexible Cam Systems," *IFToMM*, 1974, pp. 203–238.

46. Bagci, C., "Stop Designing and Testing Cam–Follower Systems Using the Rise Portions of the Displacement Programs Only," Communications of the Third World Congress for the Theory of Machines and Mechanisms, Kupari, Yugoslavia, Vol. G., Paper G–25, pp. 13–20, 1971, pp. 347–364.

47. Baker, D. A., "Cams Made from Tubing," *Am. Mach.* 48, February 28, 1918, pp. 381–382.

48. Ball, M, H., "Cams Replaced on Shafts Reamed for Taper Pins," *Am. Mach.*, 89, August 2, 1945, p. 133.

50. Ball, M. H., "Quick–Rise Cam Layout Developed to Prevent Shock in Operating a Mechanism," *Machinery*, 45, August 1939, p. 857.

51. Ball, R. C., "See–Saw Cam Motion Smoothly Converts Rotary to Linear Motion," *Product Eng.* 33, April 2, 1962, pp. 68–72.

52. Ball, W. S., "Boring–Mill Cam Follower Improves Finish, Cuts Time," *Am. Mach.*, 96, May 12, 1952, p. 176.

53. Balmer, T., "Contoured Cam Compensates for Machine Deviations," *Des. News*, 24, 1969, pp. 76–77.

54. Balzi, M. F., and Dikovskij, B. L., "Dynamic Synthesis of a Cam Mechanism" (in Russian), Issled. po dinamike masin (Swerdlowsk), 1967, pp. 87–94.

55. Baranyi, S. J., "Cams, Dynamics and Design," *Design News*, 24, 1969, pp. 108–115.

56. Baranyi, S. J., "Multiple–Harmonic Cam Profiles," ASME Paper 70–MECH, 1970, p. 59.

57. Baratta, F. L., and Bluhm, J. I., "When Will a Cam Follower Jump?" *Product Eng.* July 1954, pp. 156–159.

58. Barish, T., "Ball Bearing Used as Cam Rollers," *Product Eng.*, January 1937, pp. 2–5.

59. Barkan, P., "Calculations of High–Speed Valve Motion with Flexible Overhead Linkage," *Transactions of the SAE* 61, 1953, pp. 687–700.

60. Barkan, P., "Comment on Cam–Follower Systems Article by D. M. Mitchell," *Mech. Eng.* 72, December 1950, p. 1009.

61. Barkan, P., "High–Speed Spring–Action Cams," *Transactions of the Fifth Conference on Mechanisms*, Sponsored by *Machine Design*, 1958, pp. 64–76.

62. Barkan, P., and McGarrity, R. V., "A Spring Actuated, Cam–Follower System: Design Theory and Experimental Results," ASME Paper No. 64–Mech.–12.

63. Barkan, P., "Spring Driven Cam Systems," *Mach. Des.* 31, May 14, 1959, pp. 174–178.

64. Barlow, B. V., "Brush Up Your Knowledge on Cam Design. Linear Relationship Disc Cam," *Metalworking Prod.* 113, April 2, 1969, pp. 44–47.

65. Barlow, B. V., "The Double Disc Cam Drive," *Metalworking Prod.* 113, 1969, No. 27, pp. 37–39.

66. Barlow, B. V., "Linkage Adjusters and Effect on Geometry," *Metalworking Prod.* 113, 1919, No. 86, pp. 59–60.

67. Barlow, B. V., "Optimized Cylindrical Camtrack Section–4," *Metalworking Prod.* 113, July 30, 1969, pp. 45–47.

68. Barlow, B. V., "Two out of the Ordinary Cam Applications," *Metalworking Prod.* 113, 1969, No. 37, pp. 44–45 and 47.

69. Bartley, J. E., "Cam Design for Maximum Output at High RPM," SAE Paper, May 14, 1956, p. 7.

70. Barwell, F. T. and Roylance, B. J., "Tribological Considerations in the Design and Operation of Cams—A Review of the Situation," Rees Jones: Cams and Cam Mechanisms, pp. 99–105.

71. Bassoff, A. B., "Variable–Stroke Cam–Actuated Mechanism," *Mach.* 46, July 1940, p. 196.

72. Bauer, P., "Bewegungsgesetze für Kurvengetriebe," Wertung und Anwendung der Richtlinie VDI 2143, in der Praxis, *Antriebstechnik* 21, 1983, No. 1, pp. 18—25.

73. Baumann, W. A., "Zur Mechanik der Kurvensteuerung," *Techn. Rundschau* 54, 1962, No. 15, pp. 25, 27, and 29 ("Concerning the Mechanics of Cam Drives").

74. Baumgarten, J, R., "Preload Force Necessary to Prevent Separation of Follower from Cam," *Transactions of the Seventh Conference on Mechanisms*, Sponsored by *Machine Design*, 1962, pp. 213—218.

75. Baumgarten, J. R., "Preventing Cam–Follower Separation: Method of Determining Minimum Preset of the Follower Spring; Data Sheet," *Mach. Des.* 34, November 22, 1962, pp. 179—180.

76. Baxter, M. L., "Curvature–Acceleration Relations for Plane Cams," *ASME Trans.* 70, July 1948, pp. 483–489; Excerpts *Machine Design* 20, February 1948, pp. 159—161.

77. Beard, C. A., "Cam Mechanism Design Problems – An Engineer Designer's Viewpoint," Rees Jones: Cams and Cam Mechanisms, pp. 49—53.

78. Beckett, R. E., Pan, K. C., and Chu, S. C., "A Numerical Method for the Dynamic Analysis of Mechanical Systems in Impact" *Trans. ASME, Ser. B: J. Engng. Ind.*, 99, 1977, No. 3, pp. 665—73.

79. Beese, J. G., Dasgupta, A. K., and Peters, R. M., "Imperfections in Cam Profiles and Cam–Follower Alignment; Influence on Wear Potential," Rees Jones: Cams and Cam Mechanisms, pp. 136—140.

80. Beese, J. G. and Clark, R., "The Performance of Materials Associated with Cams," Rees Jones: Cams and Cam Mechanisms, pp. 95—98.

81. Beggs, J. S., and Beggs, R. S., "Cam and Gears Join to Stop Shock Loads," *Product Eng.* 28, September 16, 1957, pp. 84—85.

82. Beggs, J. S., and Beggs, R. S., "Planeten–Kurven–Getriebe," *VDI–Z* 99, 1957, pp. 839—40 ("Planetary Cam Mechanisms").

83. Beggs, J. S., "Planetenrad–Nockengetriebe," *VDI–Z*, 100, 1958, pp. 249—251 ("Planetary Gear–Cam Mechanisms").

84. Behr, K. G., "Kurventafeln für den Entwurf von Wälzhebelgetrieben," *Konstrukteurheft* 15, VDI–Verlag, 1958 ("Table of Curves for the Design of Rolling–Contact Mechanisms").

85. Beldon, S. T., "Cams Combine to Program Simulator," *Control Eng.* 8, March 1961, p. 165.

86. Beletskii, V. Ya., "Analytical Method of Determining the Curvature of Projected Plane Cams," (in Russian), Trudi Odessk, Tekbnol. in–ta 7, 19—26, 1955; Ref. Zb. Mekb. No. 10, 1956, Rev. 6460.

87. Belgaumkar, B. M., "Computer–Aided Design and Fabrication of Cams," *5th Int. Analogue Comput. Meet.*, Proc. Vol. 1, 2 Lausanne, Switzerland, August 28–November 2, 1967, pp. 1092—1094.

88. Beltjukov, V. P., "Dynamic Synthesis of a Band–Cam Mechanisms Connected Elastically to the Frame" (in Russian), Izv. Vyss. Ucebn. Zavedenij, Masinostr. 11, 1968, No. 5, pp. 32–35.

89. Bel'tyukov, V. P., "Algorithm for Computer Calculation of the Spasmodic Motion (Jerk) and Other Characteristics of a Roller Tappet," Sov. Eng. Res. 1, No. 10, October 1981, pp. 38–40.

90. Bel'tyukov, V. P., "Comparison of Algorithms of Computer Aided Calculation of the Jolt of a Disk Cam Roller Pusher," Izv Vyssh Uchebn Zaved Mashinostr, No. 6, 1982, pp. 27–31.

91. Benedict, C. E., Matthew, G. K., and Tesar, D., "Torque Balancing of Machines by Sub–Unit Cam Systems," Paper No. 15, O.S.U.

92. Benedict, C. E., and Smith, J. A., "Cam Design by Second Order Derivative Specification," Mach. Eng. News 8, 1971, No. 1, pp. 21–25.

93. Bennett, E., "Intermittent Drive Mechanism," Engineer 213, January 1962, pp. 191–192.

94. Bennett, N. K., "Designing Accelerated–Motion Cylindrical Cams," Mach. Des. 14, June, pp. 87–88, 86, July 1942.

95. Bennett, R. M., "Gears for Pick–Up Cams; Formula Permits Selection of Gears for Timing Pick–Up Cams on Brown & Sharpe Screw Machines," Am. Mach. 82, September 21, 1938, p. 835.

96. Bensinger, W. D., "Nocken mit Schwinghebel," MTZ 10, 1049, pp. 123–135 ("Cams with Oscillating Follower").

97. Benson, E. B., "Grinding Cam Track in Chain Cam and Gear," Tool Eng. 25, September 1950, pp. 31–32.

98. Bergmann, Von A., "Möglichkeiten zur Verringerung des Nockendrucks bei Kurvengetrieben für 8 MM–und 16 MM–Filmschaltwerke," Feinwerktechnik, No. 1, 1973, pp. 30–32 ("Possibilities for the Reduction of Compressive Stress in Cam Mechanisms for 8 and 16 mm Film Projector Systems").

99. Berry, H. P., "Abrupt Rise Cam Lobes," Am. Mach. 74, May 21, 1931, pp. 795–796.

100. Berzak, N. and Freudenstein, F., "Optimization Criteria in Polydyne Cam Design," IFToMM, Vol. 2, 1979, pp. 1303–1306.

101. Berzak, N., "Optimization of Cam–Follower Systems with Kinematic and Dynamic Constraints," Trans. ASME, J. Mech. Des. 104, 1982, No. 1, pp. 29–33.

102. Bestehorn, R., "Die Form der Steuernocken," ZVDI 63, 1919, pp. 263–266 ("The Shape of Cams").

103. Betz, W. C., "Cams for Automatic Screw Machines," Am. Mach. 76, May 5, 1932, pp. 589–590.

104. Beusch, A., "Cam Milling Attachment for the Lathe," Am. Mach. 76, September 28, 1932, p. 1043.

105. Beyer, R., "Übertragungswinkel am Zylinderkurventrieb," Maschinenbau 3, 1935, pp. 167–169 ("The Transmission Angle in Drum–Type Cam Mechanisms").

106. Beyer, R., "Der Übertragungswinkel am Zylinderkurventrieb," *Maschi-nenbau* 3, 1935, pp. 167–169 ("The Transmission Angle in Drum–Type Cam Mechanisms").

107. Beyer, R., "Das Fadengeber–Getriebe bei Nähmaschinen," *Getriebe-technik* 7, 1939, pp. 309–312 ("Thread–Feeding Cams in Sewing Machines").

108. Beyer, R., "Konstruktion von Kurbenscheibengetrieben," *Feinwerk-technik* 54, 1950, pp. 328–331 ("Design of Disk Cam Mechanism").

109. Beyer, R., "Zur Synthese ebener Kurvenscheibengetriebe," *Konstruktion* 4, 1952, pp. 208–210 ("On the Synthesis of Planar Disk Cam Mechanisms").

110. Beyer, R., "Geometrisch–kinematische Grundlagen für das Schleifen von Kurvennutteilen und das Erzeugen archimedischer Spiralen," *Industrie-Anzeiger*, 1952, pp. 1221–1223 ("Geometric–Kinematic Criteria for Grinding Techniques in the Design of Cam Grooves and Archimedean Spirals").

111. Beyer, R., "Zur Synthese der Bewegungsgesetze ebener und räumlicher Kurvengetriebe," *Konstruktion* 5, 1953, pp. 188–192 ("On the Synthesis of Kinematic and Spatial Displacement Diagrams for Cam Mechanisms").

112. Beyer, R., "Method of Curvature Determination for Curvilinear Cams," *The Engineer's Digest*, 1954, pp. 420–425.

113. Beyer, R., "Krümmungsverhältnisse der Kurven ebener Kurvengetriebe und Möglichkeiten ihrer werstattmässigen und fertigungstechnischen Auswertung," *Industrie-Anzeiger* 54, 1954, pp. 853–858 ("Criteria Concerning the Curvature of Planar Cam Mechanisms, Including Possibilities of Optimum Technological Utilization").

114. Biechele, F., "Fixture Holds Master for Face–Cam Milling," *Am. Mach.* 94, May 1, 1950, p. 106.

115. Biechele, F., "Rapid Layout Method Establishes Quick–Acting Cams," *Am. Mach.* 94, March 20, 1950, p. 123.

116. Biehler, K., "Genauigkeitsanforderungen an Kurvenscheiben im Fein-gerätebau," Internat. Wiss. Koll, TH Ilmenau 1969, No. 7, Publisher: Techn. Hochschule Ilmenau, 1970, pp. 1–11 ("Precision Requirements for Cams Used in Instruments").

117. Bilaisis, V., "Displacement Factors for Modified–Trapezoid Cam Profiles; Data Sheet," *Mach. Des.* 29, August 22, 1957, pp. 135–138.

118. Bishop, J. L. H., "Analytical Approach to Automobile Valve Gear Design," *Inst. Mech. Eng. Proc.* (Auto. Div. Pt. 4) 150, 1950–1951, pp. 150–157; *Automobile Eng.* 41, June 1951, pp. 233–238; Discussion, *Inst. Mech. Eng. Proc.* (Auto. Div. Pt. 4), 1950–1951, pp. 157–160.

119. Bisshopp, K. E., "Note on Spherical Motion," *J. Mechanisms* 4, 1969, No. 2, pp. 159–166.

120. Bittrich, W., "Erfahrungen beim Einsatz von EDVA zur Konstruktion von Kurvengetrieben in der *Büromaschinenindustrie*," *Maschinenbau-technik* 7, 1972, pp. 297–298 ("Experiences in the Use of EDP in the Construction and Design of Cam Mechanisms in Office Machine Industry").

121. Bitzel, H., "Zur Konstruktion von übertragungsgünstigen, ebenen Kurvengetrieben mit schwingendem oder umlaufendem Abtriebsglied," Diss. Univ. Stuttgart, 1969, Ref. in: *VDI–Z* 113, 1971, No. 2, p. 185 ("On the Design of Cams with Swinging and Rotating Roller Follower, Having Good Transmission Characteristics").

122. Black, D. H., and Munden, D. L., "Increasing the Rates of Fabric Production of Weft–Knitting Machinery; Part I, The Design and Performance of High–Speed Knitting Cams; Part II, An Analysis of High–Speed Knitting Cam Systems; Part III, Measurement of the Needle Forces," *J. Textile Institute* 61, July 1970, pp. 313–324, pp. 325–339, and pp. 340–348.

123. Blakey, J., and Shankely, J. A. H., "Garvin 12–in. Cam Milling Machines; a Suggested Improvement," *Eng. & Ind. Management* 5, May 19, 1921, pp. 566–568.

124. Bley, A. G., "Wälzhebelgetriebe in Messgeräten," *Feinmechanik und Präzision* 47, 1939, pp. 209–210 ("Rolling–Contact Cams in Measuring Instruments").

125. Bliss, S. C., "Applying Sine Curves to Cam Design," *Machinery* 66, January 1960, pp. 136–141.

126. Bliss, S. C., "Designing Disk Cams without Trial and Error Layouts," *Mach.* 57, October 1950, pp. 179–183.

127. Bloom, D., "The Dual (or X,Y) Cam Mechanism," Proceedings of the Oklahoma State University of Applied Mechanisms Conference, July 31–August 1, 1969, Tulsa, OK, Paper No. 30, *J. Mechanisms*, 1971.

128. Bloom, D., and Radcliffe, C. W., "The Effect of Camshaft Elasticity on the Response of Cam–Driven Systems," ASME Paper 64, Mech.–41.

129. Bloom, D., "Optimum Design of Parametric Cams for the Dual Cam Mechanism with Radial Followers," Proceedings of 2nd Applied Mechanism Conference, Oklahoma State University, Stillwater, Oklahoma, 1971.

130. Blum, J., "Problem in Precision Cam Design," *J. Res. Nat. Bur. Stand* 45, December 1950, pp. 502–504.

131. Bobancu, S., and Teodorescu, R., "Die Verbreitung des Begriffs der Exzentrizität bei ebenen Kurvengetrieben," *IFToMM*, Vol. III, 1977, I, pp. 47–56 ("On the Excentricity of Plane Cam Discs").

132. Bock, A., "Gedanken zum 'Übertragungswinkel und Vorschläge für dessen Auswertung," *VDI–Berichte* 29, 1958, pp. 158–159 ("Considerations on the Transmission Angle, Including Ideas on Its Practical Applicability").

133. Bock, A., "Konstruktion von Kurvenschaltwerken," *Maschinenbau* 4, 1936, pp. 633–637 ("Design of Cam–Ratchet Mechanisms").

134. Bock, A., "Ein Konstruktionsleitblatt für Kurvengetriebe," *Feingerätetechnik* 4, 1955, pp. 450–451 ("A Design Sheet for Cam Mechanisms").

135. Bock, A., "Richtige Form der Rollen an räumlichen Kurventrieben," *AWF–Mitteilungen*, No. 2, 1934, p. 17 ("The Correct Shape of Rollers in Spatial Cam Drive Mechanisms").

136. Bock, A., "Der systematische Aufbau der Schaltgetriebe," *Maschinen-bautechnik* 4, 1955, pp. 60–63, pp. 116–125 ("Systematic Design of Ratchett Mechanisms").

137. Bock, W. E., "Hydraulic–Electric System; Machine for Magnaflux Inspection of Crankshafts and Camshafts," *Mach. Des.* 19, December 1947, pp. 107–111.

138. Bockenmüller, E. A., and Subke, H., "Theorie und Berechnung eines Feder–Kurvenscheiben–Systems zur Erzeugung einer Kraft, die in vorgegebener Weise vom Weg abhängt," *ZAMM* 56, 1976, No. 3, pp. 145–146 ("Theory and Calculation of a Spring–Cam System Yielding a Force Which Is Dependent on the Displacement").

139. Bögelsack, G., "Praktische Anwendung und theoretische Grundlagen der Reibkurvengetriebe," *Feinwerktechnik*, No. 2, 1965, pp. 65–73 and No. 3, 1965, pp. 103–108 ("Practical Application and Theoretical Principles of Friction–Type Cams").

140. Böhme, K. R., "Potentiometer mit Rollkurvengetriebe zur Erzeugung sinusförmiger Spannungen," *VDI–Z* 91, 1949, pp. 643–644 ("Potentiometer with Cylindrical Cam Mechanism for the Generation of Sinusoidal Voltages").

141. Bona, C. F., and Ghilardi, F. G., "Influence of Tappet Rotation on Cam and Tappet Surface Deterioration," Proceedings of the Institute of Mechanical Engineers, Automobile Division 180, Part 2A, 11, 1965–1966, pp. 269–278.

142. Bondar, M. P., "Calculations for Cam Mechanisms of Automatic Precision Lathes," *Machines and Tooling*, 1966, pp. 44–45.

143. Boock, A. C., "Graphical Design of Cams," *Automotive Ind.*, May 28, 1925, p. 948.

144. Booth, A. N., "Hydraulic Contour Control for Cams," *Hydraulics & Pneumatics* 15, April 1962, pp. 96–98.

145. Borisenko, L. N., and Geronimus, J. L., "On Equivalent Mechanisms" (in Russian), Mechanika masin 7/8, 1968, No. 13–14, pp. 53–60.

146. Borun, F. L., "Cam Design Based on Law of Elastic Deformation of Homogeneous Rod: Analytic Study of Possibility to Improve Cam Design" (in Russian), *Vestnik Mashinostroeniya* 39, September 1959, pp. 41–44.

147. Bottema, O., "A Construction for the Velocity and the Acceleration of the Follower of a Cam," *J. Mechanisms*, 1966, pp. 285–289.

148. Boudreau, L., "Welded Construction Reduces Camshaft Costs," *Mach.* 44, December 1937, p. 264.

149. Bouvy, C. H., "Calculation of Proportional Cams," *Automotive and Aviation Ind.*, September 1, 1943, p. 45.

150. Bouvy, C. H., "Triple Curve Cam Gives Maximum Lift Where Space Puts Limit on Tappet Head Diameter," *Automotive Ind.* 70, June 16, 1934, p. 45.

151.   Bouvy, C. H., "Flachstösseldurchmesser und Nockenform," *ATZ* 37, December 10, 1934, pp. 575—577 ("The Relationship Between Diameter of Flat—Faced Follower and the Contour of the Cam").

152.   Bower, C. T., "Geared Disk—Cam Holder Matches Feed and Rotation for Milling," *Am. Mach.* 92, July 29, 1948, p. 110.

153.   Bowman, H. R., "Method of Developing Cam Profile," *Mach.* 32, July 1926, pp. 898—899.

154.   Boxteo, M. L., "Curvature Acceleration, Relation for Plane Cams," *Trans. ASME* 70, 1948, pp. 483—489.

155.   Bralkowicz, B. Klimowicz, T., and Swietik, M., "Changes of the Dynamic Properties of the Real Cam Profile During Its Wear," In: 8, Vol. 2, pp. 984—987.

156.   Brandenberge, H., "Ausbildung von Kurvenscheiben für Verarbeitungs-maschinen," *VDI—Tagungsheft* 1, Düsseldorf 1953, pp. 21—26 ("Development of Disk Cams for Processing Machines").

157.   Brause, A., "Neuzeitliche Fertigungsverfahren von Kurvengetrieben," *VDI—Berichte* 12, 1956, pp. 121—126 ("Modern Methods of Manufacturing Cams").

158.   Bremer, E., "Camshafts are Finish Ground to Close Tolerances; Ford Motor Company," *Steel* 93, December 25, 1933, pp. 21—22.

159.   Briana, A., "Ökonomische Serienfertigung von Kurvenscheiben mit spezieller Einrichtung," *Maschinenbautechnik* 28, 1979, No. 9, pp. 402—403 ("Economic Series Manufacturer of Cams with Special Means").

160.   Brill, J., "Unrundschleifen in der Serien- und Massenproduktion," *Werkstatt und Betrieb* 110, 1977, No. 9, pp. 623—630 ("Non—Circular Grinding in Series and Mass Production").

161.   Brittain, J. H. C., and Horsnell, R., "Prediction of Some Causes and Effects of Cam Profile Errors, *Instn. Mech. Engrs—Proc.* 182, 1967—1968, pp. 145—151 ("Computers in Internal Combustion Engine Design").

162.   Broadwin, S., "Internal Cam at Low Cost by Die Casting," *Precision Metal Molding* 10, July 1952, p. 27.

163.   Broeke, H., "Improving the Performance of Dynamically Loaded Cam Mechanisms," *Maschinenbautechnik* 14, 1965, pp. 157—163.

164.   Brokate, K., "Berechnung der Kurve eines Herzkurvengetriebes mit 2 Rollen," *Feinwerktechnik* 53, 1949, 49—53 ("Calculation of Curvature in Heart—Shaped Cam Mechanisms with Three Rollers").

165.   Brook, H., "Developing Ellipse for a Pulley—Shaft Guiding Cam," *Mach.* 30, February 1924, p. 453.

166.   Brooks, J. E., "Calculation of Feel Cam Contour," *Aeronautical Eng. R.* 15, September 1956, pp. 37—38.

167.   Brower, W. B., "Large Circular Barrel Cams Accurately Flame—Cut on Airco—DB No. 6 Oxygraph," *Iron Age* 141, June 16, 1938, pp. 31—33.

168.   Brown, H., "Fixture for Milling a Cam Slot of Constantly Increasing Radius," *Am. Mach.* 67, September 15, 1927, pp. 443—444.

169. Brown, T., "Automatic Cam Milling Fixture," *Machine & Tool Blue Book* 49, April 1953, pp. 183–188.

170. Brunell, K., "Constant Diameter Cams, Their Properties and Design Characteristics," *AMSE Trans.*, ser B 84, February 1962, pp. 161–164.

171. Buhayar, E. S., "Changing the Industrial Machinery Designer's Attitude to Cam–Driven Mechanisms," 1st ASME Design Techn. Transfer Conference, ASME 1974, pp. 161–166.

172. Buhayar, E. S., "Computerized Cam Design and Plate Cam Manufacture," *ASME Publication* 66–MECH–2, 1966.

173. Buhayar, E. S., "An Industrial Machinery Designer's Computer–Aided System for Plate Cams and Follower Linkages," Rees Jones: Cams and Cam Mechanisms, pp. 79–84.

174. Burke, E. E., "Cam Milling," *Am. Mach.* 74, May 28, 1931, p. 838.

175. Burkhardt, O. M., "Analysis of the Functions of Cam Actuated Valves," *Horseless Age* 33, January 28, 1914, pp. 176–180.

176. Burkhardt, O. H., "Characteristics of Automobile Engine Cams," *Horseless Age* 33, February 4, 1914, pp. 222–224.

177. Burow, G., "Rationelle Fertigung ruckfreier urnocken mit Hilfe lockbandgesteuerter Werkzeugmaschinen," *Maschinenbautechnik*, No. 5, 1970, pp. 226–230 ("On the Developing of Efficient Friction–Free Cams by means of Computer–Controlled Tooling Devices").

178. Cable, H. W., "Chart for Use in Designing and Milling Cams," *Mach.* 31, July 1925, pp. 852–854.

179. Cable, H. W., "Use of Dividing Head and Circular Attachment for Cam Milling," *Mach.* 32, February 1926, pp. 481–482.

180. Candee, A. H., "Formulas for Involute Curve Layouts," *Product Eng.* 19, August 1948, p. 145.

181. Candee, A. H., "Kinematics of Disk Cam and Flat Follower," *ASME Trans.* 69, pp. 709–718; Discussion October 1947, pp. 718–724.

182. Carlson, J. A., "Principles and Practice of Constant–Load Cam Design for High–Speed Operation," *Mach. Des.* 30, July 10, 1958, pp. 121–128.

183. Carlson, R. E., "Designing Cams for Four–Slides," *Tool & Manuf. Eng.* 63, August 1969, pp. 24–31; (cont. as) *Manuf. Eng. & Mgt.* 64, May 1970, pp. 49–52.

184. Carmichael, C., "Displacement Diagrams for Cam Design," *Mach. Des.* 16, July 1944, pp. 145–148.

185. Carpenter, D. E., "Cam Milling Made Sure," *Am. Mach.* 87, May 13, 1943, pp. 110–111.

186. Carter, A., "Center Shed Cams for Worsted Weaving," *Textile World*, September 8, 1917, pp. 953.

187. Carver, W. B., and Quinn, B. E., "Analytical Method of Cam Design," *Mech. Eng.* 67, August 1945, pp. 523–526.

188. Catlow, Marjorie G., and Vincent, J. J., "The Problem of Uniform Acceleration of the Shuttle in Power Looms," *J. Text. Inst. Trans.* 42, November 1951, pp. T413–T488.

189. Cazaud, R., Renout, M., and Daubertes, C., "Contribution to Study of Deterioration in Automobile Engine Cams and Tappets," *Instn. Mech. Engrs–Proc.* (Automobile Div.). 1962–1963, pp. 93–111.

190. Chamberland, H. J., "Contour Sawing, Feats, Facts, Figures," *Tool Eng.* 9, October 1940, pp. 15, 17, and 18.

191. Chapman, G. T., "Parabolic Cams Smooth Velocity Curves," *Machine Design*, February 1936, pp. 39–40.

192. Chase, H., "Automated Flame Hardening Steps Up Production of Steel Cams," *Materials & Methods* 29, April 1949, pp. 47–49.

193. Chase, H., "Rapid Machining of Plastic Parts; International Business Machines Corp.," *Mach.* 60, March 1954, pp. 202–203.

194. Chase, H., "Tumbling Nylon Cams; Removal of Flash and Sharp Edges on Typewriter Parts," *Mod. Plastics* 27, March 1959, p. 100.

195. Chase, M. H., "Cutting Barrel Cams on the Lathe," *Mach.* 22, November 1915, pp. 230–231.

196. Chen, F. Y., "An Algorithm for Computing the Contour of a Slow Speed Cam," *J. Mechanisms* 4, 1969, pp. 171–175.

197. Chen, F. Y., "Analysis and Design of Cam–Driven Mechanisms with Nonlinearities," *Trans. ASME*, Ser. B, 95, 1973, No. 3, pp. 685–694.

198. Chen, F. Y., "A Refined Algorithm for Finite–Difference Synthesis of Cam Profiles," *Mechanism and Machine Theory* 7, 1972, pp. 453–460.

199. Chen, F. Y., "A Survey of the State of the Art of Cam System Dynamics," *Mechanism and Machine Theory* 12, 1977, No. 2, pp. 201–224.

200. Chen, F. Y., "Dynamic Response of a Cam–Actuated Mechanism with Pneumatic Coupling," *Trans. ASME*, Ser. B: *J. Engng. Ind.* 99, 1977, 3, pp. 598–603.

201. Chen, F. Y., and Polvanich, N., "Dynamics of High–Speed Cam–Driven Mechanisms, Part 1: Linear System Models," *Trans. ASME*, Ser. B: *J. Engng. Ind.* 97, 1975, No. 3, pp. 769–776.

202. Chen, F. Y., and Polvanich, N., "Dynamics of High–Speed Cam–Driven Mechanisms, Part 2: Nonlinear System Models," *Trans. ASME*, Ser. B- *J. Engng. Ind.* 97, 1975, No. 3, pp. 777–784.

203. Chen, F. Y., "Kinematic Synthesis of Cam Profiles for Prescribed Acceleration by a Finite Integration Method," *Trans. ASME*, Ser. B, 95, 1973, No. 2, pp. 519–524.

204. Chen, F. Y., "Generated by an Ellipse. Part I: Quasi–Harmonic Curves," 4th OSU Applied Mechanisms Conference, 1975, pp. 31–1/31–17.

205. Cheney, Raymond, E., "High–Speed Master Cams Generated Mechanically," *Machinery*, October 1961, pp. 93–99, 148.

206. Cheney, R. E., "Production of Very Accurate High–Speed Master Cams," *Machinery (London)* 100, 1962, No. 2570, pp. 380–386.

207. Chester, H., "Three–Direction Cam," *Am. Mach.* 78, November 7, 1934, pp. 778–779.

208. Chew, M., Freudenstein, F. and Longman, R. W., "Application of Opti-
      mal Control Theory to the Synthesis of High–Speed Cam–Follower
      Systems. Part 1: Optimality Criterion," *J. Mech. Transm. Autom. Des.*
      105, No. 3, September, 1983, pp. 576–791.

209. Chiang, C. H., "Semigraphical Solution of Acceleration Problems of
      Plane Cam–Driven and Four–Bar Linkages," ASME Paper No.

210. Chicurel, R., "Cam Size Minimization by Offsetting," ASME Paper No.
      62–WA–22 for meeting November 25–30, 1962.

211. Ching–U IP, Morse, I. E. Jr., and Hinkle, R. T., "Rolling–Contact
      Mechanisms," *Mach. Des.*, July 26, 1956, pp. 75–77.

212. Chironis, N. P., "Cam–Actuated Roller Clutch Give Rapid On–Off Con-
      trol," *Product Eng.* 37, July 4, 1966, pp. 52–53.

213. Chironis, N. P., "Team of Computers and N/C Simplifies Design of
      Cams," *Product Eng.* 39, January 29, 1968, pp. 50–53.

214. Chmielewski, J., and Kotarba, M., "The Calculation of the Geometry of a
      Double–Excentric Mechanism by Means of a Computer" (in Poln.),
      Probl. projekt, hutn. i przem. maszyn. 15, 1967, No. 4, pp. 104–06.

215. Christen, G., "Ein Beitrag zur experimentellen Untersuchung von Reib-
      kurvengetrieben," *Feingerätetechnik* 18, 1960, No. 12, pp. 559–560
      ("A Contribution to the Experimental Investigation of Friction–Type
      Cam Mechanisms").

216. Christen, G., "Zur Auswahl der Übertragungsfunktion für Reibkurven-
      getriebe bei Start–Stop–Betrieb," *Feingerätetechnik* 19, 1970, No. 7,
      pp. 295–299 ("On the Choice of Transmission Function by Start–Stop
      Mechanisms").

217. Christen, G., "Zur kinematischen Analyse von Reibkurvengetrieben,"
      Feingerätetechnik 18, 1969, No. 3, pp. 115–119 ("Kinematic Analysis
      of Friction–Type Cam Mechanisms").

218. Church, J. A., and Soni, A. H., "On Harrisberger's Adjustable Trape-
      zoidal (HAT) Motion Program for Cam Design," *IFToMM*, Vol. F, pp.
      347–356.

219. Clark, A., "Turret Slide and Miller Cut Cam," *Am. Mach.* 96, March 17,
      1952, p. 154.

220. Clark, J. T., "Making Cams from a Sample," *Am. Mach.* 56, March 23,
      1922, p. 454.

221. Clark, T. E., "Crankshafts and Camshafts; Jobbing Plant, Atlas Manu-
      facturing Company," *Am. Mach.* 83, May 17, 1939, pp. 330–331.

222. Claus, Karin and Dog, M., "Automatisierung der Berechnung, Konstruk-
      tion und Fertigung von Kurvengetrieben," *Messen–Steuern–Regeln*,
      1969, No. 8, pp. 126–127 ("Automating the Calculation, Design and
      Manufacture of Cams").

223. Clickner, P., "Finishing Tool for Follower Cams," *Am. Mach.* 66, Janu-
      ary 13, 1927, p. 66.

224. Coker, E. G., and Levi, R., "Contact Pressures and Stress Distributions in
      Cams, Rollers and Wheels," *Inst. Mech. Eng. Proc.* 3, 1930, pp. 693–
      730.

225.   Cole, C., "Profile Gaging Block," *Am. Mach.* 82, August 24, 1938, p. 774.

226.   Colvin, F. H., "Crankshaft and Camshaft Work," *Am. Mach.* 57, July 27, 1922, pp. 141—143.

227.   Colvin, F. H., "Inspection Methods and Their Application," *Am. Mach.*, November 1, 1928, pp. 691—694.

228.   Colvin, F. H., "Machining of Camshafts," *Am. Mach.*, July 29, 1922, pp. 95—97.

229.   Colvin, F. H., "Making Camshafts in Two Motor Shops," *Am. Mach.* 57, August 3, 1922, pp. 186—188.

230.   Colvin, F. H., "Making Curtiss Camshafts and Connecting–Rods," *Am. Mach.* 46, May 31, 1917, pp. 939—943.

231.   Cooke, P. and Perkins, D. R., "A Computer Controlled Cam Grinding Machine," Rees Jones: Cams and Cam Mechanisms, pp. 75—78.

232.   Cormack, P., "Radial and Rotary Engine Cams," *Automobile Eng.* 15, October 1925, p. 322.

233.   Cory, C. R., "Forming Die Operated by Inside and Outside Cams," *Mach.*, May 1940, p. 115.

234.   Cotner, J. C., "Hydraulic Circuits Utilizing Cam–Operated Valves," *Product Eng.* 11, 1940, May, pp. 228—229; June, pp. 274—275; September 412—413.

235.   Cousins, R. J., "Harmonic Cams with Flat Followers," *Automobile Eng.* 11, February 1921, p. 42.

236.   Cowie, A., "The Kinematics of Contacting Surfaces," Trans. ASME, *J. Eng. Industry*, August 1968, pp. 450—454.

237.   Crabtree, J. A., "Designing Cut–Off Cams for Automatic Screw Machines," *Mach.* 26, May 1920, pp. 806—807.

238.   Cram, W. D., "Practical Approaches to Cam Design," *Mach. Des.* 28, November 1, 1956, pp. 92—103.

239.   Cram, W. D., "Cam Design," *Trans. of 3rd Conf. on Mech.*, Purdue Univ., 1956, pp. 28—29.

240.   Crane, E. V., "Beaver–Tail Stop," *Mach.* 34, February 1928, pp. 321—424.

241.   Crossley, F. R. E., "How to Modify Positioning Cams," *Mach. Des.* 32, March 3, 1960, pp. 121—126.

242.   Crossley, F. R. E., and Oledzki, A. and Szydlowski, W., "On the Modeling of Impacts of Two Elastic Bodies Having Flat and Cylindrical Surfaces with Application to Cam Mechanism," *IFToMM*, 1979, Vol. 2, pp. 1090—1092.

243.   Cuckson, J. A., "Some Unfamiliar Cam Mechanisms," *Mech. World* 131, June 1952, pp. 252—253.

244.   Culver, C. S., "Milling Profile of Small Internal Cams," *Mach.*, September 1922, p. 56.

245.   Curtis, F. W., "Cam Milling Fixture with Automatic Stop," *Mach.* 39, August 1933, pp. 783--784.

246. Curtis, F. W., "Chromium Plating Applied Automatically," *Am. Mach.* 68, May 10, 1928, pp. 765—767.

247. Dahl, J., "Cam Die Pierces Multiple Hole," *Tool & Manuf. Eng.* 45, December 1969, p. 48.

248. Danke, P. "Das dynamische Verhalten der Rolle bei Kurvengetrieben," *Industrie—Anzeiger* 90, December 24, 1968, pp. 37—40 ("Dynamic Behavior of the Roller in Cam Mechanisms").

249. Darbyshire, H., "Cam Grinding," *Automobile Eng.* 16, June 1926, p. 219.

250. Davies, J., "Turning Irregular Forms," *Am. Mach.* 55, September 1, 1921, pp. 357—358.

251. Davies, R. W., "Design of the Tangent Cam," *Machinery (London)* 101, 1962, pp. 518—520.

252. Davies, R., "Hydrodynamic Lubrication of a Cam and a Cam Follower," *ASLE, ASME, Lub. Conf.*, 54—Lub—14, Baltimore Md., 1954.

253. Davies, R. W., "A New Cam Milling and Grinding Machine," Mechanisms News, Newsheet of the Mechanisms Section of the Institution of Mechanical Engineers, No. 10, Autumn 1975, p. 3.

254. Davies, R., "Parameters Needed for Stress Calculations on Cams," *Indus. Mathematics* 11, 1961, pp. 43—55.

255. DeAngelis, A., "Generating a Cam on a Milling Machine," *Am. Mach.* 52, May 13, 1920, p. 1047.

256. DeCoursey, H., "Precision Production of Small Parts; Friden Calculating Co.," *Tool Eng.* 34, March 1955, pp. 79—80.

257. Deener, R., "Randschichthärten von Kurvenlaufbahnen," *Z. wirtsch. Fertigung* 79, 1975, No. 2, pp. 77—81 ("Hardening of Cam Tracks").

258. De Fraine, J., "The Computer Aided Design and Computer Aided Manufacturing of Cams," Brussels: Centrum voor het Wetenschappelijk en technisch Onderzoek der Metaalverwerende Nijverheid, 1982.

259. De Fraine, J., "Integration of Computer Aided Design and Computer Aided Manufacturing for Cam Driving Mechanisms," *IFToMM*, 1979, Vol. 1, pp. 122—125.

260. DeGroat, G. H., "Adjustable Cams Time Bending," *Am. Mach.* 96, June 9, 1952, pp. 155—157.

261. DeGroat, G. H., "Cam Control Cuts Contouring Costs," *Am. Mach.* 109, September 27, 1965, pp. 71—72.

262. DeGroat, G. H., "Cams Drive Small Cores in an Injection Mold; Lawrence H. Cook, Inc.," *Am. Mach.* 96, June 23, 1952, pp. 110—112.

263. Deislinger, G. W., "Grinding Pachard Crankshafts and Camshafts," *Iron Age*, September 16, 1948, pp. 162—180.

264. DeLeeuw, A. L., "Methods of Machine Tool Design," *Am. Mach.* 57, 1922, November 2, pp. 687—692; December 21, pp. 963 and 967.

265. DeLeeuw, A. L., "Milling Cams Correctly," *Am. Mach.* 73, September 25, 1930, pp. 503—505; Discussion, M. Harris, 74, 11F. 26, 1931, pp. 363—364.

266. Demans, D. D., "Cams and Spiders for Intermittent Motion," *Product Eng.* 1, May 1930, pp. 208—209.

267. Demarest, D. D., "Cams and Spiders for Intermittent Motion," *Product Eng.* May 1930, pp. 208—209.

268. Denham, A. F., "Cam Is Used Instead of Crank Train in Radial Airplane Engine," *Automotive Ind.* 54, May 27, 1926, pp. 891—893.

269. Denisov, P. S., "Question of the Kinematic Designing of Two—Dimensional Cam Mechanisms," (in Russian), Tr. Kazansk. Aviats. in—ta, Part 87, 1965, pp. 138—149, RZM No. 8, 1966, Rev. 8 A 170.

270. Denisov, P. S., "On theDetermination of the Pressure Angle of Plate Cams," (in Russian), Trudy kazanskovo Aviacionnovo Instituta, 1969, No. 105, pp. 85—88.

271. Denkemier, H., "Zur Nockenform der Ventilsteuerung beim Viertakt—Otto—Motor," *Luftwissen* 8, 1941, pp. 157—162, and pp. 181—188 ("On Cam Forms in the Valve—Control Mechanisms of Four—Stroke Internal Combustion Engines").

272. Dhande, S. G., Bhadoria, B. S., and Chakraborty, J., "A Unified Approach to the Analytical Design of Three—Dimensional Cam Mechanisms," *Trans. ASME*, Ser. B: *J. Engng. Ind.* 97, 1975, No. 1, pp. 327—333.

273. Dhande, S. G., and Chakraborty, J. "Analysis of Profiled—Follower Mechanisms," *Mechanism and Machine Theory* 11, 1976, No. 2, pp. 131—139.

274. Dhande, S. G., and Chakraborty, J., "Curvature Analysis of Surfaces in Higher Pair Contact. Part 1: An Analytical Investigation. Part 2: Application to Spatial Cam Mechanisms," *Trans. ASME*, Ser. B: *J. Engng. Ind.* 98, 1976, No. 2, pp. 397—402 and 403—409.

275. Dhande, S. G., and Sandor, G. N., "Analytical Design of Cam—Type Angular—Motion Compensators," *Trans. ASME*, Ser. B: *J. Engng. Ind.* 99, 1977, No. 2, pp. 381—387.

276. Dhande, S. G., "Kinematic Analysis of Constant—Breadth Cam–Follower Mechanisms," *J. Mech. Transm. Autom. Des.* 106, No. 2, June 1984, pp. 214—221.

277. di Benedetto, A., and Vinciguerra, A., "Kinematic Analysis of Plate Cam Profiles Not Analytically Defined," *Trans. ASME, J. Mech. Design* 104, 1982, No. 1, pp. 34—38.

278. di Benedetto, A., and Vinciguerra, A., "A New Algorithm of Kinematic Synthesis of Plate Cam Profiles for Prescribed Follower Acceleration," *IFToMM*, 1979, Vol. 1, pp. 549—552.

279. di Benedetto, A., "Some Methods of Kinematic Synthesis of Cam Profiles for Prescribed Jerk Pattern," *IFToMM*, 1979, Vol. 4, pp. 963—968.

280. Dingerkus, O., "Messungen von Bewegungen und Kräften in Gelenk—und Kurvengetrieben," *Konstruktion* 12, 1960, pp. 21—23 ("Motion and Force Measurement in Linkages and Cam Mechanisms").

281. Disteli, M., "Über Rollkurven und Rollflächen," *Z. angew. Math. Phys.* 43, 1898, pp. 1–35 ("On Cylindrical Cams and Cylindrical Surfaces").

282. Dittrich, G., "Rechnerunterstützte Konstruktion von Kurven– und Kurbelgetrieben," *Konstruktion* 33, 1981, No. 12, pp. A37–A38.

283. Dittrich, G. H., Leyendecker, W., and Zakel, H., "Systematik, Konstruktion und Fertigung räumlicher Kurvengetriebe," Forschungsberichte des Landes Nordrhein–Westfalen, No. 2833, Opladen: Westdeutscher Verlag, 1979 ("Classification, Design and Manufacture of 3–D Cams").

284. Dittrich, G., and Leyendecker, H. W., "Anwendungsgrenzen spezieller kombinierter Bewegungsgesetze bei Kurvengetrieben," *Konstruktion* 33, No. 9, 1981, pp. 337–340.

285. Dittrich, G., and Zakel, H., "Classification and Design of Three–Dimensional Cam Mechanisms," (in German), *IFToMM*, 1979, Vol. 2, pp. 1086–1089.

286. Dittrich, G., and Schopen, M., "Systematischer Einsatz von Kurbel– und Kurvengetrieben," Zur Realisierung von Bewegungsabläufen bei der Bonbonverpackung, *Verpackungs–Rundschau* 33, 1982, No. 12, pp. 1194–1197 (I) and 34, 1983, No. 4, pp. 368–374 (II).

287. Dixie, E. A., "Cam Cutting in a Jobbing Shop," *Am. Mach.* 53, December 16, 1920, pp. 1131–1132.

288. Dixie, E. A., "Finish Turning Some Heart–Shaped Cams," *Am. Mach.* 53, October 21, 1920, pp. 779–780.

289. Dixie, E. A., "Production of Accurate Cams," *Am. Mach.* 54, March 31, 1921, pp. 553–554.

290. Doerfel, R., "Bekämpfung der Unstetigkeiten bei schnellaufendem Steuernocken," *Maschinenbau* 8, 1929, pp. 729–732 ("How to Correct Irregular Behavior in High–Speed Control Cams").

291. Boerschlag, C., "Cam Valves Control Pneumatic Car Wash," *Hydraulics & Pneumatics* 19, February 1966, pp. 72–74.

292. Dog, M., "Die zentrale Kurvenkörperfertigung (ZKF) und deren geplante Realisierungsetappen," Wiss. Z. T. H. Karl–Marx–Stadt Vol. 14, 1972, No. 1, pp. 19–23 ("Centralized Cam Manufacture and Its Implimentation").

293. Dog, M., Gentzen, G., and Jacobi, P., "KUGASY – ein Rechenprogramm für die Berechnung, Konstruktion und Fertigung von Kurvengetrieben," *Maschinenbautechnik* 23, 1974, No. 12, pp. 537–538 ("KUGASY – A Program for the Calculation, Design and Manufacture of Cams").

294. Dokucaeva, E. N., "Dynamic Characteristics of a Translating Roller Follower," (in Russian), Mechanika masin Vol. 11–12, 1970, No. 25–26, p. 32.

295. Dostmann, W., and Hesse, K., "Schnellaufende Kurvenmechanismen an Kettenwirkmaschinen," *Maschinenbautechnik* 10, 1961, pp. 131–134 ("High–Speed Cam Mechanisms").

296. Downey, R. A., "Plant Engineer's Guide to Limit Switches; Cam Design," *Plant Eng.* 26, January 1972, pp. 106–108.

297.   Droke, Robert L., "Pivoted–Follower Cam Systems," *Transactions of the Fifth Conference on Mechanisms*, Sponsored by *Machine Design*, 1958, pp. 42–49.

298.   Druce, G., "Cams—The Case for the Triple Harmonic Profile," *Machine Design and Control* 7, June 1969, pp. 36–39.

299.   Druce, G., "Cam Torques Compared," *Machine Design and Control* 8, 1970, No. 3, pp. 22–25.

300.   Druce, G., and Clifton, C. J., "Development and Application of Design Data for Cam Mechanisms," *CME Chart Mech. Eng.* 29, No. 2, February 1982, pp. 43–47.

301.   Druce, G., Halton, R. P., and Warriner, D., "The Rotary Motion of Roller Cam Followers," Rees Jones: Cams and Cam Mechanisms, pp. 25–29.

302.   Druce, G., and Stride, F., "A Survey of Devices for the Generation of Cam Profiles," Rees Jones: Cams and Cam Mechanisms, pp. 69–74.

303.   Duca, C. D., and Simionescu, "The Exact Synthesis of Single–Disk Cams with Two Oscillating Rigidly Connected Roller Followers," *Mechanism and Machine Theory* 15, 1980, No. 3, pp. 213–220.

304.   Duditza, R. L., "Periodische Selbstspannung bei nichthomokinetischen kreisgeschalteten Getrieben," *Konstruktion*, No. 11, 1970, pp. 427–433 ("Periodic Auto–Kinetic Tension in Non–Homokinetic Fully Circuited Cam Systems").

305.   Dudley, Winston, M., "A New Approach to Cam Design," *Mach. Des.* 1947, pp. 143–148 and 184.

306.   Dudley, Winston, M., "New Methods in Valve Cam Design," *Trans. SAE*, January 2, 1948, pp. 19–33, 51.

307.   Dumville, J., "Cams for Building Worsted Tubes," *Textile World* 47, July 1914, pp. 398–400.

308.   Dunk, A. C., "Roll Cams: They Stop an Go on Demand," *Product Eng.* 30, January 19, 1959, pp. 68–71.

309.   Dyson, A., u.a., "Application on the Flash Temperature Concept to Cam and Tappet Wear Problems," *The Chartered Mechanical Engineer* 8, 1961, No. 4, pp. 240–241.

310.   Dzavakhyan, R. P., "Toward a Synthesis of Flat Cam Gears with Non–Uniform Rotating Cams" (in Russian), *Mashinovedenie* 5, September/October 1967, pp. 37–47.

311.   Eckerle, R., "Ein Beitrag zur Synthese schnellaufender ebener Gleit-kurvengetriebe I," *Feinwerktechnik*, No. 3, 1969, pp. 103–111 ("Remarks on the Synthesis of High–Velocity Cam Mechanisms").

312.   Eckerle, R., "Ein Beitrag zur Synthese schnellaufender ebener Gleit-kurvengetriebe II," *Feinwerktechnik*, No. 6, 1969, pp. 257–264, ("Remarks on the Synthesis of High–Velocity Cam Mechanisms, II").

313.   Eckerle, R., "Neuartige, kraftschlüssige Kurvenschaltwerke," *Feinwerktechnik* 73, 1969, No. 10, pp. 421–23 ("Novel, Force Closure Cam Ratchet Mechanisms").

314.   Edison, T. M., "Semi–Determinate Cam Problem Requiring Unusual Mathematical Treatment," *J. Fr. Inst.* 219, 1935, pp. 331–342.

315. Edmonson, J. N., "Differential Gaging Device for a Cam," *Tool Eng.* 26, June 1951, p. 41.

316. Edwards, W. V., and Wilson, R. W., "Cam Controls Deceleration of Bomarc Shelter Roof," *Ap. Hydraulics and Pneumatics* 12, October 1959, pp. 136—138.

317. Egan, E. J., "Automatic Units Speed Turning, Gaging; Detroit Automotive Plants," *Iron Age* 173, March 11, 1954, pp. 144—145.

318. Eiss, N. S., "Vibration of Cams Having Two Degrees of Freedom," *ASME Trans.*, ser. B, 36, pp. 343—349; Discussion, November 1964, pp. 349—350.

319. Ellis, M. P., "Honing Finishes on Automobile Cam," *Am. Mach.* 95, April 30, 1951, p. 114.

320. Engel, S., Nowak, H., Schirmeister, K., and Müller, J., "Fertigungsungenauigkeiten bei Kurvenscheiben, ihre dynamische Auswirkung und messtechnische Erfassung," *Maschinenbautechnik*, No. 7, 1970, pp. 344—345 ("Manufacturing Deficiencies in Disk Cams, Including their Dynamic Effects and Practicable Measurable Accuracy").

321. Engel, S., and Muller, J., "Messeinrichtungen für Kurvengetriebe," *Maschinenbautechnik* 23, 1974, No. 12, pp. 539—542 ("Measuring Devices for Cam Mechanisms").

322. Eraslan, N. F., and Guillory, J. L., "On the Characteristics of a Quasi-Elliptic Cam Profile," *J. Mechanisms* 1, 1966, pp. 43—47.

323. Eraslan, N., "Some Properties of a Family of Curves and Their Applications" (in French). 9th Congress intern. Mecan. appl., Univ. Bruxelles, 1957, pp. 88—94.

324. Erickson, H. E., "Tool for Laying Out Cams on Clamping Levers," *Am. Mach.* 83, January 11, 1939, pp. 12—13.

325. Ericson, S., "Cam Library Slashes Cost of Screw Machine Parts," *Am. Mach.* 101, August 12, 1957, pp. 101—104.

326. Erisman, R. J., "Automotive Cam Profile Synthesis and Valve Gear Dynamics from Dimensionless Analysis,"

327. Eshelman, R. H., "Multiple Spindles Take Over Production of Camshafts," *Iron Age* 184, December 24, 1959, pp. 54—55.

328. Etchells, E. B., Thomson, B. F., Robinson, G. H., and Malone, G. K., "Interrelationship of Design, Lubrication, and Metallurgy in Cam and Tappet Performance," SAE Paper No. 472, March 1, 1955, p. 35.

329. Etchells, E. B., et al., "What We Know about Cams and Tappet: based on papers by E. P. Etchells and Others," *SAEJ* 63, September 1955, pp. 56—65.

330. Eumurian, C., "Designing Cams for Analog Computers," *Transactions of the Sixth Conference on Mechanisms*, Purdue University, 1960, pp. 55—66.

331. Eumurian, C., "Designing Cams for Analog Computers," *Mach. Des.* 32, 1960, pp. 145—155.

332. Fanella, R. J., "Dynamic Analysis of Cylindrical Cam," ASME Paper No. 59-SA-3 for meeting, June 14—18, 1958, p. 8.

333. Faurote, F. L., "Equipment Makes Possible the Ford Model A; Crank-shaft and Camshaft Operations," *Am. Mach.* 68, June 28, 1928, pp. 1034–1039.

334. Faustyn, N. W., and Eastman, J., "New Ford High–Performance Engine has Single Overhead Camshaft," *SAEJ* 73, October 1965, pp. 92–93.

335. Fawcett, G. F., and Fawcett, J. N., "Comparison of Polydyne and Non Polydyne–Cams," Mechanisms 74, Cams and Cam Mechanisms, Liver-pool: *Inst. Mech. Eng.*, September 1974.

336. Fawcett, J. N., and Fawcett, G. F., "The Effect of Cam Vibration on Follower Motion," *IFToMM*, 1971, Vol. H, pp. 147–159.

337. Fazekas, G. A., "The Offset Torsion Bar," Trans. ASME, *J. Eng. Ind.*, 1970, pp. 177–180.

338. Feldinger, W., "Problems des Schnellaufes von Nockentrieben unter besonderer Berücksichtigung der Federschwingung," *Forschung auf dem Gebiete des Ingenieurwesens* 21, 1955, pp. 159–163 and 181–188 ("High–Speed Problems Encountered in Cam Mechanisms Under Special Consideration of Spring–Actuated Mechanisms").

339. Feldinger, W., "Problems of High–Speed Cam Drives and Spring Surge," *Engrs. Digest* 17, April 1956, pp. 143–145.

340. Feldman, V. Ja., "Profiling Cams with Oscillating Roller–Followers," *Russian Engineering J.* No. 5, 1961, pp. 24–25.

341. Feldmann, J. U., "Rechnergestützter Entwurf optimaler Bewegungs-gesetze für eindimensionale Kurvengetriebe," Angewandte Informatik Vol. 19, 1977, No. 4, pp. 164–170 ("Computational Layout of Opti-mum Time–Displacement Diagrams for Disc Cams").

342. Fenno, J. E., "Cam for Varying Motion of Follower," *Mach.* 35, May 1929, pp. 658–659.

343. Fenno, J. E., "Cam of Small Diameter for Imparting a Long Stroke with a Rapid Return," *Mach.* 41, December 1934, pp. 220–221.

344. Fenno, J. E., "Combination Cam and Differential Gear Movement for Chain Conveyor for Conveying Containers Through Filling Machines," *Mach.* 42, October 1935, pp. 115–116.

345. Fenno, J. E., "Combination Cam and Parallel Motion for Guiding Spindle in Square Path, Woodworking Machine," *Mach.* 42, July 1936, p. 716.

346. Fenno, J. E., "Double–Action Cam That Moves Transfer Arm in Three Places; Spring–Winding Machine," *Mach.* 42, June 1936, pp. 646–647.

347. Fenno, J. E., "Indexing Cam for Varying Stroke of Follower," *Mach.* 39, October 1932, pp. 121–122.

348. Fenno, J. E., "Interchangeable Cams," *Mach.* 78, July 18, 1934, p. 517.

349. Fenno, J. E., "Operating Two Slides in Opposite Directions with One Single–Groove Cam," *Mach.* 39, August 1933, p. 769.

350. Fenno, J. E., "Sliding Cam with Endless Groove for Reducing Cam Size and Stroke," *Mach.* 40, January 1934, pp. 276–277.

351. Fenno, J. E., "Spiral Cam for Reciprocating Motion," *Mach.* 34, May 1928, p. 663.

352. Fenno, J. E., "Switching Arrangement for Cylindrical Cam with Intersecting Grooves," *Mach*, May 1933, pp. 581–582.

353. Fenno, J. E., "Variable Rotary Motion Imparted to Epicyclic Gearing by a Cam and Rack," *Mach*. 41, October 1934, p. 91.

354. Fenton, R. G., "Cam Design—Determination of the Minimum Base Radius for Disc Cams with Reciprocating Flat Faced Followers," *Automobile Eng.*, May 1967, pp. 184–187.

355. Fenton, R. G., "Determining Minimum Cam Size," *Mach. Des.* 38, January 20, 1966, pp. 155–158.

356. Fenton, R. G., "Disc Cams," *Automobile Eng.* 58, June 1968, pp. 254–256.

357. Fenton, R. G., "Effect of Offset Tolerance on Motion Characteristics of Reciprocating Cam–Followers," *Inst. Mech. Eng. Proc.* 181, 1966–1967, pp. 331–337.

358. Fenton, R. G., "Optimum Design of Disc Cams," *IFToMM*, 1975, Vol. 4, pp. 781–782.

359. Fenton, R. G., "Reducing Noise in Cams," *Mach. Des.* April 1966, pp. 187–190.

360. Finkelnburg, H. H., "Kurventriebe für Mehrspindel–Automaten," *Maschinenbau* 6 1938, p. 650 ("Cam Mechanisms in Multiple–Spindle Automatic Machinery").

361. Finkelnburg, H. H., "Ein Beitrag zur Gestaltung von Bewegungsabläufen," *Die Werkzeugmaschine* 44, 1940, pp. 145–150 ("A Contribution Toward the Understanding of Motion").

362. Finkelnburg, H. H., "Zur Systematik der Bewegungsgesetze," *Maschinenbau* 4, 1966, pp. 695–697, 5, 1937, pp. 221–222, 425–427 ("On the Systematology of Kinematic Laws").

363. Finkelnburg, H. H., "Der Ruck," *Maschinenbau* 3, 1935, pp. 520–522 ("Sudden Change in Acceleration").

364. Finney, B., "Machining the Buick Camshaft," *Iron Age* 129, March 31, 1932, pp. 768–769.

365. Flanders, R. E., "Machining Tractor Camshafts," *Mach.*, May 1920, pp. 825–827.

366. Flocke, Karl Alexander, "Zur Konstruktion von Kurvenscheiben bei Verarbeitungsmaschinen," *VDI–Forschungsheft* 345, 1931 ("On the Design of Disk Cams in Processing Machines").

367. Fogiel, M., "Accurate Control of Mechanical Circuit Switching," *Mach. Des.* 30, January 23, 1958, pp. 145–146.

368. Folsom, C. D., "Cutting a Cam without a Milling Machine," *Am. Mach.* 53, July 1, 1920, p. 33.

369. Foote, A. B., "Improved Form of Cam for Stamp Mills," *Am. Inst. Min. E. Bull.*, 96, December 1914, pp. 2765–2766; Same, *Eng. & Min. J.* 98, December 12, 1914, p. 1046.

370. Forbes, N., "Shop–Made Cam–Milling Attachment," *Am. Mach.* 99, January 17, 1955, p. 130.

371. Forster, K. H., and Stange, H., "Berücksichtigung veränderlicher Antriebswinkelgeschwindigkeit bei der Konstruktion von Kurvenmechanismen," *Maschinenbautechnik*, No. 3, 1968, pp. 163–167 ("Considerations on Variable Rotational Velocities in the Design of Cam Mechanisms").

372. Forster, W. M., "Crank–Driven Plate Obtains Near–Uniform Velocities Through Compensating Cam," *Mach.* 63, January 1957, p. 187.

373. Forster, W. M., "Timing of Cam Changed While Machine Is in Motion," *Mach.* 63, May 1957, pp. 187–188.

374. Fowler, F. H., "Proper Preload for Cam–Follower Springs," *Product Eng.* 32, January 9, 1961, pp. 76–77.

375. Framurz, S., "Cam–Cutting Attachment Fits Any Milling Machine," *Am. Mach.* 91, September 25, 1947, p. 111.

376. Frank, C. S., "Cam Attachment Improves Core Pin Grinder," *Am. Mach./Metalworking Manuf.* 106, March 19, 1962, p. 143.

377. Freudenstein, Ferdinand, "On the Dynamics of High–Speed Cam Profiles," *Int. J. Mech. Sci.* Pergamon Press Ltd., 1960, pp. 342–349.

378. Freudenstein, F., Vitaglian, V., Woo, L. S., and Hao, C., "Dynamic Response of Mechanical Systems," IBM—Data Processing Division, New York Scientific Center, March 1969, Technical Report No. 320–2967

379. Freudenstein, F., Mayourian, M., and Maki, E. R., "Energy Efficient Cam–Follower Systems," *J. Mech. Transm. Autom. Des.* 105, No. 4, December 1983, pp. 681–685.

380. Frey, R. "Untersuchung von Steuernocken auf Geschwindigkeit und Beschleunigung des Stössels," *ATZ* 43, 1940, pp. 380–386 ("Analysis of the Influence of Control Cams on the Velocity of Acceleration in Guide Mechanisms").

381. Frey, G. F., "Cam and Crank Layout," *Product Eng.*, April 1930, p. 170.

382. Fromelt, H. A., "Methods of Cam Milling," *Am. Mach.* 86, (1) "Use of Rotary Table," April 16, 1942, pp. 321–323; (2) "Use of the Dividing Head," April 30, 1942, pp. 386–388; (3) "Use of Dummy Templates," May 16, 1942, pp. 440–442; (4) "Use of Rotary Head Milling Machine," May 28, 1942, pp. 502–504; (5) "Use of the Cam Slide," June 25, 1942, pp. 652–654; (6) "Use of Special Cam Slides," July 9, 1942, pp. 709–711.

383. Fry, R. H., "Cam Checker Tests Different Shapes," *Am. Mach.* 95, July 23, 1952, p. 154.

384. Fry, N., "Designing Computing Mechanisms, Part III, Cam Mechanisms," *Mach. Des.* 17, October 1945, p. 123.

385. Fuchs, A. M., and Zeuner, K., "Differential Cam Follower Controls Pneumatic–Hydraulic Actuator," *Control Eng.* 8, January 1961, p. 125.

386. Fuhrman, E., "Cams and Followers for High–Speed Internal Combustion Engines," *Eng. Digest* 12, October 1951, p. 340, November 1961, p. 368.

387. Furman, F. De R., "Cam Design and Construction," *Am. Mach.* 50, 51, 52, March 27, April 10, 24, May 15, June 12, September 18, October 9, 1919; January 1, March 4, April 8, May 6, 27, 1920; pp. 581–586, 685–689, 779–784, 927–931, 1123–1126, 569–574, 695–698; 21–27, pp. 493–503, 777–783, 987–994, 1129–1135.

388. Furman, F. R., *Cams Elementary and Advanced*, John Wiley and Sons, New York, 1921.

389. Gabriel, W. A., "New Process of Cutting Cams," *ASME Trans.*, 1983, p. 82.

390. Gagne, A. F., "Designing of High–Speed Cams," *Mach. Des.* 22, July 1950, pp. 108–111.

391. Galabov, V. B., "Synthesis of Contour Mechanisms with a Flexible Element," *IFToMM*, 1977, Vol. 1, 1, pp. 227–237.

392. Garret, R. E., "Tappet and Cam Wear," *Automobile Engr.*, September 1960.

393. Garret, R. E., "Force Cams," *Mach. Des.* 34, August 16, 1962, pp. 174–176.

394. Garwood, M. H., Kinker, D. R., and Mangenello, J. J., "Considerations Affecting Life of Automotive Camshafts and Tappets," SAE Paper No. 493, March 1, 1955, p. 13.

395. Gatzen, H. H., "Bestimmung der Nutz– und Störbeschleunigungen bei Kurvengetrieben," Dissertation TH Aachen, 1976.

396. Gatzen, H., and Heinzl, J., "Dynamik schnellaufender Kurvengetriebe," *IFToMM*, 1977, Vol. III, 1, pp. 268–283.

397. Gavrilas, I., Marinescu, R., and Marinescu, N., "Establishment of the Parametric Equations for Disk Cams and Their Automatic Drawing Out," *IFToMM*, 1977, Vol. III, 1, pp. 285–294.

398. Gavrilas, I., Marinescu, N., and Marinescu, R., "General Utility Program for Calculating and Designing Disk Cams," *IFToMM*, 1977, Vol. III, 1, pp. 295–306.

399. Gendzekhadze, T. N., "Some Problems in the Kinetic Designing of Three–Dimensional Cam Mechanisms" (in Russian), *Trudi Most. Aviats. In–ta*, No. 72, 1957, pp. 4–27, *Ref. Zh. Mekh.*, No. 10, 1957, Rev. 11299.

400. Gendzekhadze, T. N., "Synthesis of Spatial Cam Mechanisms with One Degree of Freedom" (in Russian), *Trudi Inst. Masbinoved.*, Akad, Nauk, SSSR 23, 89–90, 1962, pp. 111–124.

401. Gentzen, G., Rose, W., and Volmer, J., "Getriebemodelle," *Maschinenbautechnik*, No. 1, 1971, pp. 17–21 ("Cam Models").

402. Gerard, P. L., "Cams Control Mobile Platform Travel Speeds, SIMCA Auto Plant," *Hydraulics and Pneumatics* 14, August 1961, pp. 72–73.

403, Gerber, S. R., "Cam Layout for Brown and Sharpe Automatics," *Am. Mach.* 46, January 18, 1917, 107–108.

404. Gerbl, L., "Graphische Bestimmung der Geshwindigkeiten und Beschleunigungen von Ventilsteuerungen," *MTZ* 3, 1941, pp. 345–351 ("Graphic Determination of the Velocity and Acceleration of Valve-Controlled Mechanisms").

405. Gerhardt, H. W., "Camshaft Salvage Stunt," *Bus Transportation* 22, May 1943, p. 60.

406. Gernet, M. M., "Requirements Concerning Displacement Diagrams" (in Russian), *IFToMM*, 1969, pp. 135–144.

407. Geronimus, Ja. L., "Concerning the Design of Some Came Mechanisms with Piece–Wise Circular Cams," *Maschinenbau* 5, 1948, pp. 69–79.

408. Geronimus, Ja. L., "On some Problems in the Synthesis of Cam Mechanisms," *Maschinenbau* 5, 1948, 62–81.

409. Geronimus, Ja, L., "Die Ermittlung der Schablonenform nach aufgegebenem Bewegungsgesetz des Stössels," *WIT*, 1932, No. 6 ("On the Design of Master Cams Based on Applicable Kinetic Requisites and Criteria of Control Mechanisms").

410. Geschelin, J., "Essex Slashes Camshaft Costs with Electric Furnace Alloy; Proferall," *Automotive Ind.*, November 12, 1932, pp. 620–622.

411. Ghosh, A., and Yadav, R. P., "Synthesis of Cam–Follower Systems with Rolling Contact," *Mechanism and Machine Theory* 18, 1983, No. 1, pp. 49–56.

412. Gibbons, E. J., "Shedding Motions Used in Manufacture of Narrow Fabrics," *Textile World* 80, August 22, 1931, pp. 680–690.

413. Gillett, G. L., "Graphical Analysis of Circular Arc Cams," *Am. Mach.* 65, October 12, 1926, p. 671, October 28, 1926, p. 715; 66, June 16, 1927, p. 1012.

414. Giodana, F., Rognoni, V., and Ruggieri, G., "On the Influence of Measurement Errors in the Kinematic Analysis of Cams," *Mechanism and Machine Theory* 14, 1979, No. 5, pp. 327–340.

415. Giordana, F., Rognoni, V., and Ruggieri, G., "The Influence of Construction Errors in the Law of Motion of Cam Mechanisms," *Mechanism and Machine Theory* 15, 1980, No. 1, pp. 29–45.

416. Gitter, H., "Beanspruchungs–und Reibungsverhältnisse am Nockentrieb," *Maschinenbautechnik*, No. 12, 1970, pp. 643–649 ("Utilization and Friction Coefficients in Cam Mechanisms").

417. Gobel, E. F., "Bewegungsvorgänge und Massenkraft in den Triebwerken von Nähmaschinen," *Feinwerktechnik*, 1958, pp. 117–124 ("Motion Programs and Intertia Forces in the Driving Mechanisms of Sewing Machines").

418. Goldschmidt, H., "Finding Angles Between Slot and Cam," *Mach.* 35, February 1929, pp. 444–445.

419. Golorff, A., "General Considerations in Cam Design; Abstract," *SAEJ* 61, February 1953, pp. 112–113.

420. Golosman, C., and Golosman, D., "Mechanism for Controlling Cutter-Head Slide of Cam–Generating Device," *Mach.* 54, October 1947, pp. 184–86.

421. Golosman, C., "Rotating Disk–and–Pin Mechanism for Controlling Cutter–Head Slide of Automatic Cam–Generating Device," *Mach.* 52, August 1946, pp. 189–193.

422. Goodman, T. P., "Linkages vs. Cams," *Machine Design*, August 21, 1958, 102–109.

423. Goodman, T. P., "Linkages vs. Cams," *Transactions of the Fourth Conference on Mechanisms*, Sponsored by *Machine Design*, 1957, pp. 76–83.

424. Gorfinkel, A., "Calcul et trace' des cames pour les distributions de moteurs," *Genie Civil* 94, March 9, 1929, pp. 239–243.

425. Goring, E., "Die derzeitigen Herstellungsmethoden von unregelmässig gekrümmten Oberflächen, dargestellt an der Fertigung von Kurven-scheiben," *Wiss. Zeitschrift der T.H. Dresden* 9, 1959–1960, pp. 125–131 ("Current Production Methods for Irregularly Curved Surfaces, as Reflected in the Design of Disc Cams").

426. Göring, E., "Verfahren zur Herstellung von unreglemässig gekrümmten Oberflächen," *Werkstattstechnik* 52, 1962, No. 8, pp. 391–395.

427. Göring, E., "Systematische Darstellung der Bewegungsgesetze für Kurven-getriebe," *Maschinenbautechnik* 9, 1960, pp. 313–321 ("Systematic Representations of Laws of Motion for Cam Systems").

428. Gorjacko, V. I., "On the Choice of Proportions for Cam with Swinging Follower," (in Russian), Trudy Leningradskovo Politechn. Instituta, 1969, No. 309, pp. 9–15.

429. Gorr, W. B., "Twin Cams Provide Versatility; Poppet–Valve Gear Control System in Marine Engine," *Mach. Des.* 16, May 1944, pp. 101–102.

430. Götz, H., "Die Emmittlung der Geschwindigkeiten und Beschleunigungen bei Kurvenscheibengetrieben," *Werkstattstechnik* 3, 1909, p. 363 ("Computing Velocities and Accelerations in Disk Cam Mechanisms").

431. Granholm, H. W., "General Equations for Rocker Cam Design," *Design News*, November 15, 1956, pp. 137–139.

432. Granholm, H. W., "General Equations for Roller Cam Design," *Design News*, September 1, 1955, 58.

433. Grant, B., and Soni, A. H., "A Survey of Cam Manufacture Methods," *J. Mechanical Design* 101, July 1979, pp. 455–464.

434. Granville, A. E., "Fixture for Milling and End Cam," *Mach.* 39, September 1932, pp. 38–39.

435. Green, C. J., "Automatic Cycle Control of Cam Grinder," *Elec. Mfg.* 48, October 1951, pp. 118–123.

436. Green, C. J., "Metrology for Ground Cams and Shapes," *Mach.* 69, February 1963, pp. 83–93.

437. Greenleaf, W. B., "Forming and Assemblying Dies for Roll Cam," *Mach.* 28, April 1922, pp. 618–619.

438. Greim, I. A., and Arucov, I. A., "Calculation of Accuracy of Cam Mechanisms," *Feingerätetechnik* 24, 1975, No. 7, pp. 303-306.

439. Gres, W. H., "Gleichabständige Kurven in der Nachformtechnik," *Werk-stattstechnik und Maschinenbau* 39, 1949, pp. 170–172 ("Constant Distance Curves in Copying Machine Technology").

440. Griffen, R. F., "Designing Cams with Aid of Computers," *Transactions of the First Conference on Mechanisms*, December 1953, No. 12, pp. 209–215.

441. Grobe, R. E., "Analysis and Comparison of Cam Actions," *Product Engineering*, September 1936, pp. 339–340.

442. Grodzinski, P., "Cam Drive with Variable Stroke Mechanism," *Mach.* 48, December 1941, p. 144.

443. Grodzinski, P., "Cam and Eccentric Mechanisms for Crank and Lever Movements," April, pp. 503–505.

444. Grodzinski, P., "Wirtschafliche Herstellung von Kurvenscheiben," *Maschinenbau* 2, 1034, No. 4, pp. 30–33 ("Economical Production of Cam–Disks").

445. Grodzinski, P., "Improvement of Cam Mechanisms," *Mech. World* 127, February 17, 1950, p. 179.

446. Grodzinski, P., "New Clamping Device for Jigs and Fixture," *Tool Eng.* 27, August 1951, pp. 40–42.

447. Grodzinski, P., "Production of Cam Profile by Positive Mechanism," *Machinery* (*London*), May 11, 1956, pp. 683–688.

448. Grodzinski, P., "Schaltgetriebe mit Kurvensteuerung," *Maschinenbau* 6, 1927, pp. 655–657 ("Switch–Gear Mechanisms with Cam Control").

449. Groh, R. E., "Getriebefragen bei automatischen Blechverpackung-maschinen," *VDI–Berichte* 29, 1958, pp. 27–31 ("Cam Problems in Automatic Sheet Metal <Foil> Packing Machines").

450. Gronegress, H. W., "Flame Hardening of Camshafts," *Automobile Eng.* 51, August 1961, pp. 320–324.

451. Grover, F., "Wrapping Machinery," *Instn. Mech. Engrs. – Proc.* 132, 1936, pp. 345–368, and 363–-393.

452. Gruenberg, Rudolph, "Nomogram for Parabolic Cam with Rapidly Moving Follower," *Product Eng.* 26, January 1955, p. 209.

453. Gruenberg, Rudolph, "Radius of Curvature of Parabolic Cams," *Product Eng.* 27, August 1956, pp. 219–221.

454. Gruger, C. and Muller, H., "Eing Beitrag zur Untersuchung hydrody-namisch geschmierter Kurvengetriebe," Dissertation, TH Karl–Marx–Stadt, 1978.

455. Guergoze, M., "Ein Beitrag zum dynamischen Verhalten der Kurven-getriebe mit Schwinghebel," *Ingenieur–Archiv.* 51, 1982, No. 5, pp. 311–323.

456. Guertler, G., and Roth, K., "Berechnungs– und Herstellungsverfahren für Präzisions–Kurvenscheiben," *Feinwerktechnik*, No. 2, 1964, pp. 58–62 ("Calculation and Design for Precision Disk Cams").

457. Guillet, G. L., "Graphical Analysis of Circular–Arc Cams," *Am. Mach.* 65, October 21–26, 1926, pp. 671–674 and 715–717; Discussion, 66, February 24, January 16, 1927, pp. 332 and 1012–1013.

458. Gul'binas, A. S., "Angular Velocity and Angular Accelerations of the Shaft of the Guided Member of a Plane Cam–Shaft Mechanism" (in Russian), *Trudi Akad. Nauk LitSSR B* 1, 1965, pp. 185–192; Ref. Zh. Mekh. No. 8, 1965, Rev. 8 A 152.

459. Gundermann, W., "Berechnung eines Ventil–Steuernockens für einen Verbrennungsmotor," *Konstruktion* 2, February 1969, pp. 41–51 ("De-sign of Valve Cams for Internal Combustion Engine").

460. Gunderson, A. D., "New High Accuarcy Cam Contour Mill Design and Applications," *ASME* Advance Paper N 50–Sa–7 for meeting Juen 19–23, 1950, 2 supp. sheets; *Tool & Die J.* 16, August 1950, pp. 41, 44–-45, and 60–61, Machine Developed by George Gorton Co.

461. Gupta, K. C., and Wiederrich, J. L., "Development of Cam Profiles Using the Convolution Operator," *J. Mech. Transm. Autom. Des.* 105, No. 4, December 1983, pp. 654–657.

462. Gutman, A. S., "Cam Dynamics," *Mach. Des.* 23, March 1961, pp. 149–151.

463. Gutman, A. S., "Mechanical Linearizer, " *Mach. Des.* 25, May 1953, p. 175.

464. Gutman, A. S., "Synthesis of New Cam Profiles from Frequency Components," ASME Paper No. 62–WA–64, 1962.

465. Gutman, A. S., "To Avoid Vibration Try This New Cam Profile," *Product Eng.* 32, December 1961, pp. 42–48.

466. Haeussler, A. H. K., "Versatile Cam–Action Bending Die," *Mach.* 68, March 1962, pp. 128–129.

467. Hafferkamp, Harry C., "Cam Dynamics and the Digital Computer," *Transactions of the Seventh Conference on Mechanisms*, Sponsored by *Machine Design*, 1962, pp. 174–178.

468. Hagedorn, L., "Zwangläufiges Fräsen von Kurvenscheiben für form-schlüssige Kurvengetriebe," *Industrie–Anzeiger* 91, March 25, 1969, pp. 19–22 ("Positive Milling of Disk Cams for Cam Mechanisms").

469. Hain, K., "Die Kurvengetriebe," *Neue Produktion* 3, 1948, No. 7, pp. 21–22 ("Cam Mechanism").

470. Hain, K., "Wälzhebelgetriebe. Grundlagen d. Landtechnik, No. 9. 14," *Konstrukteurheft*, 1957, pp. 119–124 ("Rolling Contact Mechanisms").

471. Hain, K., "Classification of Multiple–Membered Cam Mechanisms and their Application," (in German), *Maschinenbautechnik* 9, December 1960, pp. 641–649.

472. Hain, K., "Die praktische Bedeutung der mehrgliedrigen Wälzkurven-getriebe," *Industrieblatt* 60, 1960, No. 12, pp. 749–753 ("The Practical Significance of Multi–Linked Contact Cam Mechanism").

473. Hain, K., "Systematik mehrgliedriger Kurvengetriebe und ihre Anwen-dungsmöglichkeiten," *Maschinenbautechnik* 9, 1960, pp. 641–649 ("Systematic Analysis of Multiple Cam Systems and Their Potentiali-ties").

474. Hain, K., "The Production of Large Oscillating Angles by Cam–Linkage Mechansims," *Proc. Int. Conf. for Teachers of Mechanisms*, 1961, pp. 65–90.

475. Hain, K., "Rocking Mechanisms," *Product Engineering*, September 18, 1961.

476. Hain, K., "Wälzkurvengetriebe mit zwei Wälzkurvenpaaren für höhere Koppelbewegungen," *Feinwerktechnik*, No. 7, 1965, pp. 302–305 ("Cylinder–Type Cams with Dual Cylinder Pairs for Greater Coupling Efficiency").

477. Hain, K., "Selbsttätige Getriebegruppen zur Automatisierung von Arbeitsvorgängen," *Feinwerktechnik*, September 1967, pp. 327–329 ("Automatic Cam Systems in the Automatization of Work Processes").

478. Hain, K., "Systematik und Umlauffähigkeit drei und mehrgliedriger
      Kurvengetriebe," *Konstruktion* No. 10, 1967, pp. 379–388 ("Analysis
      and Operation of Triple and Multiple Jointed Cam Systems").

479. Hain, K., "Dreigliedrige Kurvengetriebe," *Maschinenmarkt* 75, 1969, No.
      72, pp. 1601–1604.

480. Hain, K., "Mehrgliedrige Kurvengetriebe für vorgeschriebene Geschwin-
      digkeit auf gegebener Bahnkurve," *Konstruktion* 21, 1969, No. 11, pp.
      436–440.

481. Hain, K., "Aufgabenstellung für den Entwurf von Kurbel–und Kurven-
      getrieben," *VDI–Berichte*, NO. 40, Düsseldorf: VDI–Verlag, 1970, pp.
      5–13.

482. Hain, K., "Challenge: To Design Better Cams," *J. Mechanisms* 3, 1970,
      pp. 283–286.

483. Hain, K., "Dreigliedrige Funktions–Wälzkurvengetriebe für vorgeschrie-
      bene Übertragungsgüte in vorgegebenen Bewegungsgrenzen," *Konstruk-
      tion*, No. 11, 1970, pp. 434–440 ("Rolling–Contact Cam Systems for
      Pre–Determined Transmission Values Within Given Limits").

484. Hain, K., "Dreigliedrige Wälzkurvengetriebe zur Führung eines Punktes
      auf gegebener Bahnkurve," *Feinwerktechnik*, No. 9, 1970, pp. 391–396
      ("Rolling–Contact Cam Mechanisms for Tracing a Given Curve").

485. Hain, K., "Entwurf dreigliedriger Kurven–Rastgetriebe für gegebene
      Übergansfunktionen und deren Ableitungen," *Technica* 19, 1970, No. 5,
      pp. 327–334.

486. Hain, K., "Entwurf mehrgliedriger Kurvengetriebe zur Führung einer
      Koppelebene auf vorgeschriebener Bahn," *Werkstatt und Betrieb* 103,
      1970, No. 2, pp. 104–106.

487. Hain, K., "Kurvengetriebe mit Schubrolle," *Maschinenmarkt* 76, 1970,
      No. 46, pp. 1003–1005.

488. Hain, K., "Beschleunigungsgünstige Hubbewegungen mit zeitweise
      konstanter Geschwindigkeit," *Konstruktion* 23, 1971, No. 11, pp. 413–
      419.

489. Hain, K., "Entwurf übertragungsgünstigster Gegenkurven–Rastgetriebe
      mit Doppel–Schwinghebel," *Technica* 20, 1971, No. 8, pp. 691–695.

490. Hain, K., "Die Erzeugung von Bahnkurven durch dreigliedrige Kurven-
      getriebe," *Feinwerktechnik* 75, No. 7, 1971, pp. 289–294 ("The
      Plotting of Curves by Means of Cam Mechanisms").

491. Hain, K., "Möglichkeiten und Grenzen dreigliedriger Kurvengetriebe,"
      TZ Prakt. Metallbearb. 65, 1971, No. 10, pp. 480–487.

492. Hain, K., "Optimization of a Cam Mechanism to Give Good Trans-
      missibility, Maximal Output Angle of Swing and Minimal Acceleration,"
      J. Mechanisms 6, 1971, No. 4, pp. 419–434.

493. Hain, K., "Systematik und Bewertung viergliedrige Kurvengetriebe mit
      doppelt umrollter Kurvenscheibe," *Feinwerktechnik* 76, 1972, No. 8,
      pp. 387–390.

494. Hain, K., "Beschleunigungsgünstige, dreigliedrige Kurven–Schrittgetriebe,"
      TZ Prakt. Metallbearb. 67, 1973, No. 10, pp. 427–434.

495. Hain, K., "Dreigliedrige formschlüssige Kurven–Rastgetriebe mit Gleich-dick–Kurvenscheiben," *Technica* 22, 1973, No. 4, pp. 223–238, 235–240, 247–252 and 259–260.

496. Hain, K., "Kurvenscheiben mit Doppeleingriff für gegebene Bahnkurven und Geschwindigkeiten," *Feinwerktechnik* 77, 1973, No. 1, pp. 21–25.

497. Hain, K., "Kurvengetriebe mit umlaufender Doppelkurve zur Erzeugung von Bahnkurven mit vorgeschriebener Geschwindigkeit," *Mechanism and Machine Theory* 9, 1974, No. 1, pp. 7–26.

498. Hain, K., "Raumsparende übertragungsgünstige Kurvengetribe mit mehr-teiliger Kurvenscheibe," *Antriebstechnik* 13, 1974, No. 11, pp. 621–626.

499. Hain, K., "Bestimmung der Kurvenkrümmungen in Kurvengetrieben mit Schubstössel," *Maschinenmarkt* 81, 1975, No. 1.

500. Hain, K., "Geometrische Grundlagen für Rechenprogramme von Schwing-hebel–Kurvengetrieben," *Technica* 24, 1975, No. 19, pp. 1457–1470.

501. Hain, K., "Gleichdick–Kurvenschleifen–Getriebe. Erzeugung von Schwingbewegungen mit Rasten und Totlagen," *Antriebstechnik* 14, 1975, No. 11, pp. 630–636.

502. Hain, K., "Kurvengetriebe optimieren nach übertragungsgüte, Raum-bedarf und Kontaktkraft," *Maschinenmarkt* 83, 1977, No. 99, pp. 2079–2082.

503. Hain, K., "Entwurf übertragungsgünstigster Zylinderkurven–und Kegel-kurven–Getriebe," *Werkstatt und Betrieb* 111, 1978, No. 2, pp. 93–98.

504. Hain, K., "Erzeugung übertragungsgünstiger Koppelebenen–Bewegungen mit Hilfe dreigliedriger Kurvengetriebe," *Technica* 28, 1978, No. 26, pp. 2257–2270.

505. Hain, K., "Kurvengetriebe – Kostensenkung durch Mehrzweck–Kurven-scheiben. *Digest für angewandte Antriebstechnik*," Vol. 6, 1978, No. 3, pp. 33–37.

506. Hain, K., "Optimierungsfelder für Kurvengetriebe mit besonderer Berücksichtigung der Übertragungsgüte und der Scheitelkrümmung," *Feinwerktechnik & Messtechnik* 87, 1979, No. 6, pp. 283–287.

507. Hain, K., "Systematik und Bewertung der fünfgliedrigen, formschlüssigen Gegenkurvengetriebe," *Maschinenbautechnik* 28, 1979, No. 9, pp. 423–426.

508. Hain, K., "Getriebekennwerte als Mittel zum Vergleich und zur Bewertung von Getrieben," *Konstruktion* 32, 1980, No. 1, p A14.

509. Hain, K., "Drei– und viergliedrige Keilschubgetriebe mit günstigster Kräfte–und Bewegungsübertragung," *Werkstatt und Betrieb* 116, 1983, No. 2, pp. 75–80.

510. Haiquing, Yu, and Xianze, Zhan, "Evaluation of Cam Lubrication Characteristics in I.C. Engines," Neiranji Xuebao 1, No. 4, October 1983, pp. 83–-93.

511. Hale, F. W., "Cam Machining Without a Master Former; Pratt & Whitney Co.," *Tool Eng.* 35, December 1955, pp. 82–87.

512. Hale, F. W., "Milling Original Contoured Shapes with Interpolating Tracer Control," *Tool Eng.* 32, June 1954, p. 46.

513. Hall, A. E., "Cams for Small Automatic Machines," *Am. Mach.* 47, July 1926, p. 158.

514. Hall, C. L., "Grinding Heart–Shaped Cams," *Am. Mach.* 63, September 3, 1925, p. 399.

515. Hall, J. A., "Intertia Forces in Cam Mechanisms," *Product Eng.*, October 1932, pp. 393–396.

516. Hamilton, D. T., "Gear Shaper Design for Progressive Cam Cutting Operations," *Mach.* 45, September 1938, pp. 12–14.

517. Hamilton, D. T., "Generating Cams and Irregular–Shaped Work," *Am. Mach.* 49, October 24, 1918, pp. 737–740; Abstract, *ASME J.* 40, December 1918, p. 1051.

518. Hamilton, D. T., "Inspecting Cams on Gas Engine Cam–Shaft," *Mach.* 28, December 1916, pp. 307–308.

519. Hammond, E. K., "Manufacture of Accurate Camshafts," *Mach.* 28, November 1921, pp. 175–180.

520. Hannavy, A., "Spherical Cams Take Kinks out of Linking up Shafts," *Product Eng.* 37, December 5, 1966, pp. 60–62.

521. Hannula, F. W., "Designing Non–Circular Surfaces," *Mach. Des.* 23, July 1951, pp. 111–114, and 190–192.

522. Hanson, J. N., "Planar Motion of Sliding Cams by Computer Algebraic Manipulation," *Mechanism and Machine Theory* 14, 1979, No. 2, pp. 111–120.

523. Hanson, J. N., and Churchill, F. T., "Theory of Envelopes Provides New Cam Design Equations," *Product Eng.* 33, August 20, 1062, pp. 45–55.

524. Hantsch, H., "Untersuchungsergebnisse am Nockengetriebe schnellaufen-der Dieselmotoren für vershiedene Übertragungsfunktionen," *Maschinen-bautechnik* 28, 1979, No. 9, pp. 408–413.

525. Hardesty, E. C., and Stalhuth, W. E., "Adjustable–Shape Cam in Linkage Produces Several Paths of Motion," *Mach. Des.* 31, August 20, 1959, pp. 162–163.

526. Harrington, R. A., "Simplified Calculations Unshackle Designer," *Mach. Des.* 5, 1933, October, pp. 13–15; November, pp. 20–22; December, pp. 18–21.

527. Harrington, R. A., "Generation of Functions by Windup Mechanisms," *R. Sci. Instr.* 22, September 1951, pp. 701–702.

528. Harris, L. H., "Designing Harmonic Cams; Reference Sheet," *Tool & Manuf. Eng.* 45, December 1960, pp. 81–84.

529. Harrisberger, L., "Motion Programming," *Mach. Des.* 35, 1963, pp. 114–119.

530. Harrison, J. B., "Control Valve for Hydraulic Elevators Controls Speed with Helical Cam," *Mach. Des.* 16, April 1944, pp. 107–108.

531. Hart, F. D., "Coupled Effects of Preload and Damping on Dynamic Cam–Follower Separation," ASME Paper No. 64–Mech.–18.

532. Hart, F. D., Patel, B. M., and Bailey, J. R., "Mechanical Separation Phenomena in Picking Mechanisms of Fly–Shuttle Looms," *Trans. ASME,* Ser. *B. J. Engng. Ind.* 98, 1976, No. 3, pp. 835–839.

533. Hart, W. J., "Analysis of Stresses in Cam Rolls," *Product Eng.* January 1935, pp. 23–25.

534. Hartbauer, E. A., "Cam Design Chart for Constant Acceleration," *Mach. Des.* 19, August 1947, pp. 161–162.

535. Hartley, F., "Fixture for Milling Cam Surface on End of Bar," *Mach.* 43, November 1936, pp. 195–196.

536. Hartley, F., "Multiple Cam and Lever Mechanism," *Mach.* 54, November 1948, p. 165.

537. Hartmann, W., "Die Bewegungsverhältnisse von Steuergetrieben mit unrunden Scheiben," *ZVDI* 49, 1905, p. 1581 ("Motions of Control Mechanisms with Non–Circular Disk").

538. Harvey, J. J., "Computer–Assisted Cam Manufacture," *Tool & Manuf. Eng.* 55, August 1965, pp. 76–77.

539. Havely, T. W., Phalen, C. A., and Bunnell, D. G., "Influence of Lubricant and Material Variables on Cam Tappet Surface Distress," *SAE* Pas. Car Body and Mtls. Meeting, Detroit, Mich., March 2, 1954.

540. Heath, J. M., "Aluminum Disk and Felt Cover Polish Precision Cam," *Am. Mach.* 99, January 3, 1955, p. 125.

541. Hebeler, C. B., "Cycloidal–Motion Cam Systems," *Machine Design* 33, February 2, 1961, pp. 102–107.

542. Hecht, F., and Wider, E., "Gerät zum Herstellen von ruckfreien Kurvenscheiben," *VDI–Z* 101, 1959, pp. 232–235 ("Equipment for Production of Jerk–Free Cams").

543. Heeg, L., "Betrachtungen über die Ermittlung der Beschleunigungsverhältnisse ebener Kurvengetrieben," *Maschinenbautechnik* 10, 1961, pp. 216–219 ("Considerations on the Determination of Acceleration Characteristics of Plane Cam Mechanisms").

544. Heeg, L., "Betrachtungen über die Ermittlung der Beschleunigungsverhältnisse ebener Kurvengetriebe," *Maschinenbautechnik* 10, pp. 232–235 ("Equipment for Production of Jerk–Free Cams").

545. Heinrich, W., and Ristow, J., "Messeinrichtung zur Bestimmung von Geschwindigkeit und Bescheunigung an der Abtriebsschwinge eines Kurvengetriebes," *Maschinenbautechnik* 10, 1961, pp. 300–302 "Device Determine Speed and Acceleration in Output Followers of Cam Mechanisms").

546. Heinz, O., and Schlagner, B., "Problematik der Darstellung räumlicher Getriebesysteme," *Maschinenbautechnik*, No. 8, pp. 429–431 ("Problems concerning Spatial Cam Systems").

547. Heldt, P. M., "Cam Gear Design for Restricted Space," *Automotive Ind.* 82, May 15, 1940, p. 476.

548. Heldt, P. M., "Overhead Camshaft Passenger Car Engines," *Soc. Auto. Eng. J.* 19, June 1922, pp. 489–498; Same, *Automotive Ind.* 46, May 25–July 1, 1922, pp. 1109–1114, and 1158–1161; Same *Automobile Eng.* 12, December 1922, pp. 397–403; Discussion, *Soc. Auto. Eng. J.* 11, September 1922, pp. 244–247.

549. Helweg, S., "Keeping Records of Automatic Screw Machines," *Machinery* 26, March 1920, pp. 659—661.

550. Henning, J. F., "Formschlüssiges Kurven–Rast–Getriebe," *Feinwerktechnik* No. 2, 1970, pp. 61—68; Ibid., Part 2, No. 6, 1970, pp. 251—258; "Form–Closure Dwell Cams," Ibid., Part II, No. 6.

551. Henshaw, D. E., "Cam Forces in Weft Knitting," *Textile Research Journal* 38, June 1968, 592—598.

552. Herb, C. O., "Cam Slots Formed in Die–Casting by Loose Die Parts; Spider Handles for Electric Washing Machines," *Mach.* 43, November 1936, pp. 190—191.

553. Herb, C. O., "Milling Cams of Unusual Outline," *Mach.* 34, December 1927, pp. 273—277.

554. Herbst, C., "Analytische Ermittlung der günstigsten Bewegungsverhältnisse eines Schwinghebelantriebes," *Dingl. Polyt. J.*, 1908, p. 572 ("Analytical Computation of Optimum Movement Ratios in Oscillating Drive–Link Mechanisms.").

555. Herr, K., "Die Bewegungsverhältnisse an Steuernocken," *ATZ* 37, 1934, pp. 197—199, 226—227, 244—245, 403—405, and 486—487 ("Motion in Automotive Cam Mechanisms").

556. Herrmann, R., and A., "L'evolution da trace des cames en fonction des possibilites offertes par l'ordinateur," *Ingenieurs de l'Automobile* 42, 1969, No. 11, pp. 655—665.

557. Herzi, G., "Radial Layout for Precision Cams," *Am. Mach.* 99, February 28, 1955, p. 134.

558. Hickling, F. O., "Some Fixtures Used in Machining Sewing Machine Cams," *Am. Mach.* 61, November 13—20, 1924, pp. 817—818.

559. Higgins, A. P., "Tracer–Controlled Machine Speed Work at Convair," *Mach.* 62, July 1956, pp. 162—165.

560. Hildebrand, S., "Kurvengetriebe in Addiermaschinen und ihre praktischen Messungen," *Maschinenbautechnik* 6, 1957, pp. 112—117 ("Cam Mechanisms in Adding Machines and Their Practical Measurements").

561. Hildebrand, S., "Kurvengetriebe in Addiermaschinen und ihre praktischen Messunge," *VDI–Berichte* 12, 1956, pp. 41—48 ("Cam Mechanisms in Adding Machines and Their Practical Measurements").

562. Hildebrand, S., "Zur Konstruktion von Kurvenscheiben," *Maschinenbautechnik* 1, 1952, pp. 203—216 ("On the Design of Cam Mechanisms").

563. Hill. W. R., "Faceplate Cam Controls Tool for Grooving a Helical Shape," *Am. Mach.* 115, February 22, 1971, p. 89.

564. Hinkle, R. T., "Rapid and Accurate Method for Checking Cam Pressure Angles," *Machine Design* 27, July 1955, pp. 187—187.

565. Hinman, C. W., "Cam Locks for Jigs and Fixtures," *Am. Mach.* 81, June 30, 1937, pp. 573—575.

566. Hinman, C. W., "Cam Operated Lock Nut," March 1934, pp. 14—15.

567. Hinman, C. W., "Clamps and Supports Used in Drilling and Milling Tools," *Can Machy.* 45, 1934.

568. Hinman, C. W., "Standardizing Automatic Screw Machine Cams," *Am. Mach.* 77, April 26, 1933, p. 264.

569. Hinman, C. W., "Steps in Screw Machine Cam Design," *Am. Mach.* 77, November 22, 1933, p. 757.

570. Hinman, C, W., "Threading Lobes on Screw Machine Cams," *Am. Mach.* 77, November 8, 1933, p. 728.

571. Hinman, C. W., "Three–Dimension Fixture for Cutting Cams," *Am. Mach.* 72, February 6, 1930, pp. 270–271.

572. Hinman, C. W., "Use of Face Cams and Drum Type Slot Cam," February 1934, pp. 20–21.

573. Hirchenhain, A and Remmel, J., "Beitrag zur Spannungsanalyse mit Hilfe hybrider Rechenverfahren am Beispiel von Kurvenscheiben eines CYCLO–Getriebes," *Technisches Messen* 49, 1982, No. 7/8, pp. 265–269.

574. Hirschhorn, J., "Disc–Cam Curvature," *Transactions of the Fifth Conference on Mechanisms*, Sponsored by *Machine Design*, 1958, pp. 50–63.

575. Hirschhorn, J., "Disc–Cam Curvature," *Mach. Des.* 31, 1959, pp. 125–129.

576. Hirschhorn, J., "Graphical Method for Determining Pressure Angle and Minimum Base Radius in Cam Design," *Mach. Des.* 34, September 13, 1962, pp. 191–192.

577. Hirschhorn, J., "Methods for Plotting and Calculating Disc–Cam Curvature; Data Sheets," *Mach. Des.* 31, February 5, 1959, pp. 125–129.

578. Hixon, H. L., "Square Lathe Turret Supports Cam for Profile Turning," *Am. Mach.* 94, January 9, 1950, p. 130.

579. Hoffman, W., "Zeichnerische Ermittlung und Konstruktion der Kurvenbahnen für Radial–und Axialnocken," *Zeichen in Technik, Architektur, Vermessung* 12, 1974, No. 5, pp. 93–100.

580. Hollins, P. H., and Skidmore, D. W., "Method of Programming the Speed of Linear Traverse," *J. Sci. Inst.* 35, October 1958, p. 378.

581. Hollis, W. K., "Cam Rise Design Charts," *Design News*, July 21, 1958, pp. 88–89.

582. Holmes, W., "Ball–Supported Toolholder Followers Piston–Hear Cam," *Am. Mach.* 93, June 16, 1949, p. 111.

583. Holowenko, A. R., and Hall, A. S., "Cam Curvature," *Machine Design*, August 1953, pp. 170–176; "Cam Curvature, Part 2," *Machine Design*, September 1953, pp. 162–169; "Cam Curvature, Part 3," *Mach. Des.*, November 1953, pp. 148–156.

584. Holst, K. G., "Modified Gravity Curves for Quick Acting Cams," *Am. Mach.* 75, November 19, 1931, pp. 775–777.

585. Holt, J., "Don't Underrate Nylon for Mechanical Components," *Product Eng.* 24, June 1953, p. 203.

586. Holzer, H., "Wälzhebel," *VDI–Z*, 1908, pp. 2043–2051 ("Rolling Contact Cam").

587. Homewood, J., "Computing Angular Dimensions for a Cam Milling Operation," *Mach.* 51, January 1945, p. 193.

588. Honegger, J. A., "Cam–Plate with Adjustable Lobes for Transmitting an Oscillating Motion to an Arm or Lever from a Rotating Shaft," *Mach.* 42, September 1935, pp. 65–66.

589. Honegger, J. A., "Low–Cost Method of Making Temporary Cams," *Mach.* 43, October 1936, p. 113.

590. Hopkins, R. V., "Cam Profiles for Minimum Inertia Shocks," *Product Eng.* 4, May 1933, pp. 174–176.

591. Horan, R. P., "Overhead Valve Gear Problems," *SAE Trans.* 61, 1953, pp. 687–686.

592. Horn, H., "Untersuchung der an Ventilantrieben auftretenden Verschleissursachen und deren Verminderung," *Maschinenbautechnik* 3, 1954, pp. 499–506 ("Analysis of Wear and Tear in Valve–Input Mechanisms and Means of Prevention").

593. Horn, H., "Praktische Untersuchungen zur Verminderung der Verschleissursachen an kipphebelgesteuerten Ventilantriebe," *Maschinenbautechnik* 6, 1957, pp. 270–276 ("Practical Investigation for the Reduction of Causes of Wear on the Rocker–Arm–Controlled Valve Mechanism").

594. Hoschek, J., "Geometric Frictional Energy Loss of Cam Surfaces in Spherical Motion," *Zeitschrift für Angewandte Mathematik und Mechanik* 46, July 1966, pp. 315–318.

595. Hoschek, J., "Paare kongruenter Kurvenscheiben," *Mechanism and Machine Theory* 13, 1978, No. 3, pp. 281–292.

596. Hoschek, J., "Kurvengetriebe mit Kurve und Gegenkurve gleicher Gestalt," VDI–Berichte 281, Duesseldorf, pp. 65–74.

597. Hoschek, J., "Spezielle Nockengetriebe und geometrische Eigenschaften dazu verwandter Kurven," *ZAMM* 58, 1978, No. 2, pp. 75–80.

598. Hrones, J. A., "Dynamic Forces in Cam Mechanisms," *Mach. Des.* 20, January 1948, pp. 135–138.

599. Hrones, J. A., "Analysis of the Dynamic Forces in a Cam Driven System," *Trans. ASME* 1948, pp. 473–482.

600. Hrones, J. A., "Key Factors in Cam Design and Application," *Machine Design*, April 1949, pp. 127–132; May 1949, pp. 107–111, and 178; June 1949, pp. 124–126.

601. Hubert, C. P., "Cutting Plates for Welded Ship Construction; Use of Cams in Flame Cutter," *Civil Eng.* 14, March 1944, pp. 109–110.

602. Hubert, C. P., "Edge Preparation of Ship Plate for Welding; Utilization of Mechanical Cams in Conjunction with Flame–Planer," *Iron Age* 156, September 13, 1945, pp. 60–63.

603. Hubschmann, W., "Untersuchung der Reibungsverhältnisse am Rollenexzenter und Beschreibung eines Konstruktionselementes zur Vermeidung der Gleitreibung," *Konstruktion* 23, 1971, No. 10, pp. 405–407.

604. Huckert, J., "Accelerations by Vector Methods in Disk Cam Mechanisms," *Product Eng.* 19, December 1948, 128–132.

605. Hudson, F. C., "Curring Cams on a Boring Machine," *Am. Mach.* 52, June 3, 1920, p. 1203.

606. Hudson, F. C., "Machine for Cutting Face Cams," *Am. Mach.* 63, September 17, 1925, p. 481.

607. Hugk, H., Nerge, C., and Stange, H., "Untersuchung der Kontakt-Schwingungen an der Eingriffsstelle von Kurvenmechanismen," *Maschinenbautechnik*, No. 4, 1967, pp. 295–301 ("Examination of Contact Oscillations in Critical Areas of Cam Mechanisms").

608. Hugk, H., "Ergebnisse experimenteller Verschleissuntersuchungen an Kurvenmechanismen," *Maschinenbautechnik* 25, 1976, No. 10, pp. 456–462.

609. Hugk, H., "Experimentelle Untersuchungen zur Klärung des Rattermarkenproblems und des Vershcleissverhaltens bei kraftschlüssigen Kurvenmechanismen," Diss. TU Dresden, 1978.

610. Hugk, H., "Verschleissuntersuchungen mit rollbeanspruchten Prüfscheiben," *Schmierungstechnik* 10, 1979, No. 1, pp. 10–13.

611. Hugk, H., Krzeniciessa, H. and Nerge, G., "Arbeitsblatt für die Konstruktion von Mechanismen: Berechnung von Kurvenmechanismen," *Maschinenbautechnik* 28, 1979, No. 11, pp. 523–525.

612. Huhn, E., "Kinematisch exakte zeichnerische Synthese von Kurvengetrieben mit Hilfe des Übertragungswinkels," *Wissenschaftliche Zeitschift d. Hochschule für Maschinenbau*, Karl–Marx, Stadt, 3, 1961, pp. 59–81 ("Kinematically Exact Design Synthesis of Cam Mechanisms by Means of Transmission Angles").

613. Hunt, K. H., "Profiled Follower Mechanisms," *Mechanism and Machine Theory* 8, 1973, pp. 371–395.

614. Hussey, M. C., "Large Cams Milled on Twinshaft Lathe," *Am. Mach.* 95, October 29, 1951, p. 131.

615. Huszthy, L., "Analytical Examinations of Speed and Acceleration Conditions with Cam Control," *Gep 16* 4, April 1964, pp. 135–140.

616. Hutchinson, R. V., "Do We Understand the Grinding Process?" *SAE J* 42, March 1938, pp. 93–94.

617. Huther, B., "Kinematik und Kinetik der Räderkurvenschrittgetriebe," *Maschinenbautechnik*, No. 10, 1970, pp. 506–510 ("Kinematics and Kinetics of Wheel–Type Cam Movements").

618. Huther, B., "Rückkehrende Summiermechanismen," *Maschinenbautechnik*, No. 9, 1970, pp. 461–468 ("Reiterating Adding Mechanisms").

619. Huther, B., "The Sinoid as a Law of Motion for Problems with Predetermined Speeds," *Maschinenbautechnik* 15, September 1966, pp. 488–493.

620. Hyler, J. E., "Cam Applications Can Produce Variety of Cuts," *Wood-Worker* 70, March 1951, pp. 32, 34, 36, and 38; April 1951, pp. 32, 34, and 36; May 1951, pp. 46, 48, and 51.

621. Hylar, J. E., "Cam Applications Can Produce Variety of Cuts," *Wood–Worker* 70, March, 1951, pp. 32, 34, 36, and 38; April 1951, pp. 32, 34, and 36; May 1951, pp. 46. 48, and 51.

622. Hylar, J. E., "Cam Curves Aren't Tough," *Steel* 132, January 12, 1953, pp. 70–71.

623. Hylar, J. E., "Cams Their Production and Applications," *Machine and Tool Blue Book* 5, 1948, pp. 131–136; February 1948, p. 143; March 1948, p. 199; April 1948, p. 178; May 1948, p. 155.

624. Hylar, J. E., "Wooden Cams Save Money," *Tool Eng.* 40, January 1958, pp. 117–118.

625. Ingham, R. E., "Button–Pushing Job? Simple Cam Control System Improves Machine Productivity," *Plant Eng.* 13, September 1959, pp. 122–123.

626. Inglis, E. J., and Douglas, M. T., "Cammed Contour–Boring Bar Cuts Cutting Time 1/3," *Am. Mach.* 100, September 10, 1956, pp. 122–125.

627. Ip, Ching–U, Morse, E. I., Jr., and Hinkle, R. T., "How to Analyze Rolling Contact Mechanisms," *Machine Design* 28, 1956, pp. 75–77.

628. Ishihara, Y., "Numerical Control Grinder Cuts Auto Cams," *Am. Mach.* 110, September 28, 1966, pp. 112–113.

629. Iudin, V. A., Petrokas, L. V., and Kolman–Ivanov, E. E., "The Method of Construction and Application of Nomograms for the Synthesis of Plane and Cam Mechanism," *Vestn. Inj. Techn.*, 1947, pp. 167–172.

630. Jackowski, C. S., and Dubil, J, F., "Single Disk Cams with Positively Controlled Oscillating Followers," *J. Mechanisms* 2, 1967, pp. 157–184.

631. Jacob, R, J., "Indexing with Concave Barrel Cams," *Mach. Des.* 21, February 1949, pp. 93–96.

633. Jacobi, P., and Unger, W., "Projizierbare bewegliche Getriebemodelle," *Maschinenbautechnik* No. 1, 1970, pp. 8–10 ("Projectable Movable Mechanism Models").

634. Jagow, L., "Chuck–Type Driver for Camshaft," *Mach.* 26, March 1920, pp. 661–662.

635. Jahr, W., "Der Entwurf von Kurventrieben mit Rücksicht auf den Beschleunigungsverlauf," *Maschinenkonstrukteur–Betriebstechnik*, 1931, p. 30 ("Design of Cam Mechanisms with Special Regard to the Rate of Acceleration").

636. Jahr, W., "Entwicklung eines Kurvengetriebes auf neuzeitlichen Grundlagen," *Werkstatt und Betrieb* 83, 1950, pp. 54–59 ("Development of a Cam–Mechanism Based on New Fundamentals").

637. Jamoulle, Andre, "Quelques aspects particuliers de la distribution des moteurs a combustion," *R.U.M.* – Revue Universelle 111, 1968, No. 12, pp. 343–351.

638. Janie, Z., "Analytical Methods for Deriving Cam Profiles," *Tool Eng.* 31, October 1953, pp. 81–88.

639. Janie, Z., "Design Procedure for Cam Follower Springs," *Design News*, February 15, 1954, pp. 47–51.

640. Jante, A., "Über Nocken an Verbrennungsmotoren," *Maschinenbautechnik* 10, 1961, pp. 142-149 ("Concerning Cams in Combustion Engines").

641. Jante, A., "Einheitsnocken mit sprung- und knickfreiem Beschleunigungsverlauf," *Kraftfahrzeugtechnik* 8, 1958, pp. 5-12 ("Standard Cams with Jerk-Free Acceleration").

642. Jarunov, A. M., "Cams with Constant Pressure Angles" (in Russian), *Machanika masin* 11/12, 1970, No. 25/26, p. 46.

643. Jeans, H., "Designing Cam Profiles with Digital Computers," *Mach. Des.* 29, October 31, 1957, pp. 103-106.

644. Jehle, F., and Spiller, W. R., "Idiosyncrasies of Valve Mechanisms and Their Causes," *Trans. SAE* 24, 1929, pp. 133-134.

645. Jellig, W., "Precision Machines Assure Cam Accuracy; Ford Instrument Co.," *Iron Age* 173, April 15, 1954, pp. 140-142.

646. Jenner, A., "Making Cams for Automatic Screw Machines," *Am. Mach.*, January 31, 1918, pp. 189-190.

647. Jennings, J., "A New Analysis of the Tangent Cams," *Mechanical World*, 1954, pp. 150-154.

648. Jennings, J., "Calculating Springs for Cams," *Mach. (London)* 52, July 28, 1938, p. 521.

649. Jennings, J., "Calculating Springs for Cams," *Mach. (London)* 57, January 16, 1941, p. 433.

650. Jensen, P. W., "Konstruktion, Berechnung und Herstellung von Kurvenscheiben," *Technica* 6, No. 22, 1957, (Switzerland) pp. 1245-1250 ("Cam Design").

651. Jensen, P. W., "Konstruktion, Berechnung und Herstellung von Kurvenscheiben," *Technica* 6, No. 23, 1957, (Switzerland) pp. 1319-1321 ("Cam Calculations").

652. Jensen, P. W., "Konstruktion, Berechnung und Herstellung von Kurvenscheiben," *Technica* 6, No. 24, 1957, (Switzerland) pp. 1391-1393 ("Cam Manufacture").

653. Jensen, P. W., "Kurveskiven-Konstruktion, Beregning og Fremstilling," *Ingeniør-og Bygningsvaesen*, May 10, 1958, pp. 174-184 ("Cam Design sign and Manufacture").

654. Jensen, P. W., "Kurveskiven-Konstruktion, Beregning og Fremstilling," *Ingeniør-og Bygningsvaesen*, May 10, 1958, pp. 174-184 ("Cam Design and Manufacture").

655. Jensen, P. W., "Intermittent Motion from Two Synchronized Cams," *Machy.*, November 1960, pp. 134-135.

656. Jensen, P. W., "Mouvement Intermittent Communique par deux cames synchronisees," *Machine Moderne*, April 1961, pp. 47-48 ("Intermittent Movement Transmitted by Synchronized Cam Mechanisms").

657. Jensen, P. W., "Slide Motion Differential," *Mach.*, Decmeber 1961, p. 135.

658. Jensen, P. W., "Charts for Radial Translating Followers," *Design News* February 21, 1962, pp. 90-95.

659. Jensen, P. W., "Rotary Work-Transfer Device," *Machin.*, March 1962, pp. 138–139.

660. Jensen, P. W., "Differential a mouvement de coulisse," *Machine Moderne*, April 1962, p. 49 ("Differentials with Cam Movements").

661. Jensen, P. W. "How to Proportion the Smallest Possible Cam for Given Pressure Angle," *Trans. of the Seventh Conference on Mechanisms*, sponsored by *Machine Design*, 1962, pp. 202–213.

662. Jensen, P. W., "Single Closed-Track Cam Drives Glue-Transfer Mechanism," *Machy.*, February 1963, pp. 119–120.

663. Jensen, P. W., "Adjustable Indexing Mechanisms with 180-Degree Dwell," *Machinery*, July 1963, p. 120.

664. Jensen, P. W., "Concave and Convex Machining on a Shaper," *Machinery*, July 1963, p. 114.

665. Jensen, P. W., "Usinage concave et convexe sur etau-limeur," *Machinery*, July 1963, p. 114 ("Concave and Convex Cam Designs").

666. Jensen, P. W., "Cam unique a chemin de roulement ferme command un mecanism transporteur de colle," *Machine Moderne*, September 1963, pp. 45–46 ("Single and Fixed Roller Type Cams Controlling One Locking Mechanism").

667. Jensen, P. W., "Mechanisms," *Product Eng.* September 2, 1963, p. 67.

668. Jensen, P. W., "Konstruktion von Kurvenscheiben," *Technische Rundschau*, October 4, 1963, pp. 65–69 ("Design of Disk Cams").

669. Jensen, P. W., "Cam Mechanisms with Translating Followers," *Design News*, November 27, 1963, pp. 92–97.

670. Jensen, P. W., "Minimum Cam Size," *Product Eng.* March 2, 1964, pp. 69–76.

671. Jensen, P. W., "New Charts Quickly Lead to Minimum Cam Size," *Product Eng.* 35, March 2, 1964, pp. 69–76.

672. Jensen, P. W., "Machinery Mechanisms," *Product Eng.* June 8, 1964, pp. 66–74.

673. Jensen, P. W., "Cam-Controlled Differential Mechanism," *Machy.*, October 1964, p. 156–157.

674. Jensen, P. W., "Machinery Mechanisms," *Product Eng.*, October 26, 1964, pp. 108–116.

675. Jensen, P. W., "Mekanisme-Hjørnet," *Maskinindustrien*, 1964, September 1, p. 647 ("Cam Mechanisms").

676. Jensen, P. W., "Mekanisme-Hjørnet," *Maskinindustrien*, 1964, September 15, p. 688 ("Cam Mechanisms").

677. Jensen, P. W., "Mekanisme-Hjørnet," *Maskinindustrien*, 1964, October 1, p. 738 ("Cam Mechanisms").

678. Jensen, P. W., "Mekanisme-Hjørnet," *Maskinindustrien*, 1964, October 15, p. 784 ("Cam Mechanisms").

679. Jensen, P. W., "Mekanisme-Hjørnet," *Maskinindustrien*, 1964, November 1, p. 825 ("Cam Mechanisms").

680. Jensen, P. W., "Mekanisme-Hjørnet," *Maskinindustrien*, 1964, November 15, p. 873 ("Cam Mechanisms").

681. Jensen, P. W., "Mekanisme-Hjørnet," *Maskinindustrien*, 1964, December 1 ("Cam Mechanisms").

682. Jensen, P. W., "Mekanisme-Hjørnet," *Maskinindustrien*, 1964, December 15 ("Cam Mechanisms").

683. Jensen, P. W., "Mekanisme-Hjørnet," *Maskinindustrien*, January 1, 1965, p. 158 ("Planetary Cam Mechanisms").

684. Jensen, P. W., "Mechanisms for Generating Cam Curves," *Product Eng.*, March 1, 1965, pp. 41-47.

685. Jensen, P. W., "Konstruktion von Kurvenscheiben," *Antriebstechnik* 4, No. 2, pp. 43-48 ("Design of Cams").

686. Jensen, P. W., "Mekanisme-Hjørnet," *Maskinindustrien*, July 1 and 15, 1965, p. 453 ("Rolling-Contact Cams").

687. Jensen, P. W., "Mekanisme-Hjørnet," *Maskinindustrien*, August 15, 1965, p. 493 ("Constant-Diameter Cam").

688. Jensen, P. W., "Tailored Cycloid Cams," *Product Eng.*, November 8, 1965, pp. 108-111.

689. Jensen, P. W., "Mekanisme-Hjørnet," *Maskinindustrien*, December 1, 1965, p. 809 ("How to Decrease Maximum Pressure Angle").

690. Jensen, P. W., "Mekanisme-Hjørnet," *Maskinindustrien*, December 15, 1965, p. 845 ("Heart-Shaped Cam").

691. Jensen, P. W., "Mekanisme-Hjørnet," *Maskinindustrien*, January 1, 1966, p. 49 ("Wedge Cams").

692. Jensen, P. W., "Mekanisme-Hjørnet," *Maskinindustrien*, January 15, 1967, p. 57 ("Motion Converters").

693. Jensen, P. W., "Mekanisme-Hjørnet," *Maskinindustrien*, February 15, 1967, p. 125 ("Motion Converters").

694. Jensen, P. W., "Kurveskivemekanismer I," *Maskinindustrien*, August 15, 1967, pp. 563-573 ("Design of Cams I").

695. Jensen, P. W., "Kurveskivemekanismer II," *Maskinindustrien*, September 15, 1967, pp. 615-620 ("Design of Cams II").

696. Jensen, P. W., "Zur Synthese und Analyse der dreigliedrigen Kurvengetriebe VDI-Berichte No. 77, 1965 ("On the Synthesis and Analysis of Cam Mechanisms").

697. Jensen, P. W., "Mekanisme-Hjørnet," *Maskinindustrien*, December 1, 1965, p. 877 ("Screw Mechanisms").

698. Jensen, P. W., "German Methods Simplify Layout of Tailored Cycloid Cams," *Product Eng.* 36, November 8, 1965, pp. 108-111.

699. Jensen, P. W., *Cam Design and Manufacture*, The Industrial Press, New York, and The Machinery Publishing Co. Ltd., Brighton, England, 1965.

700. Jensen, P. W., "Berechnung von Krümmungen bei Kurvenscheiben," *Maschinenmarkt* 75, 1969, No. 97, pp. 2119-2122.

701. Jensen, P. W., "Mekanismeteknik" ("Kinematics and Dynamics of Mechanisms"), Teknisk Forlag, 1971.

702. Jensen, P. W., "Systematisk Mekanismekonstruktion" ("Systematic Development of Mechanisms"), Teknisk Forlag (Copenhagen), 1971.

703. Jensen, P. W., "Cam Manufacture," *Project 2*, (Copenhagen) No. 2, 1982 (in Danish), pp. 20–27.

704. Jensen, P. W., "Cam Manufacture," *Project 2*, (Copenhagen) No. 3, 1982 (in Danish), pp. 27–32.

705. Jensen, P. W., "Finding Radius of Curvature of Plate Cams Using Complex Vectors," 8th Applied Mechanism Conference, Saint Louis, Missouri, 1983.

706. Jensen, P. W., *Cams and Cam Design: Machinery's Handbook*, 22nd Edition, Industrial Press, New York, 1984, pp. 586–609.

707. Jhala, P. B., and Venkataramanan, C. G., "Optimization of Picking Cam Design Through Dynamic Studies," *IFToMM*, 1979, Vol. 2, pp. 992–995.

708. Jog, A. P., "Graphical Construction to Determine the Angular Acceleration of a Rocker for a Tangent Cam," *J. Inst. Engrs.*, India, 47, January 1967, pp. 118–122.

709. Johns, L. G., "Practical Cam Design," *Am. Mach.* 48, May 16, 1918, pp. 823–824.

710. Johnson, A., "How to Choose Short-Run Cams for Automatic Screw Machines," *Tool Eng.* 34, May 1955, pp. 81–84.

711. Johnson, A. R., "Motion Control for a Series System of 'N' Degrees of Freedom Using Numerically Derived and Evaluated Equations," ASME Paper No. 64-Mech.-7.

712. Johnson, R. C., and Nymberg, R. J., "Reliability of Numerical Control Proved by Ten Years of Precision Machining," *Mach.* 71, October 1964, pp. 137–141.

713. Johnson, R. C., "Analysis and Design of Cam Mechanisms Having a Varying Input Velocity," *Trans. of the Seventh Conference on Mechanisms*, Sponsored by *Machine Design*, 1962, pp. 190–201.

714. Johnson, R. C., "Cam Design," *Mach. Des.* 27, November 1955, pp. 195–204.

715. Johnson, R. C., "Cam Mechanisms," *Mach. Des.* 28, January 26, 1956, pp. 85–89.

716. Johnson, R. C., "Development of a High Speed Indexing Mechanisms," *Trans. of the Fourth Conference on Mechanisms*, Sponsored by *Machine Design*, 1957, pp. 39–43.

717. Johnson, R. C., "Development of a High Speed Indexing Mechanism," *Mach. Des.* 30, September 4, 1958, pp. 134–138.

718. Johnson, R. C., "The Dynamic Analysis and Design of Relatively Flexible Cam Mechanisms Having More than One Degree of Freedom," *ASME J. Eng. Ind.*, November 1959, pp. 323–331.

719. Johnson, R. C., "Dynamic Analysis of Cam Mechanism," *Trans. of the Fifth Conference on Mechanisms*, Purdue Univeristy, October 1958, pp. 21–35.

720. Johnson, R. C., "How Profile Errors Affect Cam Dynamics," *Mach. Des.* 29, February 7, 1957, pp. 105–108.

721. Johnson, R. C., "How to Design High-Speed Cam Mechanisms," *Mach. Des.* 28, January 26, 1956, pp. 85–89.

722. Johnson, R. C., "Impact Forces in Mechanisms," *Mach. Des.* 30, June 12, 1958, pp. 138-146.

723. Johnson, R. C., "Method of Finite Differences in Cam Design," *Mach. Des.* 29, November 14, 1957, pp. 159-151.

724a. Johnson, R. C., "Method of Finite Differences in Cam Design," *Mach. Des.* 27, November 1955, pp. 195-204.

724b. Johnson, R. C., "Minimizing Cam Vibrations," *Mach. Des.*, August 9, 1956, pp. 103-104.

725. Johnson, R. C., "Rapid Method for Developing Cam Profiles Having Desired Acceleration Characterists," *Mach. Des.* 28, December 13, 1956, pp. 129-132.

726. Johnson, R. C., "Simple Numerical Method for Dynamic Analysis and Design of Flexible Cam Mechanism," *Mach. Des.* 31, September 3, 1959, pp. 140-145.

727. Johnstone-Taylor, F., "Cam Profiles and Fuel Injection on Diesel Engines," *Marine Eng.* 29, November 1924, pp. 665-669.

728. Jones, A. H., "Thermit Welding of Camshafts," *Met Chem. Eng.* 13, December 1, 1915, p. 929.

729. Jones, L., and Davies, D. N. C., "Overhead Camshaft Drives," *Automobile Eng.* 56, May 1966, pp. 175-178.

730. Jonge, A. E. R. De, "The Motion of a Link Cam Over a Roller," *Trans. ASME* 75, May 1953, pp. 747-757.

731. Just, E., "Determining Wear of Tappets and Cams at Volkswagen," *Metal Prog.* 98, August 1970, pp. 110-112.

732. Kadar, L., "Bulldozer Cam Mechanism," *Mach.* 34, August 1928, p. 915.

733. Kaffine, H., "Grinding True Flat Surfaces on the Sides of Thin Cams," *Mach.* 40, 1933, pp. 157-158.

734. Kahr, J. S., "New Equations Speed Design of Circular-Arc Cams," *Product Eng.* 31, March 21, 1960, pp. 58-63.

735. Kalb, C. W., "Manufacture, Machining Molded Nylon Parts," *Tooling & Production* 18, April 1952, pp. 90-94.

736. Kaluzhnikov, A. N., "Geometrical Theory of Single-Row Cam Mechanisms," *Russian Eng. J.*, LI, 1971, pp. 33-39.

737. Kammel, E., "Angewandte Getriebetechnik im Glasmaschinenbau," *VDI-Tagungsheft* 1, 1953, pp. 170-193 ("Applied Kinematics in Glass Machine Construction").

738. Kanango, R. N., and Patnaik, N., "Improving Dynamic Characteristics of a Cam-Follower Mechanism Through Finite Difference Techniques," In: 8, Vol. 1, pp. 591-594.

739. Kanarchos, A., "Über die Anwendung von Optimierungsverfahren bei dynamischen Problemen in der rechnerunterstützten Konstruktion," *Konstruktion* 28, 1976, No. 2, pp. 53-58.

740. Kanarachos, A., "Zur Auslegung von Kurvengetrieben unter Berücksichtigung des dynamischen Verhaltens," *Konstruktion* 29, 1977, pp. 188-190.

741. Kanarachos, A., "Rechnereinsatz zur Optimierung der dynamischen Eingenschaften ungleichförmig übersetzender Getriebe," *VDI-Berichte* 281, Düsseldorf, pp. 133-140.

742. Kanzaki, K., and Itao, K., "Polydyne Cam Mechanisms for Typehead Positioning," *J. Eng. Ind.* 94, February 1972, pp. 250-254.

743. Kaplan, J., "Cam Control Gets More Out of Planetary Gear," *Product Eng.* 31, January 4, 1960, pp. 38-41.

744. Kappler, P., "Bestimmung der Kurvenformen bei kraftschlüssigen Kurvengetrieben nach dem Kräftespiel zwischen Massen-und Rückstellkraften *VDI-Z* 96, 1954, pp. 782-788 ("Determination of Curvature in Force-Closed Cam Systems Dependent on Mass and Force Potentials").

745. Kappler, P., "Allgemeingültige Kurvenberechnung," *Maschinenbautechnik* 5, 1956, pp. 134-140 ("General Cam-Calculations").

746. Karelin, N. M., "Methods for Continuous Automatic Checking of Cylindrical Components with Curvilinear Cross Sections," *Measurement Technique*, 1961, No. 11, pp. 863-868.

747. Karl, B., "2 1/2-Dimensional Fräsen unter Verwendung der Achsumschaltung," *Industrie-Anzeiger* 99, 26, November 1972, pp. 2513-1514 ("2 1/2-Dimensional Shifts Using Central Circuit Switching").

748. Kasper, L., "Cam Mechanism Designed to Control Shaft Speed," *Mach.* 44, December 1937, pp. 242-243.

749. Kasper, L., "Shaper Fixture for Machining Special Cams," *Mach.* 48, March 1942, pp. 153-154.

750. Kasper, L., "Cam Designed to Operate on Alternate Revolutions," *Mach.* 51, June 1945, pp. 173-174.

751. Kasper, L., "Cam Designed to Provide Longer Stroke Without Enlarging Operating Space," *Mach.* 51, December 1944, pp. 187-188.

752. Kasper, L., "Intermittent Feeding Mechanism Operating Two Slides from One Cam," *Machy. (London)* 82, August 29, 1946, pp. 270-271.

753. Kasper, L., "Automatic Variable-Lift Cam Mechanism," *Mach.* 53, September 1946, p. 178.

754. Kasper, L., "Oscillating Cam Made Adjustable; Drilling Machine Work-Holding Fixture," *Mach.* 54, June 1948, pp. 193-194.

755. Kasper, L., "High-Lift Cam with Low Pressure Angle; Wire Fabricating Machine," *Mach.* 56, September 1949, pp. 193-194.

756. Kasper, L., "Simple Cam Made Practical by Increasing Speed," *Mach.* 60, October 1953, pp. 207-208.

757. Kasper, L., "Reciprocating Cam with Half-Cycle Dwell," *Mach.* 61, September 1954, pp. 203-204.

758. Kasper, L., "Follower for Cam-Plate Grinding," *Am. Mach.* 99, September 26, 1955, p. 159.

759. Kasper, L., "Gear Mechanism for Varying Cam Timing," *Mach.* 63, December 1956, pp. 189-190.

760. Kasper, L., "Four-Lobed Cam Transmit Variable Motion to Follower," *Mach.* 66, November 1959, pp. 144-145.

761. Kasper, L., "Cam Produces Motion on Alternate Revolutions," *Mach.* 66, January 1960, pp. 143-144.

762. Kasper, L., "Cylindrical Cam Positions Wire Guide," *Mach.* 69, October 1962, pp. 146-147.

763. Kasper, L., "Intermittent Rotary Motion from a Uniform Reciprocating Drive," *Mach.* 69, February 1963, p. 118.

764. Kasper, L., "Cam and Link System Produces Varying Rotation Rates," *Mach.* 69, June 1963, pp. 132-133.

765. Kasper, L., "Cam Modifications for Smoother Action," *Design Eng.* 9, June 1963, pp. 52--53.

766. Kasper, L., "Tables Oust Pressure Angle Layouts," *Design Eng.* 9, September 1963, pp. 58-59.

767. Kasper, R. H., "Cam Milling Attachment," *Mach.* 34, July 1928, p. 851.

768. Kasper, R. H., "Unique Lever Motion Obtained by Toggle Link and Cams," *Mach.* 38, May 1932, p. 659.

769. Kasper, Th., "Geschränkte oder exzentrische Schubkurbel," *Maschinenbautechnik* 8, 1959, pp. 434-437 ("Offset or Eccentric Slider Crank Systems").

770. Kassamanian, A. A., and Müller, J., "Räumliches Kurvengetriebe mit Antriebsebene," *Maschienbautechnik* 26, 1977, No. 10, pp. 453-546.

771. Katona, J., "2D Output Disc Cams," *Mach. Des.* 38, May 1932, p. 659.

772. Kearney & Trecker, "Kearney & Trecker Cam-Milling Attachment," *Am. Mach.* 78, October 24, 1934, p. 752; *Mach.* 41, November 1934, pp. 182-183; *Steel*, October 29, 1934, p. 46.

773. Keller, F. L., "Simple Method of Duplicating Cam surfaces on a Surface Grinding Machine," *Mach.* 52, July 1946, p. 194.

774. Kellog, E. W., "Calculations of Accelerations in Cam-Operated Pull-Down Mechanisms," *Soc. Motion Picture Eng. J.* 45, August 1945, pp 143-155.

775. Kennison, G. F., "Charts Simplify Design of Cycloidal-Motion Cams; Data Sheet," *Mach. Des.* 28, January 12, 1956, pp. 141-142.

776. Kenny, R. H., "Cam-Operated Indexing Drives," *Assembly Eng.* 7, September 1964, pp. 40-45.

777. Kerle, H., "Ein Beitrag zur rechnerunterstützten dynamischen Synthese von Kurvengetrieben," *VDI-Berichte* 281, Düsseldorf, pp. 153-162.

778. Kerle, H., "Zur Auslegung eines schnellaufenden einfachen Kurvengetriebes unter Berücksichtigung des Antriebs," Dissertation TU Braunschweig, 1973.

779. Kerle, H., "How Effective is the Method of finite Differences as Regards Simple Cam Mechanisms," Rees Jones: Cams and Cam Mechanisms, pp. 131-135.

780. Kerle, H., "Berechnung der Eigenkreisfrequenzen einiger Schwingungsmodelle für Kurvengetriebe mit geradgeführtem Abtriebsglied," *Konstruktion* 28, 1976, No. 11, pp. 423-428.

781. Kerle, H., "Experimentelle Bewegungsuntersuchunghen an einem einfachen Kurvengetriebe," *VDI-Berichte* 195, pp. 171-177.

782.   Kerle, H., "Drehnachgiebige Kopplung eines Kurvengetriebes mit einem Elektromotor," *IFToMM*, 1974, pp. 181-202.

783.   Kerle, H., Debus, H., and Lohe, R., "Berechnung und Optimierung schnellaufender Gelenk- und Kurvengetriebe," Grafenau/Wuertt.: Fachverlag für Wirtschaft & Technik, 1981,

784.   Kersten, L., "Computer-Aided Methods to Relate Analytical and Graphical Design of Mechanisms," ASME Paper 70-MECH-77, 1977.

785.   Kiger, H., "Die Ermittlung der Geschwindigkeiten und Beschleunigungen an Nocken mit kreisförmigen Profil," *Der Ölmotor*, 1919, No. 2, p. 349 ("Calculation of Velocity and Acceleration in Cams with Circular Profile").

786.   Kim, H. R., and Newcombe, W. R., "Stochastic Error Analysis in Cam Mechanisms," *Mechanism and Machine Theory* 13, 1978, No. 6, pp. 631-641.

787.   Kim, H. R., and Newcombe, W. R., "The Effect of Cam Profile Errors and System Flexibility on Cam Mechanism Output," *Mechanism and Machine Theory* 17, 1982, No. 1, pp. 57-72.

788.   Kingsford, L. L., "Spooler Heart Cam-Maintaining Its Efficient Operation," *Textile World* 79, January 10, 1931, pp. 150-151.

789.   Kingsman, A. S., "Simplified Method of Plotting a Gravity Cam Curve," *Mach.* 39, April 1933, pp. 503-504.

790.   Kiper, G., "Möglichkeiten und Grenzen des einfachen Kurvengetriebes mit umlaufenden Gliedern für Antrieb und Abtrieb," *Maschinenbautechnik* 5, 1956, pp. 575-582 ("Possibilities and Limitations of the Simple Cam Mechanism with Rotating Link for Input and Output").

791.   Kiper, G., "Einfache Kurvengetriebe," *VDI-Forschungsheft* 461, 1957, pp. 38-40 ("Simple Cam Mechanism").

792.   Kirchdorfer, J., "Ein Beitrag zur Berechnung von Kurvengetrieben für handbetätigte Schalter," *Feinwerktechnik*, No. 2, 1968, pp. 56-63 ("On the Analysis of Cam Systems for Manually Operated Switches").

793.   Kirchhof, M., "Ausgleichsgetriebe zur Erhöhung der Leistungsfähigkeit und Betriebssicherheit von Maschinen und Geräten," *Wiss. Z. T. H. Karl-Marx-Stadt* 14, 1972, No. 1, pp. 51-59.

794.   Kirchner, Egon, "Der Kreisbogenkeil," *Tech. Zentralblatt für prakt. Metallbearbeitung* 56, 1962, pp. 82-88.

795.   Kirmse, W., and Johannsen, P., "Regelung der Werkstück-Drehfrequenz beim Nockenschleifen," *Werkstatt und Betrieb* 115, March 1982, No. 3, pp. 149-155.

796.   Kitche, H. W., "Making Helical Gear Segments Serve as Cams," *Mach.* 38, May 1932, p. 661.

797.   Klaus, R., "Wenig beachtete und wenig bekannte Kurventrieb-Schaltwerke mit kontinuierlichem Antrieb," *Werkstattstechnik* 24, 1934, pp. 629-635 ("Little Observed and Little Known Cam Actuated Mechanism with Intermittent Motion and Continuous Input").

798.   Klein, B., "Übertragungsfunktionen für optimale Bewegungsgesetze bei Kurvengetrieben," *Maschinenmarkt* 85, 1979, No. 39, pp. 756-759.

799. Klima, W., "Interlocked Cams and Pilots Sequence Cylinders," *Hydraulics & Pneumatics* 16, July 1963, pp. 80-81.

800. Klingeman, W. E., "Adjustable Cams Give Flexibility; Controlling Location of Spot Welds on Welding Machine," *Mach. Des.* 13, April 1941, pp. 35-36.

801. Kloomok, M., and Muffley, R. V., "Computers Simplify Solutions of Polynomical Cam Curves," *Product Eng.*, March 1957, pp. 196-202.

802. Kloomok, M., and Muffley, R. V., "Design of Cam Followers," *Product Eng.*, September 1956, pp. 197-201.

803. Kloomok, M., and Muffley, R. V., "Determination of Pressure Angles for Swinging-Follower Cam Systems," *Trans. ASME* 78, 1956, pp. 803-806.

804. Kloomok, M., and Muffley, R. V., "Determination of Radius of Curvature for Radial and Swinging Follower Cam Systems," ASME Paper No. 55-SA-29, 1955.

805. Kloomok, M., and Muffley, R. V., "Determination of Radius of Curvature for Radial and Swinging Follower Cam Systems," *Trans. ASME*, May 1956, pp. 795-806.

806. Kloomok, M., and Muffley, R. V., "Plate Cam Design: Evaluating Dynamic Loads," *Product Eng.*, January 1956, pp. 178-182.

807. Kloomok, M., and Muffley, R, V., "Plate Cam Design: Radisu of Curvature," *Product Eng.* 26, September 1955, pp. 186-192.

808. Kloomok, M., and Muffley, R. V., "Plate Cam Design, Pressure Angle Analysis," *Product Eng.*, May 1955, pp. 155-159.

809. Kloomok, M., and Muffley, R. V., "Plate Cam Design with Emphasis on Dynamic Effects," *Product Eng.*, February 1955, pp. 156-162, 211, 213 and 215.

810. Klyuiko, E. V., "Accuracy of Cam Mechanisms" (in Russian), Izv Vyssh Uchebn Zaved Mashinostr, No. 5, 1983, pp. 54-59.

811. Knechtel, P., "Beispiel für kurvengesteuertes Schaltwerk," *Betriebstechnik* 7, 1939, pp. 601-602 ("An Example for Cam-Controlled Intermittent Motion").

812. Knechtel, P., "Beispiel für kurvengesteuertes Schaltwerk," *Getriebetechnik* 9, 1940, p. 85 ("An Example for Cam-Controlled Intermittent Motion").

813. Knechtel, P., "Kurvengesteuertes Schaltwerk für eine Stanze," *Getriebetechnik* 9, 1940, p. 129 ("Cam-Controlled Switch Gear for a Punching Machine").

814. Knight, B. A., and Johnson, H. L., "Motion Analysis of Flexible Cam-Follower Systems," ASME Paper No. 66-MECH-3, 1966.

815. Kobrinskij, A. E., "Dynamic Stresses in Cam Mechanisms with Elastic Follower" (in Russian), *Tr. Sem. p. T. M. M.* Band VI, 1949, No. 24.

816. Kobrinskij, A. E., "On the Selection of the law of Motion of a Follower in a Cam Mechanism," *Maschinenbautechnik* 9, February 1960, pp. 93-98.

817. Kocian, G., "Large-Radius Cams Grounded by Milling-Machine Attachment," *Am. Mach.* 89, May 10, 1945, p. 119.

818. Koen, R. F., "Equations and Layout Method to Balance Grooved Cams," *Product Eng.* 34, April 1, 1963, pp. 64-69.

819. Kogan, Yu, A., "Factors in the Design of Cams for Overhead Valves," *Russian Eng. J.*, April 1965, pp. 29-31.

820. Koganov, I. A., "Machining Cams of Small Profile Radius," *Machine and Tooling*, April 1965, pp. 34-35.

821. Kolchin, N. I., "Determination of Radii of Curvature of Cams in Cam Mechanisms by Author's Method for Unroud Wheels," *Trudi Sem. teor, Mash. Mekh.* 10, 1951, pp. 16-21.

822. Koller, R., "Konstruktion und Optimierung von intermittierenden Getrieben mit Unterstützung elektronischer Datenverarbeitungsanlagen," *VDI-Berichte*, No. 167, Düsseldorf, VDI-Verlag, 1971, pp. 143-152.

823. Koller, R., Farwich, H., and Spiegels, G., "Konstruktion von Kurven-scheiben mit Unterstützung elektronischer Datenverarbeitungsanlagen und einer Bildschirmeinheit," *Industrie-Anzeiger* 30, 1972, pp. 1011-1017 ("Construction of Cam Disk Systems with the Aid of Electronic Computer Analysis and Monitor").

824. Konietschke, H., "Toleranzen an Doppelkurvengetrieben und ihr Einfluss auf das Spiel zwischen Rolle und Kurvenscheibe," *Maschinenbautechnik*, No. 5, 1972, pp. 194-198 ("Tolerances in Cam Gears and Their Importance on the Play Between Rollers and Cam Disks").

825. Koneitzko, P., "Computergesteuerte Kurvenfrässmaschine," *Werkstatt und Betrieb* 111, 1978, No. 12, pp. 807-808.

826. Konstantinoff, M., and Zenoff, P., "Isochronische Wälzkurvenpendel," Bul. Inst. Polit. Lasi 14 (18), 1968, No. 1/2, pp. 449-460.

827. Konstantinov, M. S., "Kraftausgleichende Bandkurvengetriebe mit line-arem Ausgleichsmoment," *Maschinenbautechnik*, No. 10, 1970, pp. 513-518 ("Forces Equalizing Linear Cams with Equalizing Linear Movement").

828. Kopp, F. O., "Entwicklung des Konzepts eines vielseitig verwendbaren ungleichförmig übersetzenden Kurvenverstellgetriebes," *Konstruktion* 35, 1983, No. 3, pp. 101-107.

829. Koster, M. P., "The Effects of Backlash and Shaft Flexibility on the Dynamic Behavior of a Cam Mechanism," Reese Jones: Cams and Cam Mechanisms, pp. 141-146.

830. Koster, M. P., "The Application of Dynamic Models in the Analysis of the Dynamics of Cam Followers and Cam Shafts," *IFToMM*, 1974, pp. 131-151.

831. Koster, M. P., "Digital Simulation of the Dynamics of Cam Followers and Camshafts," *IFToMM*, 1975, Vol. 4, pp. 969-974.

832. Koster, M. P., "Vibrations of Cam Mechanisms. Consequences of Their Design," London: MacMillan Press Ltd., 1974.

833. Koster, M. P., "Effect of Flexibility of Driving Shaft on the Dynamic Behavior of a Cam Mechanism," *Trans. ASME*, Ser. B: J. Eng. Ind. Vol. 97, 1975, No. 2, pp. 595-602.

834. Kosticyn, V. T., "The Friction of a Roller in the Groove of a Cylindrical Cam," *Trudi Inst. Maschinoved* 16, 1956, No. 62, p. 56.

835. Kosticyn, V. T., "Methods for the Calculation of Minimum Mass on the Cam Mechanism with Reciprocating Follower," *Tr. Sem. p.T.M.M.* Band III, 1947, No. 12.

836. Kosticyn, V. T., "Über die kleinsten Abmessungen räumlicher Nocken-mechanismen," *Maschinenbautechnik*, 1953, pp. 50-53 ("On the Minimum Proportions of Space Cam Mechanism").

837. Kosticyn, V. T., "On the Minimum Mass of a Cam Mechanism with Oscillating Follower," *Tr. Sem. p.T.M.M.* Band IX, 1950, No. 35.

838. Kosticyn, V. T., "On the Minimum Dimensions of Cam Mechanisms," *Bull. Acad. Sci. USSR Ser. Tech. Sci.*, 1948, pp. 1531-1537 (Izv. Akad. Nauk SSSR Ser. Tekh. Nauk).

839. Kosticyn, V. T., "Determination of Minimum Cam Dimensions for Straight-Line Followers," *Trudi Sem. p.T.M.M.*, 1947, pp. 23-63.

840. Kotoc, S., "Berechnung des Nockens eines elektrischen Steuermecha-nismus" (in Tschech.), Strojirenstvi 20, 1970, No. 6, pp. 330-334.

841. Koumans, P. W., "A Special Cam-Milling Machine," Rees Jones: Cams and Cam Mechanisms, pp. 65-68.

842. Koumans, P. W., "The Calculation of the Maximum Radius of Curvature of a Cam Using Special Graphs," Rees Jones: Cams and Cam Mechanisms, pp. 116-122.

843. Koumans, P. W., "Numerisch gesteuerte Maschine zum Fräsen von Kurvenkörpern," *Maschinenbautechnik* 28, 1979, No. 9, pp. 397-401.

844. Kozhevnikov, S. N., Antonyuk, E. Y., and Tkachuk, A. I., "Synthesis of a Cam-Differential Mechanism with Periodic Dwell of the Output Link," *Mechanism and Machine Theory* 9, 1974, No. 2, pp. 219-229.

845. Kraemer, O., "Bestimmung des Krümmungshalbmessers im Nocken-gipfel," *MTZ* 4, 1942, pp. 430-433 ("Determination of Radius of Curvature at the Cam Nose").

846. Krames, J., "Über die durch die aufrechte Ellipsenbewegung erzeugten Regelflächen," *Jber. dtsch. Math.-Ver. Abt.* 1, 1940, pp. 58-65 ("On the Ruled Surfaces Generated Directly Through the Eliptical Motion").

847. Krasnikov, V. F., "Some Questions Relating to the Analysis and Synthesis of Cam Mechanisms with Considerations Being Given to the Precsion of the Manufacture of Their Bars," *V. sb. Analize Sintex Mekhanizmov i Teoriya Peredach*, Moscow, Nauka, 1965, pp. 87-101; RZM No. 5, 1966, Rev. 5.A 124.

848. Krasnikov, V. F., "On Methods to Decrease the Additional Dynamic Loads in Cam-Mechanisms," (in Russian), *Mechanika masin* 11/12, 1970, No. 23/24, pp. 146-149.

849. Kreutzinger, R., "Zur Getriebesynthese räumlicher Kurven," *Getriebetechnik* 10, 1942, pp. 441-443 ("On the Mechanism Synthesis of Space Curves").

850. Krzenciessa, H., "Zur Berechnung von Kurvenmechanismen (Arbeitsblätter)," *Maschinenbautechnik* 18, 1969, No. 9, pp. 455-457; No. 11, pp. 581-587; and No. 12, pp. 663-665.

851. Krzenciessa, H., "Verschleissregistriergerät für Kurvenscheiben,"

Maschinenbautechnik, No. 8, 1970, pp. 423-428 ("Wear-Registration Mechanisms for Cam Systems").

852. Krzenciessa, H., "Das Gesetz der 7. Potenz und die biharmonische für Hubbewegungen von Kurvenmechanismen," *Maschinenbautechnik*, No. 1, 1973, pp. 8-12 ("The Law of the 7th Power and of Bi-Harmony in Rise Motions of Cam Mechanisms").

853. Krzenciessa, H., "Zur Getriebetechnik der Papierverarbeitungs-maschinen," *Maschinenbautechnik* 9, 1956, pp. 481-488 ("On Cam Technology in Paper Manufacturing Machinery").

854. Krzenciessa, H., "Verminderung der Beschleunigung an Abtriebsgliedern von Kurvenmechanismen durch Verlängerung der Hubzeit," *Maschinenbautechnik* 11, 1962, pp. 437-442 ("Reduction of Acceleration in Output Links of Cam Mechanisms by Means of Stroke Extension").

855. Kuckenberg, H., "Die Kurvenscheibe im Werkzeugbau," *Werkstatt und Betrieb* 95, 1962, No. 4, pp. 236-242 ("The Disk Cam in the Tooling Industry").

856. Kul'bachnyi, O. I., and Pimenov, V. A., "Geometric Investigation of Cam Mechanisms," *Teoriya Mashin i Mekhanizmov* 98/99, 1964, pp. 28-44.

857. Kulitzscher, P., "Die Analyse von ebenen Kurvenmechanismen auf digit-alen Rechenanlagen," *Maschinenbautechnik* No. 3, 1968, pp. 153-158 ("Analysis of Plane Cam Mechanisms by Means of Computers").

858. Kulitzscher, P., "Ein Beitrag zur Analyse und Optimierung ebener Kurvenmechanismen," Diss., T. H. Karl-Marx-Stadt, 1968.

859. Kulitzscher, P., "Kurvengetriebe mit vorgeschalteter Doppelkurbel," *Maschinenbautechnik*, No. 9, 1967, pp. 498-502 ("Cam Systems Driven by a Double Crank Mechanism").

860. Kulitzscher, P., "Die Optimierung ebener Kurvenmechanismen," *Maschinenbautechnik*, No. 5, 1969, pp. 265-268 ("Optimizing Disc Cams").

861. Kulitzscher, P., "Kurventafeln zur Auswahl der Bewegungsgesetze für Kurvenmechanismen," *Maschinenbautechnik* 19, 1970, No. 6, pp. 324-325.

862. Kulkarni, S. V., "Graphical Solutions for Velocities and Accelerations of Roller-Followers in Tangent-Cam Mechanisms," *Mach. Des.* 34, August 30, 1962, pp. 115-116.

863. Kulkarni, S. V., "Kinematic Analysis of Cam Mechanisms," *Engineers, India* 45, March 1965, pp. 139-154.

864. Kunad, G., "Die Berechnung der Hauptabmessungen von Kurven-getrieben mit gerade geführtem Eingriffsglied," *Maschinenbautechnik*, No. 11, 1967, pp. 41-51 ("Calculations of Proportions of Cam Systems with Straight-Line Follower").

865. Kunad, G., "Die genaue Bestimung der Abmessungen übertragungs-günstiger Kurvengetriebe," Teil I *Maschinenbautechnik*, October 1961, pp. 669-692; Teil II *Maschinenbautechnik*, December 1961, pp. 668-672; Teil III *Maschinenbautechnik* 12, 1963, pp. 324-327 ("Exact Determination of Proportions of Cam Mechanisms with Optimum Transmission Angles").

866. Kutzbach, K., "Zur Ordung der Kurventriebe," *Maschinenbau* 8, 1929, pp. 706-710 ("On the Classification of Cam-Mechanisms").

867. Kuzheleva, I. V., "Synthesis of Planetary Cam Mechanisms According to Prescribed Curves," *Izv Vyssh Uchebnb Zaved, Mashinostr* 1, 1971, pp. 91-95.

868. Kqakernaak, H., and Smit, J., "Minimum Vibration Cam Profiles," *J. Mech. Eng. Sci.* 10, 1968, pp. 219-227.

869. Lache, J., "Anwendung von Kurvengetrieben in Schnellflechtmashinen," *Maschinenbautechnik* 29, 1980, No. 1, pp. 29-33.

870. Lafuente, J. M., "Interactive Graphics in Data Processing-Cam Design on Graphics Console," *IBM Systems J.* 7, 1968, pp. 365-372.

871. Lai, Chien, "Design of Cam Profiles from Arbitrarily Prescribed Acceleration Curves," Chi Hsieh Kung Ch'eng Hsueh Pao, Vol. 17, No. 4, 1981, pp. 11-16.

872. Laibach, J. E., "Converted Milling Machine Makes Short Work of 3-D Cam Grooves," *Mach.* 67, March 1961, pp. 112-113.

873. Lakso, E. E., "Adjustable Cams," *Mach.* 31, October 1924, p. 133.

874. Lalime, L. J., "Surface Grinder Attachment for Finishing Cams," *Am. Mach.* 88, January 6, 1944, p. 88.

875. Lammel, G. "Beitrag zur Berechnung dynamisch hochbeanspruchter Kurvengetriebe," Dissertation T. H. Karl-Marx-Stadt, 1970.

876. Lammel, G., "Schwingungsberechnung in Kurvengetrieben," Wiss. Z. T. H. Karl-Marx-Stadt Vol. 14, 1972, No. 1, pp. 67-75.

877. Lancucki, C. J., and Cole, W. F., "Method of Scribing Cams for Program Temperature Controllers," *J. Sci. Instr.* 42, January 1965, p. 54.

878. Landmisser, F., and Matthias, K., "Beanspruchung des Werkstoffes in rollenden Rädern," *Maschinenbautechnik*, No. 2, 1971, pp. 84-90 ("Material Strain and Wear in Rolling Wheels").

879. Lankester, J. A., "Proportional Braking with Fluid Motor," *Appl. Hydraulics* 9, September 1956, pp. 88-90, 100, and 102.

880. Leaf, G. A. V., and Blackman, F., "Impact on Needles on a Knitting Cam," *Textile Res. J.* 38, June 1968, pp. 651-652.

881. Lebedeev, S. P., "The Analysis and Synthesis of Cam-Mechanisms," *Tr. Kasansk. Avaitzil Inst.* 30, 1954, pp. 3-97.

882. Ledinegg, M., "Rolldaumensteuerungen für schnellaufende Dampfmaschinen," *Maschinenbau und Wärmewirtschaft*, October 1951, pp. 162-169 ("Rolling Cam for High Speed Steam Engines").

883. Lee, W. H., "Cams for Operating Poppet Valves," *Mechanical World* III, June 12, 1942, p. 533.

884. Le Grand, R., "Box Tool Cuts Hexes or Squares," *Am. Mach.* 108, November 9, 1964, pp. 92-94.

885. Le Grand, R., "Tooling for a Cable Drum; Special Cam Generating Attachment for Lathe for Producing Landing-Gear Part; Glen L. Martin Company," *Am. Mach.* 82, December 28, 1938, pp. 1136-1138.

886. Lengyel, A., and Church, A. H., "Cycloidal Cam Charts for Maximum
      Pressure Angle; Reference Book Sheet," *Product Eng.* 22, August 1951,
      p. 183.

887. Lengyel, A., and Church, A. H., "Radial Disk Cam Design Charts for
      Maximum Pressure Angle; Reference Book Sheet," *Product Eng.* 22,
      March 1951, pp. 155, 157, and 159.

888. Lenk, E., "Der Übertragungswirkungsgrad," *VDI-Berichte* 29, 1958, pp.
      75-78 ("Optimum Transfer Effieciency").

889. Lenpartz, J., "Multiple Cam-Grinding Attachment," *Mach.* 25, November
      1918, p. 214.

890. Lenskij, M. F., "The synthesis of plane mechanisms having point contact
      according to certain criteria of quality" (in Russian), Masinovedenie,
      1969, No. 3, pp. 20-21.

891. Lenz, R., "Graphical Method of Modifying Cam Accelerations," *Mach.
      Des.* 29, September 19, 1957, pp. 168-170.

892. Lermann, P., "Ein neuartiges Messverfahren mit kontinuierlicher Ab-
      tastung ebener Kurvenprofile," *Feinwerktechnik* 74, 1970, No. 8, pp.
      352-356.

893. Lermann, Peter, "Automatische Messung und Registrierung der Krüm-
      mungs- und Neigungsverhältnisse ebener Kurvenkonturen," Diss. T. H.
      München, 1968.

894. Levitskii, J. I., "Determination of Fundamental Parameters of Two-Di-
      mensional Cam Mechanisms with a Roller Guided Member," *Sovrem.
      Prob. Teorii Mashin i Mekhanizmov*, M. Nauka, 1965, pp. 95-100, *RZM*
      No. 8, 1966, Rev. 8 A 167.

895. Levitt, A. I., "Aluminum Disk and Felt Cover Polish Precision Cam,"
      Am. Mach., January 3, 1955, p. 125.

896. Lewi, W., "Cam Actions for Injection Molds," *Plastic Technol.* 1, Febru-
      ary 1955, pp. 27-30.

897. Lietz, H., "Carbide Mills Aluminum Cams to Two Tenths," *Am. Mach.*
      102, December 1, 1958, p. 105.

898. Liima, V., "Die-Shoe Aids Punching Round Work," *Am. Mach.* 99, Dec-
      ember 19, 1955, pp. 134-135.

899. Linderoth, L. S. Jr., "Calculating Cam Profiles," *Machine Design* 23, July
      1951, pp. 115-119.

900. Lindner, K., "Untersuchungen über Störbeschleunigungen an Kurven-
      scheiben infolge von Formfehlern sowie ein neues Verfahren zum Berech-
      nen und Herstellen von Kurvenscheiben," *VDI-Forschungsheft* 477, VDI-
      Verlag, 1960 ("Examination of Irregular Acceleration in Disc Cams Due
      to Structural Defects, Including a New Procedure for Calculation and
      Manufacturing Cam Disks").

901. Lindner, K., "Berechnungsverfahren für die Form von Kurvenscheiben
      (Leistungsfähige und wirtschaftliche Getriebekonstruktion, Getriebe-
      tagung Aachen 1959)." *Konstruktion* 12, 19 October, pp. 33-34
      ("Methods of Calculation for the Form of Disk-Cams. Efficient and
      Economic Mechanism Design, Mechanism Conference Aachen 1959").

902. Lindner, K., "Vorschläge für die Berechnung und spanlose Formung von Kopiermodellen (Meisterkurven) zur Herstellung von Kurvenscheiben," *Konstruktion*, 1959, 55, and 1960, pp. 33-34 ("Suggested Techniques for the Computation and Fault-Free Design of Master Copiers for the Production of Disc Cams").

903. Lindsey, J. B., "Simple Compensating Set-Up for Milling Cams," *Mach.* 56, May 1950, pp. 190-192.

904. Liniechi, A. G., "Optimum Design of Disc Cams by Nonlinear Programming," 4th OSU Applied Mechanisms Conference, 1975, pp. 6.1-6.18.

905. Litovcenko, V. P., and Sergeev, P. V., "The Synthesis of Globoidal and Spherical Cam Mechanisms with Respect to Pressure Angles" (in Russian), Teorija mechanizmov i masin Vol. 14, 1973, pp. 90-101.

906. Litten, W. H., "Eccentric Clamps for Jigs and Fixtures," *Machy. (London)* 69, October 10, 1946, pp. 453-458; October 31, 1946, pp. 551-555; December 5, 1946, pp. 726-728.

907. Litwin, F. L., "Die Konstruktion eines Kurvengetriebes für die Wiedergabe einer gegeben Funktion," *Maschinenbautechnik*, No. 10, 1967, pp. 520-523 ("Construction of a Cam Disk System to Ascertain a Given Function").

908. Locke, L. L., "Fixture for Milling Face Cams," *Am. Mach.* 76, November 9, 1934, p. 1136.

909. Lockenvitz, E., Oliphant, J. B., Wilde, W. C., and Young, J. M., "Geared to Compute," *Automation* 2, August 1955, p. 37.

910. Lockenvitz, A. E., Oliphant, J. B., Wilde, W. C., and Young, J. M., "Non-Circular Cams and Gears," *Mach. Des.* 24, No. 5, May 1952, pp. 141-145.

911. Loeff, L., and Soni, A. H., "Optimum Sizing of Planar Cams," *IFToMM*, 1975, Vol. 4, pp. 777-780.

912. Loeser, E. H., "Bench Test of Cam and Tappet Wear in Additive Oils," *ASME Trans.* 3, October 1960, pp. 184-190.

913. Loeser, E. H., Wiquist, R. C., and Twiss, S. B., "Cams and Tappet Lubrication, IV: Radioactive Study of Sulfur in the EP Film," *ASLE Trans.* 2, October 1959, pp. 199-207.

914. Loewe, A. G. von, "Design of Valve Control Organs—Formulas Developed for Springs, Cams and Camshafts," *Automobile* 31, September 24, 1914, pp. 584-588.

915. Longstreet, James, R., "Systematic Correlation of Motions," *Transactions of the First Conference on Mechanisms*, Sponsored by *Machine Design*, 1953, pp. 43-48.

916. Longstreet, J. R., "Systematic Correlation of Motion," *Mach. Des.* 25, December 1953, p. 215.

917. Lorman, S. A., "Mechanisms with Disc Cams of Equal Width" (in Russian), Masinovedenie, 1968, No. 3, pp. 62-69.

918. Lorman, S. A., "A Method to Combine Motiom Diagrams for Cam Followers" (in Russian), Mechanika masin, 1969, No. 15, bis 16, pp. 74-82.

919.   Lotze, W., "Measuring of Flat Cam Profiles on the JENA-Made Universal Measuring Microscope as Well as Evaluation and Tolerance Simulation with the Electronic Data Processing Equipment" (Technical Univ., Dresden, West Germany), Harmann, M., Urban, D., *Jena Rev.* 16, 1971, pp. 136–141.

920.   Lotze, W., Hartmann, M., and Urban, D., "Messung ebener Kurvenprofile auf dem JENA-Universalmessmikroskop sowie Auswertung und Toleranz-simulation mittels, EDV-Anlage," *Jenaer Rundschau* 16, 1972, No. 2, pp. 136–141.

921.   Loudon, H. A., "Machining a Large Cam Drum," *Mach.* 31, August 1925, pp. 979–982.

922.   Low, B. B., "Internal-Combustion Engine Cams," *Engineering* 115, May 25, 1923, pp. 641–644.

923.   Luder, Siegfried, "Steuernocken aus Kunststoff," *Ind. Anz.* 91, 1969, No. 61, pp. 1485–1486.

924.   Lurz, F. C., "Milling Machine Cuts Slotted Cams Using Removable Fixtures on Table," *Am. Mach.*, April 8, 1948, p. 121.

925.   Maas, margaret A., "Glass-Reinforced Nylon Battle High Impact Forces on High-Speed Cams," *Design News* 24, 1969, No. 9, p. 66.

926.   Mac, T., "New Machines at Pachard used for Crankshafts and Camshafts," *Automotive Ind.* 112, May 15, 1955, pp. 70–73.

927.   Mader, O., "Konstruktion der Ventilbeschleunigung bei Füllungsände-rungen," *Dingl. Polyt. J.*, 1911, p. 17.

928.   Mahig, J., "Spring and Follower Characteristics Due to Internal Damping and Cam Actuation," ASME Paper 70-MECH-76, 1970.

929.   Maker, P., "Cam Design and Manufacture," *Transactions of the Second Conference on Mechanisms*, Purdue, University, Lafayett, 1954, pp. 4–6.

930.   Maker, P., "Cam Design and Manufacture," *Mach. Des.* 26, December 1954, pp. 188–190.

931.   Malikow, G. F., "Die Berechnung der Kurvenscheiben von Waagen mit linearer Skala," *Feinwerktechnik*, No. 9, 1960, pp. 313–315 ("Computa-tion of Disk Cam Mechanisms for Scales Having Linear Graduation").

932.   Mallina, R. F., "Analysis of Uniform-Rise and Uniform-Pressure Angle Cam Curves," *ASME Trans.*, 1929.

933.   Mallory, J. W., "Super Nerve Center Cam has 720 Exact Stations on One Square Inch; for Automatic Control of Aircraft in Flight," *Am. Mach.*, March 31, 1952, pp. 83–86.

934.   Maltendorf, M., "Kurventrieb an Druckmaschinen," *Schriftenreihe Verl. Technik Band 193*, pp. 52–61 ("Cam Mechanisms in Printing Machines").

935.   Mal'tsev, V. F., "Determination of the Shape of the Roller of Three-Di-mensional Cylindrical Cam Mechanisms," *Trudi Odesk jekhnol. in-ta 7*, 1955, pp. 47–52 (in Russian), *Ref. Zh. Mekh.*, No. 10, 1956, Rev. 6458.

936.   Malyarova, V. M., "The Dynamics of Cam Gears," (in Russian), *Trudi Sibirsk. Meatllurg, In-ta 4A* (Prikl. Mat. Mekh.), 1957, pp. 139–153; *Ref. Zb. Mekb.*, No. 5, 1959, Rev. 4728.

937. Mang, Josef, "Basics of Index Drives," *Power Transm. Des.* 23, No. 6, June 1981, pp. 79-84.

938. Manolescu, N. I., Stanescu, D., and Boncoi, G., "The Optimization of the Cams Utilized in the Construction of the Machine-Tools" (in French), *IFToMM*, 1979, Vol. 2, pp. 1295-1298.

939. Marina, M., and Cioara, T., "Über die Vorteile der Benutzung der soge-nannten Polydyne Kurventriebe in der Steuerung der Verbrennungs-motoren," *IFToMM*, 1971, Vol. B, pp. 89-101.

940. Markhauser, Anthony W., "New Method of Automotive Cam Design," ASME Paper 65-MECH-37.

941. Marklew, J. J., "Utilization of Automatics Improved by Applying NC to Cam Making," *Machinery and Production Eng.*, April 1971, pp. 580-582.

942. Markovskii, E. A., Krasnoshchekov, M. M., and Tikhonovich, V. I., "The Character of Failure of Friction Surfaces in Abrasive Wear," *Soviet Ma-terials Science* 2, March/April 1966, pp. 162-163 (Translation of Fiziko-Khmimicheskaya Mekhanika Materialov 2, 1966, pp. 222-223, by Fara-day Press, New York).

943. Marks, R. E., "Laying-Out a Variable Cam," *Am. Mach.* 66, April 14, 1927, p. 632.

944. Marsch, W., "Circular-Arc Cam with Flat Follower; Method of Drawing Displacement, Velocity and Acceleration Curves," *Engineering* 157, May 12, 1944, p. 362.

945. Marton, R. D., "Cam-Controlled Air Regulator Provides Constant Nip Pressure for Bundling Fabric on Roll," *Instruments & Control Systems* 33, May 1960, p. 816.

946. Marx, G., "Harmonischer Nocken mit Flachstössel," *VDI-Z* 87, 1943, pp. 557-561 ("Harmonic Cams with Flat Face Follower").

947. Marx G., "Getriebe mit stillstehendem Nocken," *Motortechn. Z.* 8, 1947, pp. 44-47 ("Mechanisms with Fixed Cams").

948. Mason, A. J., "Camshaft Bearings Made by Unique and Efficient Me-thods," *Mach.* 66, August 1960, pp. 107-109.

949. Matthaei, H. D., "Ein Kurvengetriebe für vier Rasten veränderlicher Dauer," *VDI-Berichte* No. 167, Düsseldorf: VDI-Verlag, 1971, pp. 153-158 ("Cam Mechanisms with Four Adjustable Dwells").

950. Matthew, G. K., and Tesar, D., "The Design of Modeled Cam Systems. Part I: Dynamic Synthesis and Chart Design for the Two-Degree-of-Freedom Model," *Trans. ASME*, Ser. B: J. Engng. Ind. Vol. 97, No. 4, pp. 1175-1180.

951. Matthew, G. K., and Tesar, D., "The Design of Modeled Cam Systems. Part II: Minimization of Motion Distortion Due to Modeling Errors," *Trans. ASME*, Ser. B: J. Engng. Ind. Vol. 97, 1975, No. 4, pp. 1181-1189.

952. Matthew, G. K., and Tesar, D., "Cam System Design: The Dynamic Synthesis and Analysis of the One-Degree-of-Freedom Model," *Mecha-nism and Machine Theory* 11, 1976, No. 4, pp. 247-257.

953. Matthew, G. K., "The Modified Polynomial Specification for Cams," *IFToMM*, 1979, Vol. 2, pp. 1299-1302.

954. Matthew, G. K., and Tesar, D., "Formalized Matrix Method for Nth Derivative Trapezoidal Motion Specifications for Cams," *IFToMM*, 1971, Vol. H, pp. 247-259.

955. Matveev, K. K., "Design of Drum-Type Cam Mechanisms Located by One Roller," *Russian Eng. J.* XLIX, 1969, pp. 32-34.

956. Matveev, K. K., "Wear on Cams in Rotating Mechanisms," *Russian Eng. J.* XLVI, 1966, pp. 35-38.

957. Matveev, K. K., "Rotating Cam Mechanisms," *Machines and Tooling*, December 1960, pp. 23-25.

958. Maul, F., "Konstruktion von Kurvenscheiben," *Reuleaux-Mitt.* 3, 1935, pp. 105-107 ("Design of Disk Cams").

959. Maul, F., "Getriebepraxis bei Verpackmaschinen," *Z-VDI* 73, 1929, p. 481 ("Practical Cam Technology in Packing Machinery").

960. Mauzitov, G. G., "Calculation of A Cam Profile for Cam Drives," *Sov. Eng. Res.* 3, No. 10, October 1983, pp. 28-29.

961. Mawson, R., "Fixture Speeds Critical Grinding of Octagonal Shaped Cams," *Iron Age* 165, February 9, 1950, pp. 91-92.

962. Mawson, R., "Improved Cam Movement," *Tool Eng.*, December 1948, p. 41.

963. Mawson, R., "One Way to Produce a Cam," *Steel* 120, April 28, 1947, p. 108.

964. Mawson, R., "Simple Milling Fixture for Profiling and Slotting Cam Blanks," *Mach.* 57, February 1951, pp. 199-200.

965. Mawson, R., "Unusual Cam Milling Fixture," *Iron Age* 163, January 27, 1949, pp. 66-67.

966. McDade, J. R., "Improvised Cam-Cutting Attachment," *Am. Mach.* 68, May 3, 1928, pp. 744-745.

967. McEvoy, S., "Cams for Brown and Sharpe Automatic Screw Machines," *Am. Mach.* 57, September 21, 1922, p. 460.

968. McGrew, J., "Research Refines the Theory of Lubrication Under Heavy Loads," *Mach. Des.* 43, April 15, 1971, pp. 88-92.

969. McIntosh, J., "Making Master Grinding Cams," *Am. Mach.* 54, January 27, 1921, pp. 125-126.

970. McKay, "The Design of Cams for High Speed Motors," *The Engineer* III, p. 614.

971. McKellar, M. R., "Overhead Camshaft Stirs New Tempest," *SAEJ* 74, November 1966, pp. 69-73.

972. McKenna, A. E., "Multiple Cams," *Rayon & Melliand* 17, February 1936, pp. 87-88.

973. McLaughlin, C., "Grinding Cams on a Lathe," *Tool Eng.* 40, April 1958, p. 83.

974. Meltendorf, M., "Kurventrieb an Druckmaschinen," Schriftenreihe Verlag Technik," *Getriebetechnische Probleme* 193, pp. 52-61 ("Operations of Cam Mechanisms in Printing Presses").

975. Meneghetti, U., and Andrisano, A. O., "On the Geometry of Cylindrical Cams," *IFToMM*, 1979, Vol. 1, pp. 595-598.

976. Meneghetti, U., "Polycentric Cams for Translating Roller Follower," *Technica Italiana* 33, December 1968, pp. 809-813 (in Italian).

977. Mercer, S., and Holowenko, A. R., "Dynamic Characteristics of Cam Forms Calculated by the Digital Computer," *ASME Paper* No. 57-A-42.

978. Merrill, C. F., "Cam Layout for Loom Harnesses," *Am. Mach.* 47, August 23, 1917, pp. 336-337.

979. Mery, R., "Cam for Jig Clamping," *Am. Mach.* 96, December 9, 1952, 145.

980. Mery, R., "Table of Cam Dimensions Simplifies Jig and Fixture Design," *Am. Mach.* 93, August 25, 1949, p. 105.

981. Meulemeester, D., de, "Formule generale pour la determination des accelerations des organes commandes par came," *Genie Civil* 98, March 21, 1931, pp. 285-288.

982. Meyer, C. F., "Jig to Mill Cam Slots," *Am. Mach.* 49, October 31, 1918, pp. 820-821.

983. Meyer, C. F., "Jig to Mill Cam Slots," *Mach.* 19, September 1912, pp. 54-55.

984. Meyer, Zur Capellen, W., Rischer, K. A., and Danke, P., "Beschleunigungsermittlung in Gelenkgetrieben, Kurventrieben und Räderkurbelgetrieben," *Industrie-Anzeiger* 84, 1962, No. 78, pp. 1873-1878, No. 96, pp. 2243-2248 ("Determining Acceleration in Links, Cams and Geared Crank Mechanisms").

985. Meyer, Zur Capellen, W., and Janssen, B., "Kraftübertragung bei ebenen Kurventrieben mit geradegeführtem Abtrieb," Part 1: *Antriebstechnik*, April 1962, pp. 8-13; Part 2: *Antriebstechnik*, July 1962, pp. 75-80 ("Force Transmission in Planar Cam Mechanisms with Translating Follower").

986. Meyer, Zur Capellen, W., "Nonogramme zur geneigten Sinuslinie," *Forschungsbericht des Landes Nordrhein - Westfalen* 772, Köln - Opladen, 1959, ("Nomograms for a Modified Cycloid").

987. Meyer, Zur Capellen, "Spherical Cams Take Kinks out of Linking Up Shafts," *Product Eng.* 1066, pp. 60-62.

988. Midha, A. M., Badlani, L., and Erdman, A. G., "Periodic Response of High Speed Cam Mechanism with Flexible Follower and Camshaft using a Closed-Form Numerical Algorithm," IFToMM, 1979, Vol. 2, pp. 1311-1314.

989. Miller, M. C., "Action of Camshaft in Full-Fashioned Hosiery Narrowing," *Textile World* 78, April 25, 1931, pp. 1880-1883.

990. Millington, W. D., "Tape-Controlled Camshaft Inspection," *Mach.* 72, January 1966, pp. 87-89.

991. Minshi, Zeng, and Cunhou, Hu, "New Progress in the Cam Design of Supercharged Diesel Engine Fuel Injection Pumps." *Neiranji Xuebao* 1, No. 4, October 1983, pp. 73-82.

992. Mirel, A., "Mechanical Memory Devices," ASME Paper No. 58-MD-1, for meeting April 14-17, 1958, p. 5.

993. Mischke, C., "Optimal Offset on Translating Follower Plate Cams," *J. Eng. Ind.* 92, February 1970, pp. 172-176.

994. Mischke, C., "Assessment of Ultimate Fidelity of an Eccentric Circular Disk Cam with Translating Flatfaced Follower as a Function Generator," *Trans. ASME*, Ser. B., Vol. 96, 1974, No. 1, pp. 256-260.

995. Mitchell, D. B., "Cam Follower Dynamics," *Mach. Des.* 22, June 1950, pp. 151-154; Same, *Mech. Eng* 72, June 1950, pp. 467-471; Discussion 72, December 1950, pp. 1009-1011.

996. Mitchell, D. B., "Tests on Dynamic Response of Cam-Follower Systems," *Mech. Engng.* 72, June 1950, pp. 467-471 ("Dynamische Versuche an Kurvenrollengebeln").

997. Moberg, I., "Loomfixer's Manual: Cotton," *Textile World* 88, December 1938, p. 85; 89, January 1939, p. 77.

998. Molian, S., "Cam Mechanism Design," *Engineering Materials and Design* 8, June 1965, pp. 396-397.

999. Molian, S., "Cam Mechanism Trajectories," *Engineering Materials and Design* 9, July 1966, pp. 76-79.

1000. Molian, S., "Use of Algebraic Polynomials in Design of Cams," *Engineer* 215, February, 1963, pp. 352-354.

1001. Molian, S., "Mechanisms with Discontinuous Transmission Ratios," *Mechanism and Machine Theory* 8, 1973, pp. 365-369.

1002. Moon, C. H., "Designing Cam Profiles; Data Sheet," *Mach. Des.* 33, April 13, 1961, pp. 179-187.

1003. Moon, C, H., "Cam Curve Synthesis," *Design News* 21, 1966, pp. 156-161.

1004. Moon, C. H., "Simplified Method for Determining Cam Radius of Curvature; Data Sheet," *Mach. Des.* 34, August 2, 1962, pp. 123-125.

1005. Moran, J. M., "Construction of Monochromator Cams for Recording Spectrophotometers," *R. Sci. Instr.* 14, October 1943, pp. 287-293.

1006. Morgan, H., "Cam-Controlled Punch Forms Accurate Tube," *Am. Mach.* 96, September 29, 1952, p. 131.

1007. Morgan, M., "Punched-Tape Control for Automatic Milling of Master Cams," *Machinery* 63, April 1957, pp. 182-184.

1008. Morris, A. W., and Simonds, H. R., "Die Casting Iron and Steel; Casoomatic Die Casting Machine," *Iron Age* 131, June 29, 1933, pp. 1028-1030.

1009. Morrison, Ralph A., "Load/Life Curves for Gear and Cam Materials," *Mach. Des.* 40, 1968, No., 18, pp. 102-108.

1010. Morrison, R. A., "Test Data Let You Develop Your Own Load/Life Curves for Gear and Cam Materials," *Mach. Des.* 40, August 1, 1968, pp. 102-108.

1011. Morrow, L. C.,"Time-Savers on the Oregon Short Line," *Am. Mach.* 66, February 3, 1927, pp. 207-210.

1012.   Morse, J. E. C., Id. and Hinkle, R. T., "How to Analyse Rolling-Contact Mechanisms for Acceleration Characteristics," *Mach. Des.* 28, July 26, 1956, pp. 75-77.

1013.   Moskowitz, David, "Equations for Calculating Heart Shaped Cams," *Design News*, July 1, 1952, p. 47.

1014.   Mountain, K. L., "Casting Camshafts in Shell Molds; General Motors Corp." *Foundry* 84, September 1956, pp. 124-128.

1015.   Müller, F. O., and Müller, J., "Zur Leistungssteigerung von Kurvengetrieben," *Technik* 10, 1955, pp. 145-155 ("How to Increase Performance of Cam Mechanisms").

1016.   Müller, J., "Ebene Kurvenkoppelgetriebe," *Maschinenbautechnik* 6, 1957, pp. 118-123 ("2-D Cam-Linkage Mechanisms").

1017.   Müller, J., "Herstellung von Kurvenschablonen im zwanglaufmechanischen Erzeugungsverfahren," *Konstruktion*, 1961, p. 71; *Maschinenbautechnik*, 1961, p. 220 ("Manufacture of Cams Using Motion of Mechanisms").

1018.   Müller, J., "Maschinelle Herstellung von Kurvenkörpern (Kurvenschablonen) im zwanglaufmechanischen Erzeugungsverfahren," *Maschinenbautechnik* 12, December 1963, pp. 657-666 ("Manufacture of Cams Using Motion of Mechanisms").

1019.   Müller, J., "Begriffe für Kurvengetriebe," *Maschinenbautechnik*, No. 9, 1969, pp. 489-493 ("General Definitions in Cam Systems").

1020.   Müller, J., "Über den Einfluss der Bearbeitungsverfahren der Kurvenkörper," *Maschinenbautechnik* 18, 1969, No. 5, pp. 269-274.

1021.   Müller, J., "Einsatz der EDV zur rationellen Kurvenkörperherstellung," *Maschinenbautechnik* 18, 1969, No. 7, pp. 387-388.

1022.   Müller, J., "Fertigungsgenauigkeiten der Abtastrollen von Kurvengetrieben," *Fertigungstechnik und Betrieb* 19, 1969, pp. 172-175.

1023.   Müller, J., "Bestimmungsgrössen an Kurvengetriebe," *Maschinenbautechnik*, No. 1, 1971, pp. 22-26 ("Numerical Values in Cam Systems").

1024.   Müller, J., "Grundbegriffe für Abweichungen an Kurvengetrieben," *Maschinenbautechnik* 23, 1974, No. 4, pp. 146-149.

1025.   Müller, J., and Buchholz, H., "Experimentelle Untersuchungen des Spieleinflusses im Ventilgetriebe," *Agrartechnik* 26, 1976, No. 3, pp. 139-140.

1026.   Müller, J., and Kassamanian, A. A., "Zur Systematik des Kurvengetriebes," *Maschinenbautechnik* 28, 1979, No. 9, pp. 427-428.

1027.   Müller, P. M., "Generation of Cycloid Curves," *Am. Mach.*, April 7, 1927, p. 597.

1028.   Mumper, A., "Cam Cutting on a Boring Mill," *Mach.* 28, March 1922, pp. 552-553.

1029.   Myatt, D. J., "Determining Cam Profiles," *Mach. Des.* 37, June 1965, pp. 174-182.

1030.   Mykiska, A., "Utilization of Cam in Feedback of Pneumatic Drive for Linearizing Control Element," *Automatizace* 11, 1968, pp. 5-8.

1031.   Nagellis, O. O., "Cam and Ratchet Intermittent Mechanism," *Mach.* 64, April 1958, p. 156.

1032.   Nahaj, T., "Fixture for Low Cost Machining of Cams," *Machine & Tool Blue Book* 53, June 1956, pp. 136-139.

1033.   Neff, A. A., "Cam-Controlled Variable-Speed Drive," *Machinery* 44, March 1938, pp. 436-437.

1034.   Neklutin, C. N., "Cams for High Speeds," *Product Eng.* 5, 1934, pp. 333-335.

1035.   Neklutin, C. N., "Designing Cams for Controlled Inertia and Vibration," *Mach. Des.* 24, June 1952, pp. 143-160.

1036.   Neklutin, C. N., "Vibration Analysis of Cams," *Transactions of the Second Conference on Mechanisms*, Purdue University, 1954, pp. 6-14.

1037.   Neklutin, C. N., "Vibration Analysis of Cams," *Machine Design* 26, December 1954, p. 190.

1038.   Neklutin, C. N., "Cam Design Tables," *Transactions of the Fifth Conference on Mechanisms*, Purdue University, October 1958, pp. 77-90.

1039.   Neklutin, C. N., "Trig-Type Cam Profiles," *Mach. Des.* 31, 1959, pp. 175-187.

1040.   Neklutin, C. N., "Springs for Cam Followers," *Mach. Des.* 33, December 7, 1961, pp. 195-200.

1041.   Neklutin, C. N., "Consider Cam Feeds for Increasing Press Speeds," *Automation* 9, March 1962, pp. 61-63.

1042.   Neklutin, C. N., "Mechanisms and Cams for Automatic Machines," New York: Elsevier Publ. Co., 1969.

1043.   Nelson, A. L., "Cam Design Considered Mathematically," *Horseless Age* 34, September 23, October 21, 1914, pp. 466-469, 501-502, 533-535, 631-633.

1044.   Nelson, A. L., "Special Cams Designed for Aluminum Engine of Premier," *Automobile* 36, April 5, 1917, pp. 689-692; Abstract, *ASME J.* 39, May 1917, p. 469.

1045.   Nerge, G., "Wurfhebelantrieb von Druckmaschinen," *VDI-Berichte* 29, 1958, pp. 157-158 ("Driving Links in Printing Presses").

1046.   Nerge, G., "Über den Stand der Analyse und der Synthese von Kurvengetrieben in den USA," *Maschinenbautechnik* 8, 1959, pp. 620-621 ("Summary Analysis and Synthesis of Cam Mechanisms in the United States").

1047.   Nerge, G., "Zur Beurteilung der Bewegungsgesetze nach ihren Beschleunigungskennwerten," *Maschinenbautechnik* 9, 1960, pp. 334-335 ("Considerations on the Laws of Movement Based on Acceleration Coefficients").

1048.   Nerge, G., "Beziehungen zwischen Antriebsmoment und Kurvenkörpergrösse bei Kurvengetrieben," *Maschinenbautechnik* 9, 1960, pp. 661-664 ("Relation Between Input Moment and Size in Cam Mechanisms").

1049.   Nerge, G., "Der Stand der Technik auf dem Gebiet der Konstruktion von Kurvenmechanismen, insbesondere für Verarbeitungsmaschinen," *Maschinenbautechnik* 10, 1961, pp. 135-141 ("Current Technology in the Construction of Cam Mechanisms, Particularly of Manufacturing Machinery").

1050. Nerge, G., and Hugk, H., "Zur Ermittlung der geneigten Sinuslinien mit kleinster Maximalbeschleunigung," *Maschinenbautechnik* 11, November 6, 1962, pp. 327-329 ("On the Computation of Modified Cycloids with Lovest Maximum Acceleration").

1051. Nerge, G., "Tafel der Kennwerte symmetrischer Bewegungsgesetze für Kurvenmechanismen," *Maschinenbautechnik* 11, November 8, 1962, pp. 433-437 ("Table of Characteristic Data of Symmetrical Laws of Movement for Cam Mechanisms").

1052. Nerge, G., "Dynamische Untersuchungen zum Verschleissverhalten der Kurvenmechanismen," *Maschinenbautechnik*, No. 2, 1967, pp. 57-59 ("Dynamic Investigations of Wear Behavior of Cam Systems").

1053. Nerge, G., "Verschleiss an Kurvenmechanismen," *Maschinenbautechnik* No. 9, 1969, p. 494 ("Wear in Cam Mechanisms").

1054. Nerge, G., "Berechnung der hydrodynamischen Tragfähigkeit der Eingriffspaarungen von Kurvenmechanismen," Wiss. Z. TU Dresden Vol. 22, 1973, No. 1, pp. 103-112.

1055. Nerge, G., "Ein einfaches grapho-analytisches Verfahren zur Ermittlung von Wertefolgen für die Fertigung von Kurvenschablonen," *Maschinenbautechnik* 23, 1974, No. 2, pp. 50-52.

1056. Nerge, G., "Zweckmässige Formgestalt von Kurvenscheibe und Rolle," *Maschinenbautechnik* 28, 1979, No. 9, p. 422.

1057. Nerge, G., "Einfluss des Rollenspiels auf das dynamische Verhalten von Nutkurvenmechanismen," *Maschinenbautechnik* 28, 1979, No. 9, pp. 419-421.

1058. Neumann, R., "Changierende Textilmaschinengetriebe mit zwei konstanten Geschwindigkeitsbereichen," *Maschinenbautechnik*, No. 3, 1971, pp. 125-129 ("Differentiating Textile Machine Mechanisms with Two Constant Speed Ranges,").

1059. Neumann, R., "Zum dynamischen Verhalten von Rastgetrieben," *Klepzig-Fachberichte* 81, 1973, No. 3, pp. 97-101.

1060. Nevin, D. A., "Standard Cams for Brown & Sharpe Automatic Screw Machines," *Am. Mach.* 56, June 8, 1922, pp. 839-841.

1061. Nevin, D. A., "Universal Cams for Automatic Screw Machines," *Mach.* 28, June 1922, p. 833.

1062. Newman, P. A., "Cam-Generating Fixture," *Mach.* 35, February 1929, pp. 443-444.

1063. Newton, J. A., "Valve-Gear Fundamentals for the Large-Engine Designer," *ASME Trans.*, 1954, pp. 137-151.

1064. Nicolay, K., "Getriebetechnische Aufgaben an Nähmaschinen," *VDI-Z* 96, 1954, pp. 363-365.

1065. Nicolay, K., "Lösung einer getriebetechnischen Aufgabe für eine Haushaltsnähmaschine," *Feinwerktechnik* 58, 1954, pp. 44-48.

1066. Nicolosi, R. V., "Elements of Cam Design," *Mach.* 62, August 1956, pp. 183-190; September 1956, pp. 179-187.

1067.   Niessen, "Einrichtung zum Drehen von Brennstoffnocken aus Rund-
         stangen," *Werkstattstechnik* 30, 1936, p. 192 ("Device for the Tooling
         of Fuel Cams from Round Bars").
1068.   Nilson, H. J., "Special Hone Aligns Machine Bearings," *Am. Mach.* 93,
         October 6, 1949, p. 131.
1069.   Nittle, K. W., "Combination Cam Controls Stock Feed of Wire-Forming
         Machine," *Mach.* 64, August 1958, pp. 114-115.
1070.   Noltimier, F., "Extending Camshaft Operating Life for Increased Engine
         Power Output," *Pet. Refiner* 22, July 1943, pp. 100-104.
1071.   Noortgate, L. van den, and Fraine, J. de, "A General Computer Aided
         Method for Designing High Speed Cams Avoiding the Dangerous Exci-
         tation of the Machine Structure," *Mechanism & Machine Theory* 12,
         1977, No. 3, pp. 237-245.
1072.   Nooss, W., "Iterative Kurvenscheiben Synthese mittels Digitalrechner,"
         Feinwerktechnik, No. 8, 1967, pp. 378-382 ("Iterative Cam Disk
         Synthesis by Means of Digital Computations").
1073.   Nooss, W., "Ein universelles Rechnerprogram für ebene Kurvenscheiben-
         getriebe," *Konstruktion* 21, 1969, No. 11, pp. 441-446.
1074.   Norton, "Norton Grinder for Radial Type Airplane Motor Cams," *Iron
         Age* 12, February 28, 1929, p. 614; *Am. Mach.* 70, March 7, 1929, pp.
         412-413; *Automotive Ind.* 60, March 2, 1929, p. 382; *Mach.* 35, March
         1929, p. 539.
1075.   Nourse, J. H., "Designing an Optimum Cam," *SAEJ* 68, November
         1960, pp. 92-94.
1076.   Nourse, J. H., Dennis, R. C., and Wook, W. M., "Recent Developments
         in Cam Design," 1960, pp. 585-613.
1077.   Nourse, J. H., "Recent Developments in Cam Profile Measurement and
         Evaluation," *SAE-Paper* 964 A for meeting January 11-15, 1965.
1078.   Nowak, H., "Experience Gained at Dimensioning Cams for Valve Con-
         trol Mechanisms," *Maschinenbautechnik* 16, November 1967, pp. 612-
         615.
1079.   Nowak, "Bewegungsgesetz für Ventiltriebe an Verbrennungsmotoren,"
         Maschinenbautechnik 18, 1969, No. 11, pp. 609-610.
1080.   Nowak, H., "Bewegungsgesetze für Ventilsteuerungen, ihre Berechnung
         und Beurteilung," *Maschinenbautechnik* 19, 1970, No. 12, pp. 650-
         655.
1081.   Nuhn, G. L., "Double-Acting Indexing Cams," *Mach.* 37, January 1931,
         p. 333.
1082.   Oertel, F., "Getriebetechnik bei Textilmaschinen," *VDI-Tagungsheft* 1,
         pp. 41-46 Düsseldorf, 1953 ("Mechansims in Textile Machinery").
1083.   Ogarva, K., "Analysis of Complex Motion in Cam Mechanisms," *Bull.
         Japan Soc. M.E.'s* 6, 1963, pp. 113-121.
1084.   Ogina, S., "Designing Characteristics of Constant Diameter Cams: Con-
         tribution to Brunell's Paper," *Bell. JSME* 10, December 1967, pp.
         1032-1038.

1085. Ogozalek, F. J., "Theory of Catendidal-Pulse Motion and its Application to High-Speed Cams," ASME Paper 66-MECH-45, 1966.

1086. O'Hanlon, E., "Cam Design," *Automobile Eng.* 60, July 1970, pp. 297-299.

1087. Okouoglu, S. A., "An Application of Polydyne Cam Design," *Transactions of the Sixth Conference on Mechanisms*, Purdue University, 1960, pp. 44-47.

1088. Oledzki, A., "Method for Obtaining Kinematic Closure of Plane Cam Mechanisms," *Archiwum Budowy Maszyn* 10, 1963, pp. 383-388.

1089. Oledzki, A., "Tests of Cam Mechanisms for Intermittent Motion of a Cinematographic Band with Reduced Time of Band Displacement," *Archiwum Budowy Maszyn* 11, 1964, pp. 354-379.

1090. Oledzki, A., "Some Dynamic Properties of Systems with Cam Mechanisms," *Mechanism and Machine Theory* 8, 1973, No. 4, pp. 543-554.

1091. Oledzki, A., "Transients in Cam-Mechanisms," *IFToMM*, 1971, Vol. G, pp. 179-188.

1092. Omstead, E. H., and Taylor, E. A., "Poppet Valve Dynamics," *Journal of Aeronautical Sciences* 6, 1938, p. 370.

1093. Onq, G. Y., "Pressure Angles in Cam Roller Followers," ASME Paper 52-SA-33 for meeting June 15-19, 1962.

1094. Oppen, H., "Nockenwellenmessmaschine, eine Anlage zum Ausmessen der Konturen von Rotations-Körpern," *Siemens* 44, July 1970, pp. 474-476 ("New Cam Profile Measuring Machine").

1095. Orrok, N. E., "Machining with Oxide Tooling," *Metal Prog.* 90, August 1966, p. 148.

1096. Orthwein, William, C., "Numerical Design of Plate Cams for a Translating Roller Follower," *Comput. Mech. Eng.* 2, No. 1, July 1983, pp. 63-70.

1097. Oshima, T., "Cam Profile Machining by Numerically Controlled Milling Machine," *Rev. Electrical Communication Laboratory* 18, March/April 1970, pp. 226-234.

1098. O'Wril, J. J., "Double Spring-Loaded Tooling Cuts Combination Cams," *Am. Mach.* 100, April 9, 1956.

1099. Page, S. C., "Wear of Valve Cams on Vertical Engines; Value of Proper Lubrication," *Power Pl. Eng.* 39, November 1935, p. 643.

1100. Pagel, B. F., "Mathematical Solution for Involute Cam Problems of Radial Follower Type," *Product Eng.* 12, February 1941, pp. 8-87.

1101. Pagel, B, F., "Custom Cams from Building Blocks," *Machine Design*, 1978, May 11, pp. 2-6.

1102. Pajares, Diaz E., "Application of Theory of Plane Curves Involved in Mechanism of One Class of Cam Profiles," *Revista de Ciencia Aplicada* 8, July/August 1954, pp. 338-343.

1103. Pan, H. H., "General Equations for Finding Cam Curves and Cutter Pitch Curves," *Mach. Des.* 29, July 11, 1957, pp. 137-140.

1104. Raquin, J. R., "Side Cam Solve Die-Design Problems; Drawings with Text," *Am. Mach.* 97, August 31, 1951, pp. 89-91.

1105. Park, J., "Cam Design," *Automobile Eng.* 15, April 1925, pp. 110-113.

1106. Parsons, A. B., "Cam-Shaft Damper," *Eng. & Min. J.* 98, October 24, 1914, p. 743.

1107. Pasin, F., "Über die Berechnung der Eigenfrequenzen von Schwingungen bei Kurvenmechanismen," *Mechanism and Machine Theory* 9, 1974, No. 2, pp. 231-238 ("On the Calculation of Vibrations in Cam Mechanisms").

1108. Pasin, F. "Über die Kinetische Stabilität der Stösselstange in Kurvengetrieben," *Mech. and Machine Theory* 18, No. 2, 1983, pp. 151-155 ("Kinetic Stability of Translating Follower in Cam Mechanisms").

1109. Pasternak, S. F., "Cam Feed Speeds Undercutting on 77-mm Shell," *Am. Mach.* 98, April 26, 1954, pp. 110-111.

1110. Patton, W. G., "Flame Hardener Treats Cams at High Speeds; Oxygen-Propane Flame-Hardening Machine," *Iron Age* 169, March 20, 1952, 98-99.

1111. Patton, W. G., "Heat Treating Facilities Provide Versatility, Rigid Control; Continental Motors Corp," *Iron Age* 172, October 29, 1953, pp. 106-108.

1112. Patton, W. G., "Selective Flame Hardening Improves Camshaft Wear Resistance," *Iron Age* 172, November 12, 1953, pp. 171-174.

1113. Pauli, J. H., "Cam Operated Fixture Simplifies Machining of Spiral Grooves," *Am. Mach.* 97, July 6, 1953, p. 157.

1114. Pechenik, J., "Special Problems Solved in Designing Corvair-Spyder Camshafts," *SAEJ* 71, April 1963, pp. 82-83.

1115. Peczkowski, J., "Wege zur Leistungserhöhung von Kurvenscheibengetrieben," *Feinwerktechnik* 78, 1974, No. 1, pp. 29-35 ("Means to Increase Speed of Cam Mechanisms").

1116. Pejsach, E. E., "The Problem of Synthesizing Optimum Motion Diagrams" (in Russian), *Mechanika masin* 7/8, 1968, No. 13/14, pp. 35-52.

1117. Pelecudi, C., "Application of Translating Cams with Sliding Tappet for the Representation of Functions," *Rev. Mecan, Appl.* 7, 1962, pp. 297-300.

1118. Pelecudi, C., and Sava, I., "On the Analysis and Synthesis of Cam Mechanisms" (in Romanian), *Constr. mas* 19, 1967, No. 8/9, pp. 506-519.

1119. Peralta, J. L. B., "Drehung von Gleichdicken in quadratischen Bohrungen. Erzeugung von speziellen Punktführungen," *Mechanism and Machine Theory* 17, 1982, No. 5, pp. 349-354.

1120. Peralta, J. L. B., "Spezielle Punktführungen durch Gleichdickwellenkörper in vorzugsweise quadratischen Bohrungen," Diss. Univ. Hannover, 1979.

1121. Perle, R. J., "Determining Switch-Lever Angle and Cam Profile for Reliable Limit-Switch Operation," *Mach. Des.* 40, November 7, 1968, pp. 185-189.

1122. Perrill, D. M., "Cam Cutting Attachment for Milling Machine," *Mach.* 24, February 1918, p. 541.

1123. Pestel, K., "Konstruktion und Dimensionierung eines Band-Wälz-hebelgetriebes," *Maschinenbautechnik* 5, 1967, pp. 275-278 ("Construction and Design of a Rolling-Contact Cam").

1124. Petersen, E., "Kombinationsgesetze für Bewegungen mit Rast-und Umkehrlagen," *Konstruktion* 28, 1976, No. 3, pp. 90-96 ("Combined Displacement Diagrams for Dwell and Rise-Return").

1125. Petrokas, L. V., "Analytical Design of Cam Mechanisms of Automatic Machines" (in Russian), *Vest. Inzh. Tekhn.*

1126. Petrokas, L. V. V.,"Calculations of Some Type of Cam-Shaft-Rod Mechanisms with Hydraulic and Elastic Mechanisms" (in Russian), *Kinamika Mashin*, Moscow, Masgiz, 1960, pp. 186-202; Ref. *Zb. Mekb.* 5, 1961, Rev. 5A, p. 149.

1127. Petru, K., "Kinematic Dependencies Suppressing Free Oscillations" (in Czech.). *Strojnicky Csopis* 17, 1966, pp. 465-477.

1128. Pfau, W., "Massenbeschleunigungsprobleme bei Lochkartenmaschinen," *VDI-Berichte* 31, 1958, pp. 49-55 ("Dynamic Forces in Key Punches").

1129. Pisano, A. P., and Freudenstein, F., "Experimental and Analytical Investigation of the Dynamic Response of a High-Speed Cam-Follower System. Part 1: Experimental Investigation," *J. Mech. Transm. Autom. Des.* 105, No. 4, December 1983, pp. 692-704.

1130. Pollitt, E. P., "Some Applications of the Cycloid in Machine Design," *ASME Trans.* E2 B, J. Engng. Industry, 4, November 1960, pp. 407-414.

1131. Porter, W. G., "Cams for Automatic Screwing Machines," *Am. Mach.* 83, August 9, 1939, p. 610.

1132. Porter, W. G., "Involute Plate Cams for Use in Screw Machines and Similar Applications," *Am. Mach.* 83, April 5, 1939, pp. 206-208.

1133. Poschl, Th. "Über eine einfache Darstellung der Beschleunigung von Steuergetrieben mit unrunden Scheiben," *Z.D. Österr. Ing. Arch.-Vereins* 65, 1912, p. 296, Zeitschift für angewandte Mathematik und Mechanik, 3, 1923, pp. 132-134; 4, 1924, pp. 241-242 ("On a Simple Presentation of Acceleration in Cam Mechanisms with Non-Circular Disks").

1134. Povcza, A., "Planetenrad-Nockengetriebe zum Ausgleich der Ungleich-förmigkeit in Kettentrieben," *VDI-Z* 101, 1959, pp. 1130-1134 ("Planetary-Wheel Cam Mechanisms to Equalize Inaccurate Performance in Chain Drives").

1135. Praetorius and Dinslage, "Betrachtungen über Nockenformen," *Motorwagen* 13, 1910, pp. 597, 623, and 736 ("Consideration on Cam Profiles").

1136. Prentis, J. M., "On the Use of Equivalent Mechanisms in the Velocity and Acceleration Analysis of Cams," *Bull. M. Eng.*, Education Manchester V., 1960.

1137. Preusser, E., "Getriebetechnische Probleme bei Nagema-Erzeugnissen," *Maschinenbautechnik*, No. 7, 1972, pp. 299-301 ("Technological Problems with Mechanisms in Nagema-Products").

1138.   Preusser, E., and Heise, G., "Machanisch-hydraulisches Verstellgetriebe," *Wiss. Z. T. H. Karl-Marx-Stadt* 14, 1972, No. 1, pp. 97-103 ("Mechanical-Hydraulic Adjustable Mechanism").

1139.   Pritchard, B. S., "Dye-Control with R-Cam and Ruler," *Optical Soc. America-J.* 42, October 1952, pp. 752-753.

1140.   Pruszynski, A. J., "Laying Out Positive Return Disk Cams," *Mach.* 33, April 1927, pp. 613-614.

1141.   Pullar, J. A., "Eccentric with Floating Liner," *Ry. Mech. Eng.* 93, September 1919, p. 516.

1142.   Purser, M. W., "Cam-Operated Air Vise," *Mach.* 54, November 1947, pp. 170-171.

1143.   Rachwal, J. F.,"Milling Internal Cams," *Tool Eng.* 39, July 1957, pp. 80-81.

1144.   Ragulskis, K. M., "Proportioning of Cams with x Min. Circumference," *Trudy Akademii nauk litovskof SRR*, serija B, Vol. 2, 1958, pp. 149-156.

1145.   Ragulskis, K. M., "On the Calculation of Cam Proportions," *Trudy Akademii nauk litovskof SSR*, serija B, Vol. 2, 1958, pp. 157-164.

1146.   Ragulskis, K. M., and Rössner, W., "Die Behandlung der Kurvengetriebe in der sowjetischen Literatur," *Maschinenbautechnik* 9, 1960, pp. 206-215 ("Status of Cam Technology in Soviet Literature").

1147.   Rahman, Z. U., and Bussell, W. H., "Iterative Method for Analyzing Oscillating Cam Follower Motion," *J. Eng. Ind.* 93 pp. 149-154; Discussion, February 1971, pp. 154-156.

1148.   Rahn, E., "Modification and Uses for Basic Types of Cams," *Product Eng.* 20, April 1949, pp. 120-121.

1149.   Rakoczi, F., "Diminishing the Dimensions of Cam Mechanisms," (in Hungarian), *Magyar Tudomanyos Akademia, Muszaki Tudomanyok Osztalyanak, Kozlemenyei* 37, 1966, pp. 57-71.

1150.   Rankers, H., "Das elektronisch berechnete Kurvengetriebe," *VDI-Nachrichten*, April 29, 1964, p. 6 ("Computer Designed Cam Systems").

1151.   Rankers, H., "Einfache und zusammengesetzte Rollkurvengetriebe mit einem Beitrag zur Rationalisierung der Konstruktionsarbeit," *Konstruktion*, No. 11, 1970, pp. 421-427 ("Pre-Calculated Construction of Cam Systems; A Contribution Toward Greater Efficiency in the Design Process").

1152.   Rankers, H., "Übertragungswinkel und Grundkreishalbmesser bei Kurvenscheiben mit zentrisch geradegeführtem Stössel," *VDI-Berichte* 29, 1958, pp. 71-74 ("Transmission Angle and Minimum Base Circle Disc Cams with Translating Follower").

1153.   Ranney, C. R., "Production Milling of Cams," *Am. Mach.* 74, April 23, 1931, pp. 49-50.

1154.   Rantesch, E. J., "Laying Out Master Cam Templets," *Mach.* 30, September 1923, pp. 49-50.

1155.   Rao, A. C., "Optimum Elastodynamic Synthesis of a Cam-Follower Train Using Stochastic-Geometric Programming," *Mechanism and Machine Theory* 15, 1980, No. 2, pp. 127-135.

1156.   Rao, S. S. and Gavane, S. S., "Analysis and Synthesis of Mechanical Error in Cam-Follower Systems," *Trans. ASME, J. Mech. Design*, 104, 1982, No. 1, pp. 52-62.

1157.   Rappaport, S., "How Direction of Rotation Affects Cam Profiles," *Mach. Des.* 28, April 5, 1956, p. 120.

1158.   Rappaport, S., "Small Indexing Mechanisms," *Machine Design* 29, April 18, 1957, pp. 161-163.

1159.   Raven, Fr. H., "Analytical Design of Disk Cams and Three-Dimensional Cams by Independent Position Equations," *Trans. ASME*, Paper No. 58-A-17, J. Appl. Mech.

1160.   Raven, F. H., "Analytical Design of Disk Cams and Three-Dimensional Cams by Independent Position Equations," *ASME Trans.* 81 E, J. Appl. Mech. 1 March 1959, pp. 18-24.

1161.   Redeer, H., "Vorrichtung zur Serienherstellung genauster Kurven," *Feinwerktechnik*, December 1951, pp. 331-332 ("Device for Series Manufacture of Accurate Cams").

1162.   Rees Jones, J., and Reeve, J. E., "Dynamic Response of Cam Curves Based on Sinusoidal Segments," Rees Jones: Cam and Cam Mechanisms, pp. 14-24.

1163.   Rees Jones, J. "Mechanisms '74," *Cam and Cam Mechanisms*, London: Mech. Engineering Publ. Ltd., 1978.

1164.   Rees Jones, J., "A Comparison of the Dynamic Performance of Tuned and Non-Tuned Multiharmonic Cams," *IFToMM*, 1977, Vol. III, pp. 553-564.

1165.   Reeser, W., "Textile Machinery Casting," *Precision Metal Molding* 15, August 1957, pp. 46-47.

1166.   Reeve, J. E., "Component Design–Cam Indexing Mechanisms: Robust, Accurate, Versatile," *Electromechanical Design*, February 1972, pp. 24-29.

1167.   Rehwalk, W., "CAD-Berichte: DISKO-Digitale Simulation von Kurven- und Koppelgetrieben," Karlsruhe: Kernforschungszentrum, 1979.

1168.   Reid, N. E., "Two Methods for Determining the Minimum Radius of Curvature of Disk Cams," ASME Paper No. 64-MECH-20.

1169.   Reid, R. R., and Stromback, D. E., "Mechanical Computing Mechanisms," *Product Eng.* 11, September 1949, pp. 119-120.

1170.   Reid-Green, Keith S., and Marder, William Z., "Numerically Controlled Machining of Cams," *RCA Eng.* 26, No. 6, May-June 1981, pp. 80-83.

1171.   Reinholtz, D. F., Dhande, S. G., and Sandor, G. N., "Kinematic Analysis of Planar Higher Pair Mechanisms," *Mechanism and Machine Theory* 13, 1978, No. 6, pp. 619-629.

1172.   Reiser, G. C., "Beryllium Copper Cams Give Good Wear Life," *Precision Metal Molding* 14, May 1956, pp. 42, 44, and 68-69.

1173. Resetov, L. N., "Approximate Profile Synthesis of Cams with Swinging or Translating Roller Follower," *Tr. Sem. p.T.M.M.* LVIII, 1950, No. 31.

1174. Resetov, L. N., "Cylindrical and Straight-Line Moving Cams with Translating Follower," *Tr. Sem. p.T.M.M.* VII, 1949, No. 28.

1175. Resetov, L. N., "Cylindrical and Straight-Line Moving Cams with Swinging Roller Follwer," *Tr. Sem. p.T.M.M.* X, 1951, No. 40.

1176. Resetov, L. N., "Design of Disk Cam of Least Dimensions," (in Russian), *Vopr teorii mekhanizmov i mashin, Moscow, Mashiqiz,* 1955, pp. 21–24; Ref. Zh. Mekh. 1956, Ref. 3437.

1177. Reswick, J. B., and Wright, J., "Velocity-Displacement Transfer Mechanism Eases Materials-Handling Problems," *Control Eng.* 4, December 1957, pp. 76–79.

1178. Rhys, C. O., "Designing of Harmonic Cams," *Mach.* 29, August 1923, pp. 934–935.

1179. Richards, W., "Cam Design-General Principles and Classification of Cams," *Machinery* (London), May 7, 1936, p. 169.

1180. Richards, W., "Cam Design, Constant Velocity Motion Cam," *Machinery* (London), July 16, 1936, p. 473.

1181. Richards, W., "Cam Design - the Tangent Cam," *Machinery* (London), December 24, 1936, p. 381.

1182. Richards, W., "Cam Design, the Design of Uniformly Accelerated and Retarded Motion Cam," *Machinery* (London), August 11, 1938, p. 573.

1183. Richards, W., "Cam Design of Uniformly Increasing Acceleration and Retardation Cam," *Machinery* (London), December 22, 1938, p. 365.

1184. Richards, W., "High Speed Cam Design," *Machinery* (London), August 17, 1939, p. 621; and October 26, 1939, p. 105.

1185. Richards, W., "The Harmonic Motion Cam," *Machinery* (London), February 1, 1930, p. 481.

1186. Richards, W., "The Harmonic Motion Cam of Equilateral Triangular Form," *Machinery* (London), November 28, 1940, p. 231.

1187. Richards, W., "The Harmonic Motion Cam with Flat-Footed Followers," *Machinery* (London), October 17, 1940, p. 63.

1188. Richards, W., "The Triangular Harmonic Motion Cam with Square-Frame Follower," *Machinery* (London), January 9, 1941, p. 393.

1189. Richert, H., "Programmsystem zur betriebsinternen Berechnung, Optimierung und Fertigung von Kurvengetrieben mittels kleinerer EDV-Anlagen," *Konstruktion,* No. 10, 1971, pp. 380–387 ("Programmation of Calculations, Optimization and Design of Cam Systems by Means of EDV Installations").

1190. Riftin, L. P., "Analytic Design of Computing Cams with Two Degrees-of-Freedom," (in Russian), *Trudi Sem. Teor. Mash. Mekh.* 8, 1950, pp. 5–30.

1191. Ringwald, M., "Nockenform und Ventilbeschleunigung mit besonderer Berücksichtigung der Verbrennungsmotoren," *ZVDI* 71, 1927, p. 47 ("Cam Shape and Valve Action Acceleration with Particular Emphasis on Combustion Engines").

1192. Riopelle, E. F., "Important Factors in Designing Camshaft Chain Drives," *Automotive Ind.* 99, October 15, 1948, pp. 36-39.

1193. Roban, I., and Tempes, I., "High-Speed Cams" (in English) *Bluefinul Institutului Politehnic Gheorghe GheorghiuDej Bucuresti* 30, January/ February 1968, pp. 67-80.

1194. Rock, A. L., and Copper, J. D. Jr., "Precision Photoengraving of Caron Steel Machine Parts," *Indus. Photography* 7, May 1958, pp. 36-39.

1195. Roemer, W. C., "Attachment for Cutting Cams and Profiles," *Am. Mach.* 47, September 20, 1917, pp. 513-515.

1196. Rössner, E., "Calculation of Proportional Cams for Internal Combustion Engines," *Automotive & Aviation Ind.* 88, May 1, 1943, pp. 38-40; Discussion (Calculation of Proportional Cams), 89, September 1, 1943, pp. 45+.

1197. Rössner, W., "Güte der Kraft-und Bewegungübertragung," *VDI-Berichte* 29, 1958, pp. 65-70 ("Quality of Force and Movement Transmission").

1198. Rogers, A. S., "Undulator: A Modified Reciprocator for Ornamental Lathes," *Sci. Am. S.* 76, July 5, 1913, pp. 14-16.

1199. Rogers, M. D., and Schaffer, R. R., "Kinematic Effects of Cam Profile Errors," ASME Paper No. 63-WA-296, November 1963.

1200. Roggenbuck, R. A., "Designing the Cam Profile for Low Vibration at High Speeds," *Trans. SAE* 61, 1963, pp. 701-705.

1201. Rohrich, D., "Analytische Darstellung von Flankenprofilen zur Übertragung von Drehbewegungen," *Maschinenbautechnik* 19, 1970, No. 4, pp. 189-196.

1202. Rose, W., "Ein Verfahren zur Kraftmessung an Kurvengelenken," *Maschinenbautechnik*, No. 11, 1969, pp. 601-604 ("A Method to Measure Force in Cam Joints").

1203. Ross, H. R., "Percentage Charts for Motion Analysis; Reference Book Sheet," *Product Eng.* 19, 1948, pp. 171 and 173.

1204. Roth, K., Franke, J., and Simonek, R., "Algorithmisches Auswahlverfahren zur Konstruktion mit Katalogen," *Feinwerktechnik*, No. 8, 1971, pp. 337-345 ("Algorithmic Selection Process for Design through Catalogues").

1205. Roth, K., "Die Kennlinie von einfachen und zusammengesetzten Reibsystemen," *Feinwerktechnik*, No. 4, 1960, pp. 133-142 ("Analysis of Characteristics Between Simple and Composite Friction Systems").

1206. Rothbart, H. A., "Basic Cam Systems," *Mach. Des.* 28, May 31, 1956, pp. 133-136.

1207. Rothbart, H. A., "Basic Factors in Cam Design," *Mach. Des.*, October 18, 1956, pp. 107-113.

1208. Rothbart, H. A., "Basic Factors in Cam Design," *Transactions of the Third Conference on Mechanisms*, 1956, pp. 21-27, Sponsored by *Machine Design*.

1209. Rothbart, H. A., "Cam Dynamics," *Mach. Des.* 28, March 8, 1956, pp. 110-107.

1210. Rothbart, H. A., "Cam Pressure Angles," *Product Eng.* 28, January 1957, pp. 193-195.

1211. Rothbart, H. A., "Cams in Control Systems," *Control Eng.* 8, November 1961, pp. 97-101.

1212. Rothbart, H. A., "Cam Torque Curves," *Mach. Des.* 31, July 23, 1959, pp. 127-129.

1213. Rothbart, H. A., "Cam Torque Curves," *Trans. Fifth Conference on Mechanisms*, Purdue Univ., October 1958, pp. 36-41.

1214. Rothbart, H. A., "Catalog of Equivalent Mechanisms for Cams; Data Sheet," *Mach. Des.* 30, March 20, 1958, pp. 175-180.

1215. Rothbart, H. A., "Equivalent Mechanisms for Cams for Simplified Graphical Analysis of Complex Cam Follower Motions," *Trans. Fourth Conference on Mechanisms*, Purdue Univ., 1957, pp. 25-30.

1216. Rothbart, H. A., "Here's Your Guide to Cam-Inspection Equipment," *Am. Mach.* 102, October 6, 1958, pp. 97-100.

1217. Rothbart, H. A., "Mechanical Control Cams," *Control Eng.* 7, June 1960, pp. 118-122; *Product Eng.* 32, September 4, 1961, pp. 408-411.

1218. Rothbart, H. A., "New Developments in Design and Application of Cams," *Proc. 1957 ASME Design Engng. Conference*, May 1957, pp. 24-29.

1219. Rothbart, H. A., "A Numerical Method for Determining Cam-Follower Response," *ASME-Trans.* 57-F-17, September 23, 1957, pp. 1-4.

1220. Rothbart, H. A., "Design and Application of Cams," *Design News* 12, 1957.

1221. Rothbart, H. A., "Neue analytische und experimentelle Untersuchungs-methoden für den günstigsten Entwurf von Kurvengetrieben unter Berücksichtigung von Belastung, Elastizität und geforderter Drehzahl," *Technica* 11, 1962, No. 17, pp. 1215-1219; No. 20, pp. 1531-1534 ("New Analytic and Experimental Investigative Methods for Optimum Design of Cam Systems with Regard to Load, Elasticity and Minimum RPM Required").

1222. Rothbart, H. A., and Hain, K., "Die Kurvengetriebe und ihre Behandlung im Schrifttum des englischen Sprachgebietes," *Konstruktion* 11, 1959, pp. 360-363 ("Survey of Cam Literature in English").

1223. Rothbart, H. A., "Dynamik in Kurvengetrieben," *Konstruktion* 12, 1960, pp. 36-38.

1224. Rothbart, H. A., "Which Way to Make a Cam," *Product Eng.* 29, March 3, 1958, pp. 67-71.

1225. Rotinjan, L. A , "Dynamic Synthesis of Spherical Cam Mechanisms," *Tr. Sem. p.T.M.M.* X, 1951, p. 40.

1126. Rouillion, L., "Cams, Layout and Drafting," Rev. Ed., Henley, 1928.

1127. Rubinovich, D. Kh., "Synthesis of an Automatic Control System for a Cam Grinding Process," *Sov. Eng. Res.* 3, No. 3, March 1983, pp. 41-45.

1128. Rudbeck, P. E., "Design of Swing Cams," *Product Eng.* 12, August 1941, p. 438.

1129. Rückert, H., "Numerische Methoden zur Bestimmung der Laufeigen-schaften ebener Kurvengetriebe unter besonderer Berücksichtigung des Einflusses der Rast-in-Rast-Bewegungsgesetze," Dissertation T. H. Darmstadt, 1974.

1230. Rückert, H., and Risse, H., "Schwungradberechnung für Kurven-und Kurbelgetriebe bei überwiegend dynamischer Beanspruchung," *Konstruktion* 27, 1975, No. 3, pp. 96-99.

1231. Rückert, H., "Bewegungsgesetze für Kurvengetriebe mit Schwing-bewegung," *Antriebstechnik* 15, 1976, No. 3, pp. 121-124.

1232. Rückert, H., "Anwendungsmöglichkeiten kurvengesteuerter Schritt-getriebe und ihre Grenzen," *Maschinenmarkt* 83, 1977, No. 79, pp. 1565-1567.

1233. Rumsey, R. D., "Redesign for Higher Speed," *Mach. Des.* 23, April 1951, pp. 123-130.

1234. Rumyantsev, A. V., "The Accurate Manufacture of Cams," *Russian Eng. J.* XLVII, 1967, pp. 61-64.

1235. Rumyantsev, A. V., "Wear and Life of Cam Mechanisms," *Russian Eng. J.* XLIX, 1969, pp. 24-26.

1236. Runge, E. A., "Checking a Camshaft with the Dial Indicator," *Mach.* 20, August 1914, pp. 1067-1068.

1237. Rzepecki, J., "Optimaler Radius der Kipphebelkuppe bei Ventil-steuerung," *Techn. Rundsch.*, 1957, p. 6.

1238. Sakmann, W., "Beitrag zur programmierten Berechnung schnellaufender Kurvenscheiben," *Konstruktion* 18, July 1966, pp. 278-283 ("Contribution to Computer Programming for Calculating High-Speed Cams").

1239. Samuels, W., "How Motion of a Mechanism May Be Analyzed Geometri-cally," *Automotive Ind.* 59, September 22-29, 1928, pp. 406-408+, and 442-445+.

1240. Sanders, F. J., "Burnishing with Tungsten Carbide," *Mach.* 37, May 1931, pp. 674-676.

1241. Sanders, M. A., "Cam Design and Analysis for High Speeds and Torques," *Product Eng.* 18, March 1947, pp. 84-87.

1242. Sanders, M. A., "When Plastic Cams Are Better," *Product Eng.* 31, August 8, 1969, pp. 44-47.

1243. Sanderson, A. E., "Minimum Cam Size as Determined by Pressure Angle," *Product Eng.* 27, July 1956, pp. 141-143.

1244. Sandig, P., and Wege, W., "Optimierung der dynamischen Eigenschaften eines Kurvengetriebes," *Wiss. Z. T. H. Karl-Marx-Stadt* 14, 1972, No. 1, pp. 105-115.

1245. Sandler, B. Z., "The Influence of Cam Profile Errors on Follower Motion," *IFToMM*, 1974, pp. 153-179.

1246. Sankar, T. S., and Osman, M. O. M., "Dynamic Accuracy of Hybrid Profiling Mechanisms in Cam Manufacturing," *Trans. ASME*, J. Mech. Design. Vol. 101, 1979, No. 1, pp. 108-117.

1247. Saronov, S. K., "The Calculation of Cams with Straight-Line Follower," (in Russian), *Vestnik masinostroenija* 49, 1969, No. 10, pp. 16-18.

1248. Sarring, E. J., "Torque Compensation for Cam Systems," *Transaction of the Seventh Conference on Mechanisms*, Sponsored by *Machine Design*, 1962, pp. 179-185.

1249. Sassano, E. C., "Engineering Cams for their Production in Limited Quantities," *Tooling & Production* 21, July 1957, pp. 92-94.

1250. Sassen, B., "Cam Indexing Motion," *Am. Mach.* 64, May 6, 1926, pp. 705-707.

1251. Sauer, L., "Spirale oder Gerade als Hinter-Drehkurve," *Werkstatt Betr.* 106, February 1972, pp. 83-89 ("Spiral or Straight Line as Relieving Curve").

1252. Sauer, R., "Hubbeschleunigung für die geneigte Sinuslinie," *Maschinenbau*, 1938, pp. 37-39 ("Acceleration of the Modified Cycloid").

1253. Savage, M., "Cam Sizing Simplified," *Mach. Des.* 39, October 26, 1967, pp. 181-185.

1254. Savvin, E. A., "Dynamic Synthesis of Motion Diagrams" (in Russian), *Mechanika masin* 11/12, 1970, No. 23/24, pp. 46-55.

1255. Scheinman, J., "Cam Motion Indicator, A New Teaching Aid," *J. Eng. Educ.* 41, October 1950, pp. 111-114.

1256. Schelling, H., "Steigerung der Leistung an bestehenden Verpackungsmaschinen," *VDI-Berichte* 29, 1958, pp. 33-37 ("Toward Greater Efficiency in Existing Material Packing Machines").

1257. Schenker, W., "Zur Bestimmung und Beurteilung des Ventilerhebungsverlaufes und der Kraftwirkungen in Ventilsteuerungen," *Dingl, Polyt. J.*, 1927, p. 327 ("On the Determination and Analysis of Valve Gear Adjustment and Kinematic Effects in Valve Control Mechanisms").

1258. Schimpf, H., "Betrachtungen über Nockenformen," *Motorwagen*, 1970, p. 778.

1259. Schirmeister, K., "Beurteilung von Bewegungsgesetzen für Kurvengetriebe im Hinblick auf Schwingungserregung," *Maschinenbautechnik*, No. 1, 1969, pp. 46-52 ("Analysis of the Laws of Motion in Cam Mechanisms with Regard to Vibrations").

1260. Schirmeister, K., "Schwingungsdämpfung in Kurvengetrieben durch filternde Eingriffsglieder," *Maschinenbautechnik*, No. 1, 1968, pp. 43-48 ("Reduction of Oscillation in Cam Mechanisms through Absorbtion Members").

1261. Schläfke, K., "Flachstösseldurchmesser und Nockenform," *ATZ*, 1934, pp. 575-577 ("Diameter of Straight-Line Follower and Cam Shape").

1262. Schläfke, K., "Über Beziehungen zwischen Form, Spektrum und Herstellungsgenauigkeit von Steuernocken," *Luftfahrtforschung* 19, 1942, pp. 353-357 ("On the Relation Between Form, Spectral Analysis and Precision in the Manufacture of Cams").

1263. Schläfke, K., "Zur harmonischen Analyse von Nockenkurven," *Luftfahrtforschung*, 17, 1940, pp. 87-88 ("Toward a Harmonic Aanlysis of Motion Diagrams").

1264. Schlippe, "Ein originelles neues Verfahren zur Herstellung von Werkzeugen unregelmässiger Form," *Werkzeug* 10, 1932, pp. 1-4.

1265. Schmidt, E., "Continuous Cam Curves," *Machine Design* 32, 1960, pp. 127-132.

1266. Schmidt, O., "Graphische Konstruktion und Untersuchung eines harmonischen Nockens mit Flachstössel," *Autom-tech. Z.* 40, 1937, pp. 363-364.

1267. Schnarbach, K., "Federanordnungen an kraftschlüssigen periodischen Getrieben," *Werkstatttechnik Maschinenbau* 3, 1955, pp. 102-105 ("Arrangements of Springs in Force Closure Cam Systems").

1268. Schnarbach, K., "Kurvengetriebe und Koppelgetriebe als Steuer-Organe an Verarbeitungsmaschinen," *VDI-Z* 93, 1951, pp. 223-228 ("Cam Drives and Link Drives for the Control of Processing Machines" (in German), *ZVDI* 93, March 1951, pp. 223-228.

1269. Schnarbach, K., "Nachformfräsen von Kurventrägern nach offenen Bezugsformen und selbsttätiges Fräsen ohne Bezugsform," *Werkstatt und Betrieb* 42, 1952, pp. 46-50 ("Cutting of Control Cam with or Without Master Cam Profiles").

1270. Schnarbach, K., "Konstruktion der Weg-Zeit-Kurven für Schaltwerke mit Bewegungsüberlagerung," *Maschinenbautechnik* 5, 1956, pp. 313-319 ("Design of Time-Displacement Diagrams for Ratchet Mechanisms with Superposed Motion").

1271. Schnarbach, K., Bedeutung und Anwendung des Übertragungswinkels an Kurventrieben," *ZVDI* 91, 1949, pp. 603-606 ("Importance and Application of the Transmission Angle in Cam Mechanisms").

1272. Schnarbach, K., "Netztafeln für grösste Geschwindigkeit und grösste Beschleunigung bei Kurventrieben mit parabolischem und sinoidischem Bewegungsgesetz," *Reuleaux-Mittleilungen* 6, 1938, pp. 373-375 ("Nomograms for Maximum Velocity and Maximum Acceleration in Cam Mechanisms with Parabolic and Sinusoidal Motion").

1273. Schnarbach, K., "Über die Gestaltung der Kurventriebe und die Fertigung der Steuerkurven," *VDI-Berichte* 12, 1956, pp. 107-119 ("On the Design of Cam Systems and Cam Manufacture").

1274. Schnarbach, K., "Über die Leistungen der Verarbeitungmachinen in Abhängigkeit von den technologischen Arbeitsfolgen und den dabei benutzten Getrieben," *VDI-Berichte* 5, 1955, pp. 119-128 ("On the Efficiency Performance of Wrapping Machines in Relation to Technological Work Stages and Types of Cams Used").

1275. Schnarbach, K., "Zur Synthese des Kreisbogengesetzes," *Konstruktion* 3, 1951, pp. 82-89 ("On the Synthesis of the Laws Governing Circular Arcs").

1276. Schnarbach, K., "Umlaufgetriebe mit Bewegungsüberlagerung für ungleichförmig umlaufenden Abtrieb," *Techn. Mitt.* 53, 1960, pp. 262-270 ("Rotating Mechanisms with Superposed Motion for Transmission with Varying Velocity Output Ratios").

1277. Schnarbach, K., "Einige Probleme aus den Verpackmaschinenbau," *Z-VDI* 95, 1953, pp. 704-705 ("Select Problems Encountered in the Design of Packing Machines").

1278. Schnarbach, K., "Einfache Getriebe mit Bewegungsüberlagerung für ungleichförmig umlaufenden Abtrieb," *Konstruktion* 12, 1960, pp. 427-435 ("Simple Mechanisms with Superposed Motion for Transmissions with Varying Velocity Output").

1279. Schnarbahc, K., "Über die Ausbildung von Stangenköpfen, Rollen und Gelenkbolzen bei Kurven und bei Gelenkgetrieben," *VDI-Tagungsheft* 1, 1953, pp. 117-120.

1280. Schoefer, E. A., "Cast Alloy Fixture Has Long Life in Thermal Schock Service," *Metal Prod.* 73, June 1958, pp. 106-108.

1281. Scholes, G. E., "Experiments on Cams and Poppet Valves," *Automobile Eng.* 12, May 1922, pp. 151-157.

1282. Schrader, E. W., "Cycloidal Cam Reducer Gives Higher Efficiencies," *Design News* 16, 1961, pp. 6-7.

1283. Schreck, H., "Kinematics of Cams Calculated by Graphical Methods," *Transactions of the ASME* 48, 1926, p. 979.

1284. Schrick, P., "Das dynamische Verhalten von Ventilsteuerungen an Verbrennungsmotoren," *MTZ* 31, February 1970, pp. 61-63 ("Dynamic Characteristics of Valve Gear of Internal Combustion Engines").

1285. Schröder, "Form und Fertigung erhabener Steuernocken," *Werkstatt und Betrieb* 81, 1948, p. 288 ("Design and Construction of Convex Cam Mechanisms").

1286. Schubert, F., "Cam Charts Eliminate Cut and Try Methods," *Mach. Des.*, May 1936, pp. 33-34.

1287. Schubert, K. P., "Cams Multiply Low Pressure Input Forces," *Hydraulics & Pneumatics* 16, November 1963, pp. 86-87.

1288. Schuh, W., "Oszillographische Untersuchung der Ventilsteuerung von Viertaktmotoren," *Industrie-Elektronik*, Mitt. Elektrospezial Hamburg 4, 1956, pp. 3-7, Referat in *VDI-Z* 100, 1958, pp. 771-772.

1289. Schultz, H., "Wirtschaftliches Messen von Kurvenprofilen," *Feinwerktechnik* 11, 1969, pp. 478-480 ("Economical Measurement of Curve Profiles").

1290. Schultz, W., und Kulitzcher, P., "Zur Federauslegung bei kraftschlüssigen Kurvenmechanismen," *Maschinenbautechnik* 9, 1969, pp. 495-498 ("Interpreting Finite Power Cam Systems").

1291. Sejnin, G. M., "Grinding Archimedean Spirals by a Tangential Method," *Machines & Tooling* 40, 1969, No. 7, pp. 49-50.

1292. Semenov, M. V., "Determination of the Basic Parameters of Cam Mechanisms Actuated by Forces at Various Angles" (in Russian), *Teoriya Mashin i Mekhanizmov* 96/97, 1963, pp. 49-57.

1293. Serebrennikov, M. G., "Profile of a Cam Mechanism with a Flat Follower" (in Russian), *Inzhener, Sbornik, Akad, Nauk SSR* 12, 1952, pp. 23-36.

1294. Sergeev, P. V., "An Analytical Method for Dynamic Synthesis of Cam Mechanisms" (in Russian), *Mechanika Masin* 7/8, 1968, No. 12-14, pp. 97-111.

1295. Sergeev, P. V., "Calculating Cam-Lever Mechanisms," *Russian Eng. J.* XLIX, 1969, pp. 38-40.

1296. Sergeev, P. V., "The Dynamic Synthesis of Cam Mechanisms for Prescribed Mean Life" (in Russian), *Mechanika masin* 9/10, 1969, No. 19/20, pp. 69-79.

1297. Sergeev, P. V., "Synthesis of Cam Mechanisms by Using the Criterion of Permissible Wear," *Russian Eng. J.* XLVII, 1968, pp. 30-32.

1298. Sergeev, P. V., "Synthesis of Cam-Rocker Mechanisms with Respect to Contact Strength of the Profile," *Russian Eng. J.* L, 1970, pp. 47-50.

1299. Sergeev, P. V., "A Synthesis of Sliding Cam Mechanisms," *Russian Eng. J.* XLVI, 1966, pp. 43-45, 12, 1966, pp. 23-25.

1300. Sergeev, V. I., "On the Investigation of Errors in Cam Profiles" (in Russian), *Trudi Semin, tochnosti mekh, i mashin-ta mashinoved, Akad, Nauk, USSR* 7, 1954, pp. 73-82; Rev. No. 699. Ref. Zh. Mekh. 1956.

1301. Server, F., "Fixture for Form-Milling Flat Cam," *Mach.* 23, December 1916, pp. 340-341.

1302. Shaffer, B. W., "Method of Finite Differences; Another Way to Differentiate Curves," *Product Eng.* 29, October 13, 1958, pp. 77-79.

1303. Shaffer, B. W., and Krause, I., "Refinement of Finte Differences Calculations in Kinematic Analysis," *ASME Trans. B* 82, pp. 337-380; Discussion, November 1960, p. 381.

1304. Shafran, P. F., "Cam Milling Fixture," *Mach.* 33, September 1926, pp. 55-56.

1305. Shafran, P. F., "Fixture for Cutting Cams," *Am. Mach.* 75, October 8, 1931, pp. 569-570.

1306. Shafran, P. F., "Machining Cam Arcs," *Am. Mach.* 75, August 27, 1931, p. 352.

1307. Shanks, E. C., "Graphical Analysis of Cam Roller Mountings," *Product Eng.*, July 1933, pp. 251-253.

1308. Sharonov, S. K., "Roller-Guide Cam Mechanisms," *Russian Eng. J.* XLIX, 1969, pp. 18-21.

1309. Sharonov, S. K., "The Effect of Geometric and Kinematic Parameters of a Cam Mechanism on the Wear of the Cam Cross Section" (in Russian), *Teoriya Mashin i Maknarizmoy* No. 101-102, 1964, pp. 113-124.

1310. Shaw, F. W., "Dwell Cams of Uniform Diameter," *Mech. World*, October 4, 1935, pp. 329-330.

1311. Shaw, F. W., "Equations for the Design of Involute Cams," *Product Eng.* 4, April 1933, pp. 131-134.

1312. Sheffield, C. A., "Reversed Grinders for Master Cams," *Materials & Methods* 22, November 1945, p. 1468.

1313. Sheino, L. G., "Cam Mechanisms with Small Inertia Forces in the Cam-Follower System," *Russian Eng. J.*, 1961, pp. 25-28.

1314. Sheldon, E., "Cutting an Irregular Recess on the Milling Machine,"
       Am. Mach. 49, October 24, 1918, pp. 755-756.
1315. Sheldon, E., "Ingenious Device for Making a Peculiar Cam," *Am.
       Mach.* 59, November 22, 1923, pp. 773-775.
1316. Sheldon, E., "Pair of Cams of Unusual Forms," *Am. Mach.* 51, July
       10, 1919, pp. 67-69.
1317. Shepler, P. R., "Cycloidal Chord Cross-Over Cams, Tables," *Design
       News*, July 15, 1954, pp. 53-56.
1318. Sherman, W. F., "Packard Uses Induction Process to Case Harden Cam-
       Shafts; Tocco Process," *Iron Age* 141, March 31, 1938, pp. 34-35.
1319. Shipetin, S. I., "Grinding Involute Cams and Spirals," *Machines and
       Tooling* 32, December 1961, pp. 22-23.
1320. Shorter, W. H., "Cam Design Analysis," *Product Eng.* 11, May 1940,
       pp. 223-227; June 1940, pp. 264-265.
1321. Shoup, T. E., J. Berku, L. S. Fletcher, "The Kinetothermal Analysis
       and Synthesis of Cam Systems," *Mechanism and Machine Theory* 11,
       1976, No. 5, pp. 321-329.
1322. Shraer, A. B., "A Summarizing Type of Mechanism to Accomplish a
       Intermittent Motion," (in Russian), *Mekhanika Mashin* 5-6, 1967,
       pp. 81-90.
1323. Sie, S. L. "Kurvenscheiben mit Flachstössel," *Ind.-Anz.* 90, 1968, No.
       70, pp. 1596-1597; No. 87, pp. 1939-1940.
1324. Simon, F., "Herzkurvengetriebe zur Rückstellung von Zeigern und
       Zahlen-Rollen," *Feinwerktechnik* 4, 1964, pp. 164-168 ("Heart-
       Shaped Cam Systems with Readjustment of Indicators and Rollers").
       183.
1326. Simon, H., "Cam with Interchangeable Lobes," *Am. Mach.* 64, No-
       vember 18, 1926, pp. 840-841.
1327. Simon, H., "Diagram-Model for Designing Screw Machine Cams," *Mach.*
       33, December 1926, pp. 281-286.
1328. Simon, H., "Plate Cam Blanking Die," *Mach.* 31, May 1925, pp. 727-
       729.
1329. Simon, H., "Templet for Cam-Roller Center Path," *Am. Mach.* 66,
       April 28, 1927, pp. 703-704.
1330. Simond, W. A., "Metric Scale Aids in Designing Cams," *Mach.* 38,
       August 1932, p. 909.
1331. Simond, W. A., "Straight-Line Movement Applied to a Cam Follower,"
       Mach. 38, February 1932, p. 419.
1332. Simpson, C., "Cams to Any Contour," *Engineering* 208, July 25, 1969,
       pp. 92-93.
1333. Sims, A. L., "Cutting Cam Grooves by Lathe," *J. Sci. Instr.* 38, Sep-
       tember 1961, p. 379.
1334. Sims, R. C., "Cams without Masters," *Tool Eng.* 37, September 1956,
       pp. 85-89.
1335. Singler, R. L., "Adjustable Swivel Plate Simplified Grinding of Cams
       for Automatics," *Am. Mach.* 89, February 15, 1945, p. 135.

1336. Sipe, J. P., "Uniformly Accelerated and Decelerated Motion Cams," *Design News*, January 1, 1955, p. 45.

1337. Sipetin, S. I., "Grinding Involute Cams and Spirals," *Machines and Tooling* 32, 1961, pp. 22-23.

1338. Slaymaker, P. K., "Analyzing Circular-Arc Cams," *Machine Design* 20, July 1948, pp. 129-131.

1339. Small, G. O., "New Material for Speed Cams; Combination of Molding and Laminating Phenolic," *Mod. Plastics* 17, December 1939, pp. 30-31.

1340. Smirnov, L. P., "Design of Cams with Optimal Pressure Angle," (in Russian), *Masinovedenie* 1, 1971, pp. 47-50.

1341. Smith, C. F., "Development of Special Cams," *Am. Mach.* 61, October 16, 1924, p. 627.

1342. Smith, D. L., and A. H. Soni, "Vereinfachte Konstruktion und Prüfung von Kurvenscheiben," *Forschung im Ingenieurwesen* 45, 1979, No. 1, p. 11-15.

1343. Smith, J. A., "Cams–Local Deformations Between Contacting Surfaces," *Proceedings of Oklahoma State University Applied Mechanisms Conference*, July 31, August 1, 1969, Tulsa, Oklahoma, Paper No. 29.

1344. Smith, K. D., "Simple Transition Curves for List Cams," *Machy.* (London), 111, December 13, 1967, pp. 1242-1243.

1345. Sobkowiak, J. C., "Cam-Turning Attachment," *Mach.* 66, July 1960, pp. 142-143.

1346. Soderholm, Lars, "Cam-Drive System Gives Multiple Spindles Individual Feed Control," *De᠎n News* 23, 1968, No. 22, pp. 40-41.

1347. Soderholm. Lars, "Cams Convert Opposed-Piston Travel into Rotary Motion," *Design News* 24, 1969, No. 4, p. 72.

1348. Sörgel, I., "Gestaltungsrichtlinien für kleine, schnellaufende Kurvengetriebe," *Feingeratetechnik* 30, No. 12, 1981, pp. 542-545.

1349. Sommer, G. M., "Analytical Determination of Radial Cam Profiles," *Tool Eng.* 20, May 1948, pp. 17-22.

1350. Soni, A. H., and I. P. J. Lee, "A Survey of Techniques in Cam Dynamics," *Proceedings of the 4th Applied Mechanisms Conference* 28, 1975, pp. 32.1-32.

1351. Sörgel, I., "Dynamische Messungen an kleinen, schnellaufenden Kurvengetrieben," *Feingerätetechnik* 27, 1978, No. 6, pp. 215-216.

1352. Sörgel, I., "Untersuchungen zum Betriebsverhalten ebener schnelllaufender Kurvengetriebe," *Maschinenbautechnik* 28, 1979, No. 9, pp 404-408.

1353. Spicer, C., "Mandrel for Machining Cams," *Tool Eng.* 40, January 1958, p. 79.

1354. Spiegel, K., "Schmierungsverhalten und Reibungskräfte in Kurvengetrieben mit Flachstösseln," *VDI-Berichte* 127, VDI-Verlag, Düsseldorf, pp. 29-33.

1355. Spotts, M. F., "Straight-Line Follower Motion Obtained with Circular Arc Cams," *Product Eng.* 21, August 1950, pp. 110-114.

1356. Stal, H. P., "Vergleich von Kurven- und Kurbelgetrieben," *VDI-Berichte* 140, Düsseldorf' VDI-Verlag, 1970, pp. 15-23.

1357. Stanescu, D., and G. Boncoi, "Contributions to the Analytical Optimal Synthesis of Cam Mechanism Disks with Cam Follower Oscillatory are Employed with Automatized Machine Tools which Use the Electronic Computer Hewlett Packard," *IFToMM*, 1979, Vol. 2, pp. 1287-1290.

1358. Stanescu, D. and G. Boncoi, "Contributions to the Technological Synthesis of Cams Mechanism Disk Used with Automatized Machine Tools which Use the Electronic Computer Hewlett Packard" (in French), *IFToMM*, 1979, Vol. 2, pp. 1291-1294.

1359. Stange, H., "Elektrische Nachbildung von Kurvenmechanismen," *Maschinenbautechnik* 8, 1971, pp. 385-390 ("Electric Reproduction of Cam Mechanisms").

1360. Stange, H., and K. H. Foster, "Veränderung symmetrischer Bewegungs-gesetze bei ungleichförmiger Antriebswinkelgeschwindigkeit," Wiss. Z. T. U., Dresden 17, 1968, No. 1, pp. 93-98.

1361. Stankevich, V. A., "Device for Grinding Cam Profiles," *Chem. Pet. Eng.* 1, January/February 1969, p. 154.

1362. Stanley, F. A., "Adjustable Trimming and Shaving Dies," *Am. Mach.* 52, January 22, 1920, pp. 177-180.

1363. Stanley, F. A., "Press Tools for a Toothed Cam," *Am. Mach.* 51, December 25, 1919, pp. 1075-1077.

1364. Stanton, R. F. V., "Making Cams for Automatic Cigar and Cigarette Machinery," *Mach.* 57, December 1950, pp. 141-147.

1365. Staples, C. F., "Cams for Clamping Work in Jigs and Fixtures," *Am. Mach.* 79, August 28, 1935, p. 627.

1366. Stau, C. H., "Die Herstellung von Werkstücken mit nicht kreisrunden Querschnitten auf der Drehbank," *Werkstatt und Betrieb* 91, May 1958, pp. 265-269.

1367. Stau, C. H., "Das Unrunddrehen," *Werkstattstechnik* 65, 1975, No. 5, pp. 263-269.

1368. Stedman, G. E., "Cams Control 7-Spindle Milling Machines," *Tool Eng.* 12, September 1943, pp. 67-69.

1369. Steel, A. M., "Cams Make Billet Cutter Operation Safe," *Hydraulics & Pneumatics*, November 1961, pp. 92-93.

1370. Stefanides, E. J., "Stacked Followers Give Two-Axis Motion from Single Cam," *Design News* 24, 1969, No. 16, p. 48.

1371. Steiert, H. P. and H. Kramer, "Computersimulation von Kurven-getrieben. Eine Neue Methode der Kurvenauslegung," *VDI-Z* 121, No. 19, pp. 955-961.

1372. Stern, B. J., "Fixture for Milling an Eccentric Cam-Curve," *Am. Mach.* 66, February 3, 1927, p. 218.

1373. Steuart, W. C., "Cam Milling Attachment," *Mach.* 27, February 1921, p. 574.

1374. Stewart, R. T., "Cam Eliminates Shock in Rack Movement," *Mach.* 66, February 1960, pp. 142-143.

1375. Stewart, R. T., "Enlarging Tool Speeds Layout of Master Cams," *Mach.* 65, December 1958, p. 163.

1376. Stocker, W. M., "Long Helical Cams Cut on Gear Shaper," *Am. Mach.* 93, October 20, 1940, pp. 100-101.

1377. Stocker, W. M., "Magic Touch: Cam Hydraulic System Grinds Burs Automatically," *Am. Mach.* 101, March 11, 1957, pp. 126-127.

1378. Stockwell, A. J., "Cam-Actuated Dies Promote Economy," *Tool Eng.* 33, July 1954, pp. 67-70.

1379. Stoddart, D. A., "Polydyne Cam Design," *Mach. Des.* 25, January 1953, p. 121; February 1953, pp. 145-146; March 1953, pp. 149-164.

1380. Stoner, P., "Grinding Accurate Cam Surfaces," *Machinery*, 51, September 1944, p. 163.

1381. Storck, F., "Erzeugung von veränderlichen Arbeitsräumen durch gleichförmige Bewegungen," Dissertation T. U. Hannover, 1973.

1382. Stockmann, P., "Praktische Anwendung von CAD und CAM am Beispiel der Fertigung von Steuerkurven," *Werkstatt und Betrieb* 109, 1976, No. 8, pp. 417-419.

1383. Strassburg, J. L., "Fixture for Inspecting Camshafts," *Am. Mach.* 57, July 27, 1922, p. 149.

1384. Strasser, F., "Design for Compound Bending Dies," *Steel Processing* 42, October 1956, pp. 581-584.

1385. Strasser, F., "Die with Cams That Operate Punches in Opposite Directions," *Mach.* 60, March 1954, pp. 210-211.

1386. Strasser, F., "Fifteen Ideas for Cam Mechanisms," *Product Eng.* 30, August 3, 1959, pp. 56-57.

1387. Strasser, F., "Nine More Ways to Change Straight-Line Direction," *Product Eng.* 31, March 7, 1960, p. 59.

1388. Strasser, F., "Side Cams on Assembly Die Speed Switch Assembly," *Am. Mach.* 98, August 16, 1954, p. 145.

1389. Strauchmann, H., and Patzelt, H., "Optimierung eines 4-Gliedrigen Schwinganlagenantriebes für Offset-Bogenrotationsdruckmaschinen," *Maschinenbautechnik* 10, pp. 483-487 ("Optimization of a Four-Unit Swinging-Type Cam System for Off-Set Folio Rotating Printing Machines or Presses").

1390. Straumbel, M., "Contribution to Calculation and Affecting Vibrations of Cam-Mechanisms," *Motortechnische Zeitschrift* 27, October, 1966, pp. 403-410.

1391. Straubel, Max, "Beitrag zur Erfassung und Beeinflussung des Schwingungsverhaltens von Nockengetrieben," *MTZ*, October 1966, pp. 403-410.

1392. Strombeck, G, M., "Design of Cams for Form Turning Shells," *Am. Mach.* 46, March 1, 1917, pp. 381-382.

1393. Stubner, K., "Centrifugal Cam Action," *Mach. Des.* 31, June 25, 1959, p. 106.

1394. Sugihara, S., "Constant Pressure Cam," *Soc. Mech. Engrs. J.* (Japan), 36, June 1933, pp. 405-408.

1395. Sukhanov, Yu S., "Computer-Aided Design and Machining of Cams," *Sov. Eng. Res.* 3, No. 11, November 1983, pp. 103-104.

1396. Svenson, E., "Automatic Does Contour Grinding; South Water Machinery Co.," *Am. Mach.* 99, December 5, 1955, pp. 156-157.

1397. Svenson, I. S., "The Influence of Cam Curve Derivative Steps on Cam Dynamical Properties," *Trans. Chalmers Univ. Tech.* 255, 1961, p. 92.

1398. Svidlo, I., "Designing Cams for Automatic Screw Machines," *Am. Mach.* 59, November 29, 1923, pp. 812-813.

1399. Swidlo, I. A., "Designing Cams for Automatic Screw Machines," *Mach.* 42, June/July 1936, pp. 641-644, 706-708.

1400. Tainov, A. I.,"Hyperboloid System Cam Mechanisms," (in Russian), *Sb. Nauchn. Trudi Belorusak Politekhn. In-ta* 57, 1957, pp. 3-14; Ref. Zh. Mekh. No. 7, 1958; Rev. p. 7398.

1401. Takano, M. and S. Toyama, "Dynamics of Indexing Cam Mechanism and Speed-Up of Its Motion," *IFToMM*, 1979, Vol. 2, pp. 1408-1411.

1402. Talbourdet, G. J., "Intermittent Mechanisms," *Machine Design* 20, November 1948, p. 159.

1403. Talbourdet, G. J., "Laminated Phenolics on Cams," *Product Eng.* 10, September 1939, p. 369.

1404. Talbourdet, G. J., "A Progress Report on the Surface Endurance Limits of Engineering Materials," *ASME, ASLE, 54-LUB-14,* Baltimore, Maryland, October 18, 1954.

1405. Talbourdet, G. J., "Surface Endurance Limits for Gear and Cam Materials," *Product Eng.* 26, Mid-October 1955, pp. 2-5.

1406. Tangerman, E. J., "Back to Cams Again! Bryant Internal Automatic Grinder," *Am. Mach.* 93, June 2, 1949, p. 105.

1407. Tartakowskii, I. I., "On the Contour of a Cam with Low Contact Pressure" (in Russian), *Mechanika masin* 41, 1973, pp. 3-12.

1408. Tartakowskii, I. I., "Optimal Synthesis of Cam Mechanisms" (in Russian), *Izv Vyssh Uchebn Zaved Mashinostr,* No. 12, 1982, pp. 31-34.

1409. Taylor, D, F., "Automatic Tank Transmissions Demand Close Tolerances," *Machinery* 58, April 1952, pp. 141-148.

1410. Terauchi, Yoshio, and El-Shakery, Sabry A., "Computer-Aided Method for Optimum Design of Plate Cam Size Avoiding Undercutting and Separation Phenomena - 1. Design Procedure and Cycloidal Cam Design Results," *Mech. and Mach. Theory* 18, No. 2, 1983, pp. 157-163.

1411. Terhorst, T. B., "Burner Tip Designed to Flame Harden Cam Shafts," *Gas Age* 87, March 13, 1941, pp. 33-34+; Same, *Iron Age* 147, April 24, 1941, pp. 60-62.

1412. Terry, M., "Application of Double Wedge to Die Work," *Mach.* 22, May 1916, p. 604.

1413. Terry, M., "Dynamics of the Mushroom Type Valve Lifter," *Mach.* 29, July 1914, pp. 954-957.

1414. Tesar, D., and G. K. Matthew, "The Design of Modelled Cam Systems," Rees Jones: Cams and Cam Mechanisms, pp. 30-42.

1415. Tesar, D., and G. K. Matthew, "Dynamic Distortion in Cam Systems," *Mach. Des.* 48, 1976, No. 7, pp. 186-191.

1416. Tesar, D., and G. K. Matthew, *The Dynamic Synthesis, Analysis, and Design of Modeled Cam Systems*, Lexington Book, 1976.

1417. Thiele, K., "Nockenberechnung und Herstellung von Steuerstreifen für die Urnockenfertigung mit der Rechenanlage National-Elliott 503," *Maschinenbautechnik* 18, 1969, No. 5, p. 243.

1418. Thoren, T. R., H. H. Englemann, and D. A. Stoddart, "Cam Design as Related to Valve Train Dynamics," *Trans. SAE* 6, January 1952, pp. 1-13.

1419. Thornton, H., "Milling Fixture for Plate Cams," *Am. Mach.* 77, December 6, 1933, p. 790.

1420. Thumin, A. D., "Cam Design by Computer," *Mach. Des.* 43, September 2, 1971, pp. 81-84.

1421. Thumin, C., "Constant Torque Power Cams," *Product Eng. Hdbk p F.*, 1955, pp. 16-20.

1422. Thumin, C., "Constant Torque Power Cams," *Product Eng.* 25, February 1954, pp. 180-186.

1423. Timpner, F, F., "Applications of CPC to Automatic Engineering Problems," *Indus. Mathematics* 3, 1952, pp. 57-61.

1424. Tishen, M. M., "Profilieren der Nocken für die Brennstoffzuführung bei Flugmotoren," *Tr. Sem. p.T.M.M.* I, 1947, No. 4.

1425. Tishen, M. M., "The Profiling of the Gas Distribution Cams of Aviation Motors" (in Russian), *Trudi Sem. teor, Mash. Mekh.* 1, 1947, pp. 217-239.

1426. Tishen, M. M., "The Design of Cam Mechanisms with a Rocking Crank Follower" (in Russian), *Akad. Nauk SSSR Trudi Sem. Teorii Mash. Mekh.* 2, 1947, pp. 105-110.

1427. Tolke, J., "Eine Charakterisierung sphärischer Trochoidenbewegungen vermöge eigentlich rollgleitender Hüllkurvenpaare," *ZAMM* 58, 1978, No. 2, pp. 114-115,

1428. Tolle, O., "On the Dynamics of Cam Drives" (in German), *Motortech. Z.* 13, December 1952, pp. 288-289.

1429. Tomlin, J. H., "Punch-Press Tooling Works from Cams Makes Cams; Telephone-Exchange Contact Cams Made for Bell Telephone," *Am. Mach.* 94, September 18, 1950, pp. 106-108.

1430. Toropygin, E. I., "The Important Property of the Elliptical Shaped Cam" (in Russian), *Mechanika masin* 11/12, 1970, No. 25/26, p. 49.

1431. Town, H. E., "Cam Design and Operation," *Engineering Materials and Design*, February 1962, pp. 90-96.

1432. Towner, H. W., and D. O. Blackketter, "Cam Shape Synthesis," *Instr. & Control Systems* 43, September 1970, pp. 109-110.

1433. Tränker, G., "Kurvengetriebe oder Kurbelgetriebe," *VDI-Berichte* 12, 1956, pp. 49-53 ("Cams or Linkages").

1434.   Tseitlin, G. E., and Dashevskaya, I. S., "Analytical Determination of the Radii of Curvature of Cams," *Sov. Eng. Res.* 2, No. 10, October 1982, pp. 43-44.

1435.   Tsfas, B. S., "Acceleration of Roller Follower Shaft in Cam Mechanism with Plane Translatory or Cylindrical Rotary Cam; Relations Are Derived for Determining Roll Acceleration," *Vysshikh Uhcebnykh Zavedenii, Mashinostroenie* 3, 1967, pp. 10-14.

1436.   Turkish, M. C., "Inspection of Cam Contour by the Electronic Method," *Electronics* 23, October 1950, pp. 74-77.

1437.   Turkish, M. C.,"The Relationship of Valve Spring Design to Valve Gear Dynamics and Hydraulic Lifter Pump-Up," *Trans. SAE* 61, 1953, pp. 706-714.

1438.   Turnbull, D. C., "Sixteen Shafts from One Motor; Cleaning Machine for Cast Camshafts," *Mach. Des.* 16, April 1944, pp. 109-110.

1439.   Turnbull, S. R., "Design of Planetary Cams for Short Period Events," *J. Mech. Eng. Sci.* 23, No. 1, February 1981, pp. 13-19.

1440.   Tutunaru, D., and Tr. Demian, "Construction of Cam Profiles with the Aid of a Generating Curve," (in Russian), *Trudi Instr. Mashinoved., Akad. Nauk SSSR* 22, 1961, pp. 85-86, 38-50.

1441.   Uhden, I., "Kurventafeln zur Bestimmung der kleinsten Krümmungsradien an Kurvenscheiben," *Feinwerktechnik* 7, 1971, pp. 313-319 ("Nomograms to Find Minimum Radius of Curvature of Disk Cams").

1442.   Uhing, J., "Lichtelektrische Steuerung für Kurvenkörper-Herstellung," *VDI-Berichte* 14, 1956, pp. 99-101.

1443.   Ulbricht, W., "Die Berechnung der Steuerkurven bei Einspindelautomaten," *Werstatt und Betrieb* 83, 1950, pp. 261-265 ("Calculation of Control Cams in One-Spindle Automatic Dispensers").

1444.   Unterberger, Richard, and Lermann, Peter, "Automatische Messung und Registrierung der Krümmungs- und Neigungsverhältnisse ebener Kurvenkonturen," *Konstruktion* 21, 1969, No. 7, pp. 267-271.

1445.   Unterberger, Richard, "Kurvenanalyse mit einem Bogenhöhentaster," *Ind.-Anz.* 93, 1971, No. 54, pp. 1321-1325.

1446.   Unterberger, Richard, "Some New Ideas for Measuring Curves," *Mechanism and Machine Theory* 9, 1974, No. 3/4, pp. 411-420.

1447.   Urin, D. M., "Standard Tables and Diagrams for Cam Design," *Russian Eng. J.* 45, 1965, pp. 32-36 (Translation of Vestnik Mashinostroeniya 45, 1965, pp. 32-36 by Production Engineering Research Association of Great Britain).

1448.   Vaclavik, M., "Beitrag zur Synthese des Nadelstangenantriebes einer Überwendlich-Nähmaschine," *IFToMM*, 1977, Vol. 111.2, pp. 631-642.

1449.   Vaclavik, M. and Z. Koloc, "Berechnung, Fertigung und Qualitätskontrolle von Kurvenkörpern," *Maschinenbautechnik* 26, 1977, No. 7, pp. 292-296.

1450.   Vaclavik, M. and Z. Koloc, "Programm für die Fertigung und Kontrolle von Kurvenkörpern auf einer NC Maschine mit analytisch nicht festgelegten, sondern z.B messtechnisch bestimmten Konturen," *Maschinenbautechnik* 26, 1977, No. 12, pp. 556-558.

1451. Vail, D., "Development of Proferall Cast Camshafts," *SAEJ* 39, July 1936, pp. 288-291; Abstract, *Iron Age* 137, April 30, 1936, pp. 47-48; Discussion, *SAEJ* 39, July 1936, pp. 291-292.

1452. Vakhtel, V. Yu., et al, "Load Determination for a Balancing-Mechanism Drive," *Russian Eng. J.* LI, 1971, pp. 16-19.

1453. Vandemann, J. E., and J. A. Wood, "Modifying Starwheel Mechanisms," *Machine Design*, April 1953, pp. 255-261.

1454. Van Fossen, C. H., "Interchangeable Cams," *Am. Mach.* 65, December 9, 1926, p. 962.

1455. Varnum, E. C., "Circular Nomogram Theory and Construction Technique," *Prod. Eng.* Annual handbook of Prod. Design for 1953, pp. 28-30.

1456. Varga, O. H., "Valve Actuation; Cams with Flat-Footed Followers," *Automobile Eng.* 36, February 1946, pp. 75-76.

1457. Veldkamp, G. R., "On Mating Profiles in Three-Link Rolling Cam Drives," Conference Mechanisms 72,' London, 1973, pp. 140-142.

1458. Veretennikov, E. A., and Skomorokhov, G. Ya., "Way of Increasing Cam Profile Accuracy," *Sov. Eng. Res.* 3, No. 6, 1983, pp. 84-86.

1459. Viall, E., "Machining Side Cams on a Gear Cutter," *Am. Mach.* 55, July 7, 1921, p. 28.

1460. Viall, E., "Milling Peripheral Cams on a Lathe," *Am. Mach.* 55, July 28, 1921, p. 148.

1461. Virabov, R. V., and E. E. Shchekut'ev, "Design of Gear-Lever and Cam-Gear-Lever Mechanisms with Periodic Rotation," *Russian Eng. J.* LII, 1972, pp. 3-6.

1462. Vogel, A., "Neues Bewegungsgesetz für Umlaufnocken," *Reuleaux-Mitteilungen* 6, 1938, pp. 309-313 ("New Displacement Diagram for Rotating Cams").

1463. Vogel, A., "Umlaufnocken für den Kolbenantrieb von Brennstoff-Einspritzpumpen," *Reuleaux-Mitteilungen* 6, 1938, pp. 541-543 ("Rotating Cams for Piston Drive of Fuel Injection Pumps").

1464. Vogel, A., "Günstige Krümmungshalbmesser bei harmonischen Nocken mit Pilzstössel," *Getriebetechnik* 8, 1940, p. 177 ("Optimum Radius of Curvature in Harmonic Cams with Flat Faced Follower").

1465. Vogt, W., "Developing Cam System Charts for Optimum Performance," *Tool Eng.* 44, March 1960, pp. 115-124.

1466. Volmer, J., R. Brock and G. Uhlig, "Kurvengetriebe für Doppelhub-bewegungen," Wiss. Zeitschrift der TH *Karl-Marx-Stadt* 21, 1979, No. 3, pp. 323-338.

1467. Volmer, J., R. Brock and G. Uhlig, "Übertragungsfunktionen für Kurvengetriebe mit Doppelhubbewegungen," *Maschinenbautechnik* 28, 1979, No. 9, pp. 414-418.

1468. Volmer, J., and E. Huhn, "Kurvengetriebe mit hydrodynamisch geschmierten Gleitelementen," *Maschinenbautechnik* 25, 1976, No. 10, pp. 453-455.

1469. Vorob'ev, E. I., "Question of the Wear-Stability and the Designing of Cam Mechanisms" (in Russian), *Analiz i Sintez Mashin-Avtomatov, Moscow, Nauka*, 1965, pp. 49-61; Ref. Zh., Mekh. No. 12, 1965, Rev. 12 A 149.

1470. Virgazov, A.,"The Radius of Curvature of the Outer and Inner Profiles of a Disk Cam" (in Russian), *Intl. Conf. on Mechanisms and Machines at Varna, Bulgaria*, 1965, II, p. 131.

1471. Vul'fson, I. I., "Selection of a Rational Value of Phase Angle of a Cam Mechanism Subject to Certain Supplementary Conditions" (in Russian), *Trudi., Inst. Mashinoved., Akad. Nauk SSSR* 22, 1961, pp. 85-86, 14-29.

1472. Vul'fson, I. I.,"The Calculation of Vibrations of the Driver of a Cam Mechanism" (in Russian), *Mechanika masin*, 1967, No. 9, pp. 39-54.

1473. Waech, R. R., "Camshaft Keys Fitted in a Jig," *Am. Mach.* 88, May 25, 1944, p. 116.

1474. Waitkus, V., "Single Cam Action Performs Four Different Functions," *Mach.* 39, April 1933, pp. 517-518.

1475. Wallace, C. F., "Long Milling Machine Controlled by Helical Cams," *Tool Engr.* 34, March 1955, pp. 84-85.

1476. Waller, J.,"Screws Permit Minute Adjustments on Cams," *Am. Mach.* 108, March 2, 1964, p. 102.

1477. Walls, F. J., "Cast Camshafts and Crankshafts Possess Many Advantages," *Foundry* 65, March 1937, pp. 28-30, pp. 83-84; April 1937, pp. 60+; Same, *SAEJ* 41, July 1937, pp. 284-290; Excerpts, *Mech. Eng.* 59, November 1937, pp. 857-858.

1478. Watson, H. C., and E. E. Milkins, "The Design of Camshafts for Racing Engines," Reese Jones: Cams and Cam Mechanisms, pp. 54-59.

1479. Watson, J. S.,"Overhead Valves and Silent Chain Front-End Drives," *Automotive Ind.* 55, October 14, 1926, pp. 659-660.

1480. Way, S., "Pitting Due to Rolling Contract," *Transactions of the ASME*, 1935, p. 249.

1481. Weber, T., "Cam Development by Evolute Analysis," *Machine Design* 28, February 9, 1956, pp. 117-118.

1482. Weber, T., "Cam Dynamics via Filter Theory," *Mach. Des.* 32, 1960, pp. 160-165.

1483. Weber, T., "Cams by Computer," *Mech. Eng.* 90, October 1968, pp. 56-57.

1484. Weber, T., "Computer Design and No Machining Make Better Cams," *Mach.* 76, January 1970, pp. 70-72.

1485. Weber, T., "Continuous-Path NC Machining of Computer-Designed Cams," *Mach. and Prod. Eng.* 116, 1970, No. 2993, pp. 453-456.

1486. Weber, T., "Data Processing Gives Cam Designers Increased Latitude," *Tooling and Production* 34, November 1968, pp. 83-85.

1487. Weber, T., "Filter Theory Applied to Cam Dynamics," *Transactions Sixth Conference on Mechanisms*, Purdue University, 1960, pp. 48-54.

1488. Weber, T., "N/C Speedup in the Machining of Cams," *Mauf, Eng. and Mgt.* 66, June 1971, pp. 28-30.

1489. Weber, T. Jr., "Recent Advances in Cam Design," *Instruments and Control Systems* 41, November 1968, pp. 69-71.

1490. Weichelt, H., "Zur Massynthese der Kurvengetriebe," *Maschinenbautechnik*, October 1963, pp. 545-552 ("On the Synthesis of Cam Mechanisms").

1491. Weiler, L., "Neuzeitliche Nockenschalter und Nockenschaltertechnik," *Siemens Zeits.* 31, January 1957, pp. 30-39.

1492. Weise, H., "Kinematik der Filmschaltwerke," *Feinwerktechnik*, 1949, pp. 97-104, and 137-144.

1493. Weise, H., "Bewegungsverhältnisse an Filmschaltgetrieben," *VDI-Berichte* 5, 1955, pp. 99-106 ("Kinematic Criteria in Cinematographic Mechanisms").

1494. Weise, H., "Getriebe in photographischen und kinematograpischen Geräten," *VDI-Berichte* 12, 1956, pp. 131-137 ("Cams in Photographic and Kinematographical Equipment").

1495. Weishaupt, M., "Anwendungsbeispiele für die Einheitsnocken mit sprung-und knickfreiem Beschleunigungsverlauf," *Kraftfahrzeugtechnik* 10, 1960, pp. 168-170 ("Practical Applications of Standard Cams with No Sudden Change in Acceleration").

1496. Wenrich, H. E., "Each Warp Stop Motion Has Some Point of Superiority," *Textile World* 98, December 1948, pp. 125, 202, 204, 206 and 208.

1497. White, F. B., "Constant Pressure-Angle Cam with Roller Follower," *Machinery* (London) 69, July 11, 1946, pp. 47-51.

1498. White, T. S., "Only Usable Equations Satisfy the Busy Designer," *Mach. Des.*, August 1936, pp. 36-38.

1499. Wick, C. H., "Sixteen Cams Cut at One Time on a Special Gear Shaper," *Machinery* 56, April 1950, pp. 168-171.

1500. Wickman, Ltd., "Single Spindle 2-5/8 In. Automatic," *Engineer* 213, June 8, 1962, p. 1015.

1501. Wiederrich, J. L., and B. Roth, "Design of Low Vibration Cam Profiles," Reese Jones: Cams and Cam Mechanisms, pp. 3-8.

1502. Wiederrich, J. L., and B. Roth, "Dynamic Synthesis of Cams Using Finite Trigonometric Series," *Trans. ASME*, Ser. B: J. Engrg. Ind., vol. 97, 1975, No. 1, pp. 287-293.

1503. Wiederrich, J. L., "Residual Vibration Criteria Applied to Multiple Degree of Freedom Cam Followers," *J. Mach. Des. Trans. ASME* 103, No. 4, October 1981, pp. 702-705.

1504. Wildt, P., "Zwangläufige Triebkurvenherstellung," *VDI-Tagungsheft* 1, Getriebetechnik, VDI-Verlag, 1953, pp. 11-20 ("Mechanism-Based Cam Design and Manufacture").

1505. Williams, D. N., "Sometimes the Old Ways Are Best; Cam-Operated Machines Tool Belies Notion Newer is Better," *Iron Age* 207, April 29, 1971, p. 39.

1506.   Wilson, J. W., "Closer Numerical Control via Cam Compensator," *Mach.* 69, March 1963, pp. 114-115.

1507.   Wilson, J. F., "Instrumentation to Control Strain Rate for Materials Undergoing Plastic Flow," *Exptl. Mech.* 4, January 1964, pp. 11-14.

1508.   Wilson, R. A.,"Cutting a Cam of the Lathe," *Am. Mach.* 47, November 22, 1917, p. 912.

1509.   Wittman, E., "Cam Cutting Attachment for the Drill Press," *Am. Mach.* 81, June 30, 1937, p. 578.

1510.   Wittman, E., "Cam Cutting Attachment for the Screw Machine," *Am. Mach.*, June 6, 1934, p. 420.

1511.   Wittmann, E., "Cam-Milling Attachment for the Bench Lathe," *Am. Mach.* 77, September 13, 1933, p. 604.

1512.   Wolford, J. C., and L. Kersten, "On the Optimum Value for a Modified Trapezoidal Indexing Cam," 4th OSU Applied Mechanisms Conference, 1975, pp. 30.1-30.17.

1513.   Wood, F. W., "Simple Types of Cams for Actuating Limit Switches," *Machine Design* 31, March 19, 1959, pp. 189-190.

1514.   Wood, J. K., "Heat Treatment of Steel Cams," *Am. Mach.* 62, January 8, 1925, pp. 47-49.

1515.   Woodcock, G. "Better Cam with Investment Castings; George Woodcock and Son Ltd., *Engineering* 193, June 15, 1962, p. 771.

1516.   Woodcock, W. L., "Magazine Loading Speeds," *Am. Mach.* 91, May 8, 1947, pp. 108-110.

1517.   Wotipka, J. L., "Identification of Failing Mechanisms Through Vibration Analysis" (IBM General Systems Div., Rochester, Minn.) *ASME Paper* 71-Vibr-90 for meeting September 8-10, 1971, p. 12.

1518.   Wright, J. P., "Non-Cylindrical Contours; How You Can Grind Them," *Am. Mach.* 98, May 10, 1954, pp. 137-156.

1519.   Wright, M., "Some Examples of Cam Cutting," *Am. Mach.* 57, August 24, 1922, pp. 306-307.

1520.   Wroble, W., "Nine Cams Generally Used in Mechanical Design," *Design News*, March 1950, pp. 37-41.

1521.   Wunderlich, W., "Contributions to the Geometry of Cam Mechanisms with Oscillating Followers," *J. Mechanisms* 1, 1971, pp. 1-20.

1522.   Yamada, T., U. Katsumi and M. Tanaka, "Development of Constant Peripheral Speed Cam Profile Grinder," *Bull. Japan Society of Precision Engineering* 12, 1978, No. 3, pp. 157-158.

1523.   Yanev, T. K., and M. Y. Milkov, "Power-Geometric Synthesis of Two-Stage Cam Mechanisms with Plane Oscillating Followers.. (in Russian), *Intl. Conf. on Mechanisms and Machines* at Varna, Bulgaria, II, p. 116.

1524.   Yang, Chihou, "Theory of a Method of Generating Cam Profiles," Chi Hsieh Kung ch:eng Hsueh Pao, Vol. 19, No. 4, December 1983, pp. 31-39.

1525.   Yates, K. P., and G. Sinclair, "Cam for Introducing Periodic Functions into Mechanical Drives," *R. Sci. Instr.* 18, June 1947, pp. 454-455.

1526. Yavne, R. O., "High Accuracy Contour Cams," *Product Eng.* 19, August 1948, pp. 134-136.

1527. Young, G. J., "Repairing Broken Cam Shafts," *Eng. Min. J.* 110, December 4, 1920, pp. 1092-1093.

1528. Young, V. C., "Consideration in Valve Gear Design," *Soc. Automotive Engrs.-Trans.* 1, July 1947, pp. 359-365.

1529. Zagar, F., "Cam Compensator for Jig Borer Lead-Screws," *Mach.* 70, November 1963, p. 128.

1530. Zagar, F., "Cam Grinding on a Cylindrical Grinder," *Mac.* 52, February 1946, pp. 180-181.

1531. Zakel, H., "Der Pressungswinkel bei räumlichen Kurvengetrieben," *VDI-Berichte* 321, Düsseldorf, pp. 31-56.

1532. Zbigniew, Jania, "Graphical Method of Determining Cam Shaft Torque," *Design News*, February 15, 1953. p. 71.

1533. Zegalov, L. I., "A Method of Geometric Loci for the Design of the Smallest Cam for a Given Plane Follower," 6, 1949, pp. 69-71.

1534. Zigo, M., "A General Numerical Procedure for the Calculation of Cam Profiles from Arbitrarily Specified Acceleration Curves," *J. Mechanisms* 2, 1967, pp. 407-414.

1535. Zigo, M., "A Method of Specifying the Dimensions of Cams," *J. Mechanisms* 4, 1969, pp. 283-289.

1536. Zimmerli, R. B., "How About Nylon for Wear Parts," *Am. Mach.* 97, June 8, 1953, pp. 150-152.

1537. Zinov'ev, V, A.,"Analytic Method for the Kinematic Study of Cam Mechanisms," (in Russian), *Trudi Sem. teor, Mach. Mekh.* 15, 1955, pp. 38-45.

1538. Zinoviev, V. A., "Kinematische Untersuchung an Nockengetrieben nach einer analytischen Methode" *Arb. Inst. Maschinenkunde* XV, 1955. p. 58.

1539. Zivkovic, Z., "Verbesserung der Laufeigenschaften von Kurvenscheiben mit stationärem Geschwindigkeitsverlauf," *Ind-Anzeiger* 93, January 26, 1971, pp. 153-156 ("Improvement of the Operating Characteristics of Disk Cams with Constant Velocity Ratio").

1540. "Adjustable Cam Compensates for Errors; Patent Assigned to Fairchild Aviation Corp.," device developed primarily for compensating radio compasses, *Machine Design* 15, September 1943, pp. 172+.

1541. "Adjustable-Cam Times," *Engineer* 219, January 8, 1965, p. 107.

1542. "Adjustable Cams Shoe to Drum Clearance Remains the Same Throughout the Life of the Linings," *Diesel Equip. Supt.* 48, February 1970, pp. 62-63.

1543. "Alloy Cast-Irons; Midland Motor Cylinder Co., Ltd.," *Automobile Eng. Extra* 31, November 6, 1941, pp. 383-384.

1544. "Angle Between Cams of a Four-CylinderEngine," *Mach.* 24, March 1918, pp. 650-651.

1545. "Appareil a' fraiser les cames, se montant sur le tour Monarch-Keller," *Genie Civil* 105, September 1, 1934, p. 199 (in French).

1546. "Assembling Camshaft Bushings in Cylinder Blocks; Packard Motor Car Co.," *Mach.* 44, September 1937, p. 33.

1547. "Attachment for Rowbottom Cam Cutting Machine," *Am. Mach.* 67, December 15, 1927, p. 958.

1548. "Auto Camshaft Regrinder Solves Problem Using Hardsurfacing," *Welding J.* 45, October 1966, p. 844.

1549. "Automatic Camshaft-Grinding Machine," *Engineering* 167, June 24, 1949, p. 584.

1550. "Automatic Screw Machine with Interchangeable Cam Units for Quick Change in Set-Ups," *Mach.* 59, September 1952, pp. 218-219.

1551. "Automatic Telephone Trigger Dial," *Engineer* 189, March 24, 1950, p. 370.

1552. "Barrel Cams in Actuator Produce Exotic Motions," *Product Eng.* 42, March 29, 1971, pp. 51-52.

1553. "Bearings Cut from Steel Tubing," *Iron Age* 197, January 27, 1966, p. 75.

1554. "Bewegungsgesetze für Kurvengetriebe," VDI 2143, 1, Theoretische Grundlagen, Düsseldorf, VDI-Verlag, 1980.

1555. "Bouncing Ball" Method of Indication Inspects Irregular Cam Contours," *Machine & Tool Blue Book* 50, October 1955, pp. 164-167.

1556. "British Developments in Cam Turning; Schrivener Turning Lathe for Automotive Camshafts," *Iron Age* 155, February 8, 1945, p. 47.

1557. "British Thomson-Houston Electro-Pneumatic Camshaft Control for Multiple Use Operations," *Engineering* 144, October 8, 1937, pp. 400-401.

1558. "Bronze Cams and Iron Followers in Matched Sets by Powder Metalurgy," *Precision Metal Molding* 10, January 1952, pp. 35,37-40.

1559. "Cam Action Dies," *Iron Age* 171, February 26, 1953, pp. 146-147.

1560. "Cam and Memory Rollers Can Take Up Rotational Shock," *Product Eng.*, March 1971, p. 62.

1561. "Cam and Switch Unit Controls Air-Hydraulic System," *Automation* 9, February 1962, p. 85.

1562. "Cam Automatically Clamps, Unclamps Parts," *Mach. Des.* 28, November 15, 1956, p. 118.

1563. "Cam Calculator," *Tool Eng.* 30, June 1953, p. 139.

1564. "Cam-Controlled Sewing Machine Sews Complex Stitches; Elna Corp. of America, Inc.," *Machine Design* 25, March 1953, p. 140.

1565. "Cam Curves in Common Use; Reference-Book Sheet," *Am. Mach.* 66, May 5, 1927, p. 747.

1566. "Cam-Derived Feul Injection Cuts Down on Fuel Emissions," *Product Eng.* 42, February 15, 1971, p. 62.

1567. "Cam Ejector for High Speed Stamping," *Manuf. Eng. & Mgt.* 68, April 1972, p. 15.

1568. "Cam-Controlled Planet Alters Gear-Train Output," *Machine Design* 37, 1965, p. 167.

1569. "Cam Generator Programmed from Engineering Drawings," *Mach.* 69, June 1963, pp. 148+.

1570. "Cam-Grinding," *Aircraft Production* 15, July 1953, pp. 258-261.

1571. "Cam Locks Wires in New Connector Strip," *Electronics* 32, October 9, 1959, pp. 90-91.

1572. "Cam Mechanism Reduces Lost Motion," *Product Eng.* 27, February 1956, p. 157.

1573. "Cam Mechanism Tailors Spring-Squeeze Rate," *Mach. Des.* 39, August 3, 1967, p. 107.

1574. "Cam Mechanism with Variable Quick-Drop Adjustment," *Mach.* 51, September 1944, p. 182.

1575. "Cam-Operated Deceleration Valves Control Braking of Heavy Members," *Mach. Des.* 31, April 30, 1959, pp. 114-115.

1576. "Cam Operated Nonsynchronous Assembly," *Manuf. Eng. & Mgt.* 67, December 1971, p. 9.

1577. "Cam Profile Milling Fixture," *Mach.* 51, August 1945, . 179.

1578. "Cam Sets Relief Valve Actuation Point," *Product Eng.*, December 7, 1964, p. 112.

1579. "Cam Sprags Are Broached at 100 per hr. Automatically," *Iron Age* 177, February 23, 1956, pp. 166-117.

1580. "Cam Varies Spindle Speed," *Product Eng.* 29, January 29, 1958, p. 66.

1581. "Cam and Gears Team Up in Programmed Motion," *Product Eng.* 70, April 21, 1969, pp. 168-170.

1582. "Cams and Memory Rollers Can Take Up Rotational Shock," *Product Eng.* 42, March 29, 1971, p. 62.

1583. "Cams are Cut with Perfect Helix Direct from Engineering Computations," *Am. Mach.* 94, January 9, 1950, pp. 150-151; *Tool Eng.* 24, February 1950, p. 63.

1584. "Cams by Computer," *Mech. Eng.*, October 1968, pp. 56-57.

1585. "Cams Control Machine Functions in Contour Welder," *Tool & Manuf. Eng.* 47, September 1961, p. 104.

1586. "Cams Correct Registration Errors in Automatic Typesetter," *Control Eng.* 16, June 1969, p. 92.

1587. "Cams Cure Curves in Welding; Contour Welder Produced by Expert Welding Machine Co., *Product Eng.* 32, July 10, 1961, pp. 66-67.

1588. "Cams Drive Steel-Squeezing Dies," *Product Eng.* 29, December 8, 1958, p. 95.

1589. "Cams Make Billet Cutter Operation Safe," *A. M. Steel* 14, November 1961, pp. 92-93.

1590. "Cams Rock Platen in Functionally Styled Press," *Product Eng.* 28, June 1957, pp. 152-154.

1591. "Cams Sequence Pump Motions; Shirley Developments Ltd.," *Product Eng.* 27, October 1956, p. 154.

1592.   "Cams Time Induction Heating Cycles; Ford Motor Co.," *Iron Age* 175, Juen 2, 1955, p. 99.

1593.   "Cam's The Thing—Pressed From Iron Powder and Infiltrated with Brass," *Precision Metal Molding* 11, August 1953, pp. 40-41.

1594.   "Camshaft- and Profile-Turning Lathe," *Engineering* 158, December 15, 1944, pp. 40-41.

1595.   "Camshaft; Chevrolet Manufacturing Methods," *Am. Mach.* 82, June 29, 1938, pp. 566-568.

1596.   "Camshaft Grinding; Naxos-Union Semi-Automatic Machine," *Automobile Eng.* 29, September 1939, pp. 333-334.

1597.   "Camshaft in Head Cuts Parts and Weight; *Product Eng.* 33, May 14, 1962, pp. 100-101.

1598.   "Camshaft Inspection Machine," *Engineer* 223, March 31, 1967, p. 494.

1599.   "Camshaft Milling Machine," *Engineer* 221, April 15, 1966, p. 591.

1600.   "Camshaft Production," *Automobile Eng.* 19, March 1929, pp. 104-106.

1601.   "Camshaft Tester Saves Time and Operator Fatigue," *Tooling and Production*, June 1971, pp. 50-51.

1602.   "Camshaft Turning on Multiple-Load Lathe," *Engineering* 160, August 17, 1945, p. 126.

1603.   "Cast Camshafts; Development and Application of Monikrom High Duty Cast Iron," *Automobile Eng.* 40, September/October 1950, pp. 345-347.

1604.   "Celoron Develops Spoke-Type Gear to Reduce Noise and Temperature," *Automotive Ind.* 66, May 14, 1932, pp. 723-724.

1605.   "Checking Cam Pressure Angles," *Mach. Des.* 29, September 5, 1967, p. 106.

1606.   "Churchill, Cam Grinding Machine," *Automobile Eng.* 15, October 1925, p. 340.

1607.   "Churchill, Cam Grinding Machines," *Automobile Eng.* 21, February 1931, pp. 64-65.

1608.   "Cogged Belts the Key to Overhead Camshafts," *Engineering* 203, May 19, 1967, pp. 820-821.

1609.   "Collins Tailer-Made Cam Profiles," *Mach. Des.* 26, October 1954, p. 161.

1610.   "Combination Cam and Screw Mechanism," *Product Eng.* 18, September 1948, pp. 122-123.

1611.   "Combination Cam Dies Assure Economical Production," 18, September 1948, pp. 122-123.

1612.   "Compound Cam Mechanism; French Netmaking Machine," *Mach.* 45, May 1939, pp. 625-626.

1613.   "Controls Cam Mechanism Clearance; Patent Assigned to Thompson Products, Inc.," *Mach. Des.* 14, July 1942, pp. 126+.

1614.   "Counterweight + Cam Work to Sell in Bags," *Product Eng.*, June 10, 1963, p. 113.

1615.   "Crank and Cam Reclamation," *Diesel Equip. Supt.* 41, July 1963, pp. 28-30.

1616. "Curd Nube Cam Cutting Machine Built in Three Sizes," *Iron Age* 132, October 12, 1933, p. 34.

1617. "Custom Cams Speed Drawing," *Am. Mach.* 111, January 30, 1967. pp. 86-87.

1618. "Cutting Cams as They Are Used," *Am. Mach./Metalworking Manuf.* 107, April 29, 1963, pp. 47-49.

1619. "Dashpot Control of Torsion-Spring Release," *Machine Design* 31, October 1, 1959, p. 113.

1920. "Deburring Pump Drive Gears; Roto-Finish Barrelling Operations," *Automobile Eng.* 49, December 1959, pp. 531-532.

1621. "Der heutige Stand des Nockendrehens," *Werkzeugmaschine* 46, 1942, p. 43 ("Current Technology in Cam Tooling").

1622. "Design for Differential Copy Milling Machine for Prodcuing 3-Dimensional Cams," *Machinery* (London) 85, July 30, 1954, pp. 250-251.

1623. "Design of Cams; Reference-Book Sheet," *Am. Mach.* 66 April 7, 1927, p. 596.

1624. "Design of Disc Cams with Swinging Flat-Faced Followers: Calculation Methods," *ESDU Data Items* 80022, November 1980, Amend A August 1981, 34p.

1625. "Design of Disc Cams with Swinging Roller Followers,: Calculation Methods," *ESDU Data Items* 70921 K, December 1979, Amend A August 1981, 34p.

1626. "Design of Disc Cams with Translating Flat-Faced Followers,: Calculation Methods," *ESDU Data Items* 70923 K December 1979, Amend A August 1981, 28p.

1627. "Design of Disc Cams with Translating Roller Followers: Calculation Methods," *ESDU Data Items* 79007 K, June 1979, Amend A August 1981, 33p.

1628. "Design of Mushroom Cams," *Horseless Age* 37, March 15, May 1, 1916, p. 164.

1629. "Device for Maintaining Cutting Angle of Tool when Turning Cams," *Machinery* 37, November 1930, pp. 188-189.

1630. "Developement of Master Cams for Grinding Machines," *Machinery* (London) 42, June 29, 1933, pp. 371-375.

1631. "Die Casting...Answer for Short Run of Special Shape," *Precision Metal Molding* 11, March 1953, pp. 38-39.

1632. "Digital Readout Aids Cam Milling," *Tooling & Production* 34, August 1968, pp. 91-92.

1633. "Discs Welded to a Pipe Make Low-Cost Cam," *Machine Design* 31, April 30, 1959. p. 113.

1634. "Double Cam Slide Boosts Punch Travel," *Am. Mach.* 110, December 5, 1966, p. 138.

1635. "Double Cam Increase Camshaft Grinding Accuracy," *Mach.* 72, February 1966, p. 93.

1636. "Double Crankshafts Replace Cam Action in High Speed Knitting Machine," *Product Eng.* 18, May 1947, pp. 90-92.

1637.   "Doveltailed Cam Shunts Reaction Force," *Machine Design* 39, February 2, 1967, p. 136.

1638.   "Dreheinrichtung für kreisbogenförmige Querschnitte," *ZVDI* 86, 1942, p. 689 ("Lathe Arrangement for Arc-Shaped Cross Section").

1639.   "Duplicator Control Cuts Milling Time of Screw-Machine Cams," *Steel* 108, February 17, 1941, p. 90.

1640.   "Ein Hilfsmittel zur Fertigung von Kurvenscheiben," *Industrie-Anzeiger* 104, 1982, No. 1, pp. 29-33.

1641.   "Electronically Controlled Cam Milling Machine," *Machinery* (London) 84, January 22, 1954, pp. 193-196.

1642.   "Electronically Controlled Mechanism for Milling Truing-Diamond Guide Cam," *Mach.* 44, June 1938, pp. 677-678.

1643.   "Electrochemical Machining Teams Up with Numerical Control," *Am. Mach.* 110, June 20, 1966, pp. 128-131.

1644.   "Electrolytic Grind Triples Cam Output," *Am. Mach.* 100, December 17, 1956, p. 120.

1645.   "Entire Cam Contour Checked in 40 Seconds, Optical Gaging," *Am. Mach.* 97, August 31, 1953, p. 140.

1646.   "Estimation of Basic Dimensions of Disc Cams with Swinging Followers," *ESDU Data Items* 83008 K, May 1983, 31p.

1647.   "Estimation of Basic Dimension of Disc Cams with Translating Followers," *ESDU Data Items* 82023 K, September 1982, 24p.

1648.   "Exhaust Cam Moved Forward," *Power Pl. Eng.* 25, March 1, 1921, p. 293.

1649.   "Extended Locking Cam Cuts Geneva Chatter," *Mach. Des.*, September 1971, p. 98.

1650.   "Factors in the Life of Cams and Tappets," *Engineering* 195, January 18, 1963, p. 69.

1651.   "Ferris Wheel Machine Doubles Shaft Runs," *Steel* 146, April 4, 1960, p. 157.

1652.   "Finish Grinding Cam Aperture," *Machinery* (London) 82, 859-860.

1653.   "Finishing Airplane Engine Cams," *Mach.* 30, September 1923, pp. 39-40.

1654.   "Fixture for Milling Face-Cam Profiles," *Mech.* 52, November 1945, p. 179.

1655.   "Ford Improves Quality and Cuts Cost by Casting Camshafts," *Iron Age* 136, August 15, 1935, pp. 22-24.

1656.   "Ford Overhead Camshaft Engines," *Automobile Eng.* 60, December 1970, pp. 494-497.

1657.   "Free-Transfer Line Turns Automotive Camshafts," *Automation* 17, June 1970, p. 14.

1658.   "Free-Wheel for Cam Rotor Units," *Fluid Power International*, November 1971, pp. 33-34.

1659.   "Frew Cam Milling Machine," *Mach.* 42, May 1936, p. 624.

1660.   "Gaging Cams Optically Saves Time and Money," *Tool Eng.*, November 1953, pp. 63-64.

1661. "Gaging Fixture for the Relation of the Cams to Keyway," *Automotive Ind.* 43, September 2, 1920, p. 457.

1662. "Gearless, Variable Speed Reducer," *S. African Min. & Eng. J.* 72, October 5, 1962, pp. 732-734.

1663. "Gear with Straight Teeth Rotates Table in Precise Steps," *Product Eng.* 40, April 21, 1969, p. 170.

1664. "Grinder Reproduces Three-Dimensional Cams," *Tool Eng.* 42, January 1959, p. 61.

1665. "Grinding Shaper Cross-Feed Cams," *Mach.* 27, March 1921, pp. 653-654.

1666. "Hard-Facing Cams," *Iron Age* 149, April 23, 1942, p. 35.

1667. "Herbert Automatic Camshaft Grinding Machine," *Engineer* 187, June 3, 1949, pp. 617-618.

1668. "Herstellen von Kurven und unregelmässige Formen auf der Hinterdrehbank," *Werkstattstechnik* 27, 1933, p. 206 ("Manufacture of Cams and Irregular Shapes on a Lathe Bed").

1669. "Herstellung unregelmässiger Flächen auf der Zahnradstossmaschine," *Werkstattstechnik* 39, 1949, p. 280 ("Design of Non-Uniform Surfaces on a Gear Shaper").

1670. "Hey Cam Milling Machine," *Automobile Eng.* 30, September 1940, pp. 273-274.

1671. "High Pressure Compressor Pistons Driven by Keystone," *Prodcut Eng.* 23, April 1952, pp. 140-141.

1672. "High Speed Cam Drives Indexing Table; Ferguson Machine Corp.," *Product Eng.* 28, January 1957, p. 155.

1673. "Hydraulically-Loaded Cam Milling Machine," *Engineer* 142, October 8, 1926, p. 388.

1674. "Höhere Effektivität durch Ausnutzung der Hubtoleranzen," *Maschinenbautechnik* 9, 1969, pp. 455-457.

1675. "Impulse Angle Controls Drive Speed," *Machine Design* 40, 1968, No. 19, p. 176.

1676. "Induction Hardening of Camshafts; Packard Motor Car Co.," *Steel* 102, May 2, 1938, pp. 56-57.

1677. "Induction Heat Treatment Machine for Camshafts," *Engineering* 183, May 10, 1957, p. 593.

1678. "Ingenious Machining of Small Part," *Mill & Factory* 55, September 1954, pp. 118-121.

1679. "Intermittent Movements and Mechanisms," *Product Eng.*, September 1941, pp. 488-489.

1680. "Introduction to the Design of Cams Using Radial Followers," *ESDU Data Items* 66009 K, August 1966, 15p.

1681. "Inshaw Overhead Camshaft Drive," *Automobile Eng.* 18, October 1928, pp. 382-383.

1682. Investment Cast Gears and Cams in Custombuilt Machines," *Precision Metal Molding* 10, September 1952, pp. 36-38.

1683.   "Job Spot Plated in Day vs. Two Week Conventionally; Warner &
        Swasey Co., *Steel* 151, October 1, 1962, p. 44.

1684.   "Keller Duplex Cam Grinding Machine," *Am. Mach.* 65, October,
        1926, p. 691.

1685.   "Kurvengetriebe und ihre praktische Anwendung," KDT (Hrsg.), Vor-
        träge der KDT Informationstagung Kurvengetriebe am 31, 1, 1974 an
        der TU Dresden, Karl-Marx-Stadt: Forschungszentrum des Werkzeug-
        maschinenbaus, 1974.

1686.   "Kulissenfräsmaschine," *Werkzeugmaschine* 46, 1942, p. 96 ("Cross-
        head Guide Cutters").

1687.   "Landis Attachment for Grinding of Cams," *Automotive Ind.* 43,
        Novmeber 4, 1920, pp. 921-922; *Iron Age* 106, November 4, 1920,
        p. 1204; *Mach.*, November 1920, pp. 292-294.

1688.   "Landis Camshaft Grinding Machine," *Automobile Eng.* 20, December
        1930, pp. 483-484.

1689.   "Lees-Bradner Milling Machine Produces Disc- and Face-Type Cams to
        Nine In. in Diameter," *Am. Mach./Metalworking Manuf.* 107, June 24,
        1963, p. 81.

1690.   "Les Cames: Exemple de Calcul," Prat Ind. Mecan, 51 (56), 1968, No.
        6, pp. 189-195.

1691.   "Leyland Diesel Reverts to Fixed-Head Design," *Engineer* 226, August
        30, 1968, pp. 311-314.

1692.   "Linkages Gain on Cams," *Product Eng.* 28, December 23, 1957, p.
        75.

1693.   "Longer Service Life Obtained from Powdered Metal Parts," *Precision
        Metal Molding* 11, August 1953, p. 68.

1694.   "Machinable Carbide Used for Coining Steel," *Engineering* 192,
        December 15, 1961, p. 779.

1695.   "Machine Camshafts on Tracer Lathes," *Mach.* 74, May 1968, pp. 84-
        85.

1696.   "Machining; 3-D Cam-Cutters Work to + or -0.002 In.," *Iron Age* 172,
        August 27, 1953, p. 119.

1697.   "Making Master Cams for Finishing Gas Turbine Blades," *Machy
        (London)* 86, February 11, 1955, p. 2204.

1698.   "Making Precision Points in Space," *Am. Mach.* 102, September 22,
        1958, pp. 100-101.

1699.   "Manufacture of Camshafts," *Automobile Eng.* 16, October 1926, pp.
        374-375.

1700.   "Master Cams Control Cutters; Turn Miller, Developed by Gisholt
        Machine Co., " *Product Eng.* 18, October 1947, pp. 98-99.

1701.   "Mechanism for Making Quick Change in Angular Positions of Cams,"
        Mach. 41, July 1935, p. 667.

1702.   "Milling Cam-Turning Lathe," *Automobile Eng.* 17, September 1972,
        p. 338.

1703.   "Milling Cams for a Textile Machine," *Mach.* 29, March 1923, p. 555.

1704. "Milling Cams for Printing Presses," *Mach.* 30, November 1923, pp. 212-213.

1705. "Milling Cams Without a Master or Lay-Out; Disk Cams for Use on Brown & Sharpe Automatic Screw Machines," *Mach.* 54, February 1948, pp. 178-179.

1706. "Miniature Bearings as Cam Followers," *Engineering* 199, May 7, 1965, p. 616.

1707. "Minicomputer Simplifies Cam-Cutting Operations," *Automation*, November 1970, p. 13.

1708. "Modern Foundry Methods; Making Cylinder Blocks, Brake Drums, and Cas Camshafts at the Birmid Works," *Automobile Eng.*, February 1938, pp. 49-50.

1709. "Modular Camshaft Design," *Mach. Des.* 30, January 23, 1958, p. 117.

1710. "Molded Multi-Cam, Modular Switch Improve Timer for Appliances," *Product Eng.* 41, April 13, 1970, pp. 66-67.

1711. "Monarch Cam-Milling Machine Attachment," *Am. Mach.* 78, February 14, 1934, p. 166; *Iron Age* 133, February 15, 1934, p. 36; *Automotive Ind.* 70, March 3, 1934, p. 184; *Mach.* 40, March 1934, p. 432.

1712. "Multiple Grinding of Automotive Shafts Jumps Production Rate," *Steel* 123, August 9, 1948, p. 101.

1713. "Multiple Spindle Bar Machine Eliminates Interchangeable Cams," *Product Eng.* 17, August 1946, p. 100.

1714. "Muscular Cams Lift Six Million Pound Load," *Mach. Des.* 37, April 1, 1965, p. 129.

1715. "New Cam Grinding Device of British Design," *Mach.* 37, April 1931, pp. 592-593.

1716. "New Flex-O-Timer Cycle Controller-Process Coordinator," *Taylor Technology* 2, Autumn 1949, pp. 22-25.

1717. "New Self-Contained Machine Introduces Improvements in Cam Making; Helix-Master," *Mill & Factory* 46, June 1950, p. 163.

1718. "New Tools, Automated, Make '55 Cars; Pontiac Camshaft Production Up 25%," *Am. Mach.* 98, November 8, 1954, pp. 128-130.

1719. "Nockendreheinrichtung," *Werkstattstechnik* 30, 1936, p. 440 ("Special Lathe for Cams").

1720. "Nockenform-Messmaschine mit elektronischer Auswertung," *Werkstatt and Betrieb* 108, 1975, No. 3, p. 140.

1721. "Nonlinear Cam Tailors Controller Gain," *Machine Design* 41, October 30, 1969, p. 104.

1722. "Norton Automatic Cam Grinding Machine," *Automotive Ind.* 65, October 17, 1931, pp. 604-605; *Iron Age* 128, October 15, 1931, p. 1007; *Mach.* 38, November 1931, pp. 218-219; *Steel* 89, October 26, 1931, pp. 49-50.

1723. "Norton Automatic Indexing Cam Grinding Attachement," *Am. Mach.* 67, September 1, 1927, p. 369.

1724. "Norton New Automatic Cam Grinder," *Automobile Eng.* 22, May 1932, pp. 236-237.

1725. "Oilgear 15-Ton Camshaft Bushing Press," *Am. Mach.* 80, October 21, 1936, p. 898; *Automotive Ind.* 75, September 19, 1936, p. 393.

1726. "Opeations on the Ford V-8 Camshaft," *Am. Mach.* 76, August 3, 1932, pp. 904-907.

1727. "Optical Inspection of Irregular Cam Conturs," *Can Metals* 16, September 1953, pp. 66 and 68; *Tool Engr.* 31, November 1953, pp. 63-64; *Western Machy. & Steel World* 44, October 1953, pp. 94-95.

1728. "Optics Check Cam Profile," Tool & Manuf. Eng. 54, April 1965, pp. 42-43.

1729. "Overhead Camshaft Pontiac Engine," *Engineer* 220, July 30, 1965, p. 211.

1730. "Overhead Camshaft Six Cylinder Pontiac Tempest Engine," *Engineer* 221, June 24, 1966, pp. 999-1000.

1731. "Packard Adopts Tocco Hardening Process for Camshafts," *Mach.* 44, April 1938, pp. 524-525.

1732. "Packard Axle Shafts Carburized in Gas-Fired Furnaces," *Iron Age* 135, May 2, 1935, pp. 24-25.

1733. "Packard Camshafts Induction Hardened," *Iron Age* 141, January 13, 1938, p. 37.

1734. "Part Feed and Memory Systems Synchronized by Single Barrel Cam," *Design News* 9, April 15, 1954.

1735. "Plastic Cams Control Automatic Sewing Machine," *Product Eng.* 24, March 1953, pp. 708-709.

1736. "Pontiac Overhead Camshaft Engine," *Automobile Eng.* 56, September 1966, pp. 382-390.

1737. "Power or Silence Decided by Cams," *Automobile* 30, May 7, 1914, pp. 960-061.

1738. "Precision Cam Design Methods and Analyses Computation of Machine Settings," *Product Eng.* 16, December 1945, pp. 845-848.

1739. "Precision Grinding Machines; Cam Shaft Grinding," *Engineer* 154, December 16, 1932, pp. 612-613.

1740. "Producing High Accuracy Cams," *Machinery* 89, July 13, 1956, pp. 100-102; July 20, 1956, pp. 156-162.

1741. "Production of Calculating and Ticket-Issuing Machines," *Machy (London)* 92, May 23, 1958, pp. 1192-1203; June 27, 1958, pp. 1492-1500; 93, July 9, 1958, pp. 60-76; July 30, 1958, pp. 232-242.

1742. "Production of Cash Registers and Accounting Machines," *Machinery* (London) 93, September, 24, 1958, pp. 692-704; October 8, 1958, pp. 1156-1168.

1743. "Producto Automatic Cam Milling Machine," *Mach.* 30, May 1933, pp. 610-611.

1744. "Producto-Matic No. 4, Thread and Cam Milling Machine," *Am. Mach.* 78, April 11, 1934, p. 296.

1745. "Profile Turning; Development for Maching Cam Forms," *Automobile Eng.* 34, December 1944, p. 546.

1746. "Rapid Maching of Multi-Cylinder Camshafts," *Automotive Ind.* 42, February 12, 1920, pp. 462–463.

1747. "Rapid Milling of Cam Surfaces on Railroad Steam Couplings," *Mach.* 58, February 1952, pp. 187–188.

1748. "Rechenprogramme für die Auslegung von Kurvengetrieben (Dokumentation)," Forschungszentrum des Werkzeugmaschinenbaus, Karl-Marx-Stadt (Hrsg.), *Karl-Marx-Stadt,* 1974.

1749. "Reciprocating Engine with Half as Many Parts; Novel Use of a Cam," *Power Ind.* 74, October 1958, p. 13.

1750. "Recoil of 20mm Gun Gets Lost in Cam Mechanism; Aircraft Gun," *Product Eng.* 30, October 1958, p. 13.

1751. "Recommended Practice for the Carburizing and Heat-Treatment of Camshafts," *Am. Mach.* 64, January 14, 1926, p. 46; *Automotive Ind.* 54, January 21, 1926, pp. 111–112; *Mach.* 32, February 1926, pp. 496–497.

1752. "Recording Camshaft Comparator Features Reproducibility of Results," *Tool Eng.*, pp. 92–93.

1753. "Redesigned Garvin 12-Inch Cam Cutting Machine," *Automotive Ind.* 42, March 4, 1920, p. 609.

1754. "Reo Camshaft Machining Practice has Interesting Features," *Automotive Ind.* 59, October 6, 1928, pp. 490–491.

1755. "Reversed Grinder Set-Up for Forming Master Cams," *Mach.* 51, November 1944, p. 168.

1756. "Revolver Principle Weaves Four-Color Cloth," *Product Eng.* 30, August 31, 1959, p. 37.

1757. "Roller Cam is Adjustable for Container Filling Machines; Patent Assigned to Food Machinery Corp.," *Mach. Des.* 17, May 1945, p. 150.

1758. "Rolling-Contact Cams Vary Suspension Spring Rate," *Mach. Des.* 37, February 4, 1965, p. 23.

1759. "Rowbottom Model 300 Universal Cam Milling Machine," *Am. Mach.* 75, August 20, 1931, pp. 321–322.

1760. "Saved 86% on Part Cost; Stratos Div., Fairchild Engine & Airplane Corp." *Steel* 136, February 14, 1955, p. 106.

1761. "Screw Machine Work; Pullouts for Drills," *Am. Mach.* 77, 78, October 11, 1933, February 28, 1934.

1762. "Scrivener Multi-Tool Cam Turning Machine," *Automobile Eng.* 27, May 1937, p. 183.

1763. "Selection of DRD Cam Laws," *ESDE Data Items(* 82006 K, March 1982, 26p.

1764. "Selective Hardening Automobile Camshafts," *Steel* 124, January 10, 1949, p. 62.

1765. "Sh-h-h! Nylon at Work; How It's Used to Cut Noise and Power Requirements; Nylon Cams, Abbott Machine Co. Automatic Traveling Spindle Winders," *Plant Eng.* 12, November 1958, p. 93.

1766. "Shop Made Cam Milling Attachment," *Am. Mach.*, July 17, 1955, p. 130.

1767. "Shop Modified Diesinker Tracer-Grinds Cams," *Iron Age* 177, April 26, 1956, p. 108.

1768. "Shorter Process Camshaft-Hardening Machine," *Engineering* 152, September 5, 1941, pp. 186-187.

1769. "Sintered Nylon Switch Cam Follower Replaces Metal Parts," *Electro-Tech.* 67, February 1961, p. 148.

1770. "Sinusoidal Acceleration Cams-Calculation Methods for Cam Profiles Using Radial Followers and No Operning or Closing Ramps," *ESDU Data Items* 66026 K, August 1966, Amend A July 1970, 39p.

1771. "Slidable Camshafts-In This Case Arranged for Compression Relief Only," *Automobile* 31, October 29, 1914, p. 807.

1772. "Smooth Links Blend New Machine to Current Production Line," *Iron Age* 185, February 11, 1960, pp. 150-151.

1773. "Soldered-On Wire Cams," *Mach. Des.* 28, August 9, 1956, p. 81.

1774. "Solenoid-Operated Escapement Provides Positive Control for Difficult Products in Pushbutton Vending Machine," *Product Eng.* 34, June 10, 1963, p. 113.

1775. "Some Flame Hardening Machines," *Engineer* 172, July 4, 1941, pp. 6-7.

1776. "Sonderschleifmaschine für Nocken," *Werkzeugmaschine* 36, 1932, pp. 40-41 ("Special Tool Grinders for Cams").

1777. "Special Cam-Cutting Machine for Fairchild Electric Gun Sight," *Aeronautical Eng. R.* 3, November 1944, p. 161; *Steel* 117, July 2, 1945, p. 159.

1778. "Special Methods in a Jobbing Crankshaft and Camshaft Shop; Atlas Mfg. Co.," *Mach.* 46, November 1939, pp. 185-188.

1779. "Special-Shape Cams Reduce Speed Smoothly; CONtorq, Made by Fairbairn, Lawson Combe and Barbour Ltd.," *Product Eng.* 31, December 26, 1960. p. 47.

1780. "Spring-Loaded Cams are Easily Adjusted to Drive Motor," *Product Eng.* 41, March 30, 1970, p. 61.

1781. "Spring Pressure for Constant Acceleration Cams," *Horseless Age* 37, February 1, 1916, p. 119.

1782. "Stacked Cams Drive Carousel Projector," *Machine Design* 34, April 12, 1962, pp. 128-129.

1783. "Staff Report from Product Engineering: Linkages Gain on Cams," *Product Eng.*, December 1957, p. 75.

1784. "Stamp Dispenser Counts on Cam Wheels," *Machine Design* 33, August 3, 1961, pp. 100-101.

1785. "Stationary Cam Programs Merry-Go-Round Capper," *Machine Design* 31, November 26, 1959, pp. 138-139.

1786. "Stroke of Oscillating Rotating Shaft is Varied While Turning," *Product Eng.* 39, August 26, 1938, p. 52.

1787. "Surface Hardening; Recent Developments of the Shorter Process," *Automobile Eng.* 29, January 1939, pp. 16-18.

1788. "Tailor Made Cam Profiles," *Machine Design*, October 1954, p. 161.

1789.   "Tangential Cams," *Automotive Ind.* 42, February 5, 1920, pp. 422–423.

1790.   "Tappet and Cam Wear," *Automobile Eng.* 50, September 1960, p. 351.

1791.   "Temperature Controls Cams Follower," *Machine Design* 22, June 1950, pp. 128-129.

1792.   "Testing Camshafts; A New Optical Instrument for Precision Work," *Automobile Eng.* 22, October 1932, p. 467.

1793.   "3-D Cams by Numerical Control; Bendix Aviation Cuts Cam Output Time 90%," *Am. Mach.* 99, November 7, 1955, pp. 172-173.

1794.   "Time and Costs Slashed by Digital Cam Cutter. It's Eight-Times Faster than Other NC Tools," *Iron Age* 206, September 3, 1970, p. 69.

1795.   "Timing Belt and Chain Drives for Cars," *Engineer* 225, January 5, 1968, p. 28-29,

1796.   "Timing Drives; Some Notes on American and English Practice," *Automobile Eng.* 30, January 1940, pp. 11-13.

1797.   "Tiny Internal Cam Miller Devised," *Iron Age* 154, November 9, 1944, p. 61.

1798.   "Tolerances Get Fast Check; Cam Ring is Used in Torque Converters," *Steel*, July 13, 1953, p. 145.

1799.   "Transfer Line Turns for the Better," *Iron Age* 205, May 7, 1970, p.

1800.   "Triumph Engine for Saab 99," *Engineer* 224, December 1, 1967, pp. 734-735.

1801.   "Turning Cams on a Lathe," *Mach.* 33, September 26, 1927, pp. 51-52 and 371-372.

1802.   "Turret Cams Control Rotary Capper," *Product Eng.* 24, October 1953, pp. 156-157.

1803.   "Unicams for B & S Automatic Screw Machines," *Am. Mach.* 76, September 28, 1932, pp. 1048-1049.

1804.   "Unique Lathe Turns Cams at Single Operation," *Iron Tr. R.* 68, February 24, 1921, p. 564.

1805.   "Unique Mechanism Reloads Camshafts in Nine Seconds; Seneca Falls Machine Co.," *Mach.* 60, April 1954, pp. 224-228.

1806.   "Unit Pares Cam Cutting Time 50%," *Steel* 152, May 13, 1963, p. 155.

1807.   "Use Self-Lubricated Sinterings Where You Can't Lubricate," *Precision Metal Molding* 11, July 1953, pp. 30-31.

1808.   "Vauxhall Viva Gets OHC Engine," *Engineer* 225, June 28, 1968, p. 1009.

1809.   "Vinco Recording Camshaft Comparator Features Reproducibility of Results,"*Tool Eng.* 34, January 1955, pp. 92-93.

1810.   "Vorrichtungen zum Kurvenfräsen," *Maschienbau der Betrieb* 13, 1934, p. 371 ("Equipment Used for Cam Tooling").

1811.   "Vorrichtung zur Unwandlung einer kontinuierlichen Drehbewegung in eine absatzweise hin-und hergehende Drehbewegung," Texas Instruments Deutschland GmbH (Patentinhaber), *Konstruktion* 29, 1977, No. 2, pp. 77-78.

1812.   "Wear Resistance; Fescol and Lemet Methods of Chromium-Plating Cylinders, Crankshafts and Camshafts," *Automobile Eng.* 26, August 1936, pp. 819-820.

1813.   "Wear Testing of Cam Rolls," *Iron Age* 133, June 14, 1934, pp. 16-17.

1814.   "Wider Use of Cast Alloy for Crankshafts and Camshafts," *Automotive Ind.* 76, May 15, 1937, pp. 742-743.

1815.   "Wirtschaftliches Fertigen von Trochoidenformen im Abwälzverfahren," Fa. Burr - Ludwigsburger Maschinebau GmbH (Hrsg.), Ref. in *VDI-Z* 116, 1974, No. 1, pp. 20-21.

1816.   "Zeiss Optical Device Checks Accuracy of Cam," *Iron Age* 128, September 17, 1931, p. 769; *Automotive Ind.* 65, September 5, 1931, p. 367.

1817.   "Zuführeinrichtung mit Kurvenscheibensteuerung," Siemens AG, *VDI-Z* 119, 1977, No. 14, p. 736, also in *Werkstattstechnik* 67, 1977, No. 5, p. 296.

References of which the author has made special use (see Bibliography):

205, 238, 366, 427, 523, 598, 599, 635, 769, 854, 986, 1017, 1047, 1049, 1050, 1051, 1379, 1418, 1494, and 1504.

Index

Acceleration determination, 142–145
  in the circular-arc straight-line cams,
    126–134
  in the crossed slide-crank, 242–244
  in the four-bar linkage, 207–219
  in the inverted crossed slide-crank,
    235–242
  in the modified cycloid, 42–45
  in the slider-crank, 216–219
Accuracy of cams, 186
Arrangement of springs, 196

Calculation of stresses, 153–157
Cam accuracy, 186
Cam-actuated intermittent worm-drive
    mechanism, 190–192
Cam-actuated toggle mechanism for
    operating pressure pad, 198
Cam-and-rack movement for
    increasing the throw of a lever,
    200, 201
Cam checking, 169
Cam
  closed-track, 3, 6
  conical, 6
Cam copying machines, 169
Cam displacement curves
  comparison of, 30
  types of, 10–30

Cam followers, 9, 10
Cam material, 152
Cam, open track, 6
Cam profile calculation
  for offset translating roller follower,
    65–68
  for radial translating roller follower,
    57–65
  for swinging centric flat-faced
    follower, 72, 73
  for swinging eccentric flat-faced
    follower, 73
  for swinging roller follower, 68–71
  for translating flat-faced follower, 70,
    71
Cam profile determination
  for flat-faced translating follower,
    53–55
  for offset translating roller follower,
    49, 50
  for oscillating flat-faced follower, 56
  for radial translating roller follower,
    47, 48
  for swinging roller follower, 50–52
Cam profile, layout of a circular-arc
    straight-line, 126
Cam size
  for swinging roller follower, 85–89
  for translating follower, 76–79

Cams with constant diameter, 134–137
Cams with swinging roller followers, globidal, 7
Cam used in a sewing machine, 7
Cam with swinging roller follower and closed-track, spherical, 7
Cam with swinging roller follower and open track, spherical, 7
Checking of cams, 169
Circular-arc straight-line cam profile, layout of a, 126
Circular-arc straight-line cams, determination of velocities and accelerations in, 125–133
Circular cam mechanism proportioning, 106–116
Circular cams with swinging roller follower, proportioning of, 117–119
Circular cams with translating roller follower, proportioning of, 119
Closed-track cam, 3
Closed track conical cam, 6
Comparison of cam displacement curves, 30
Computer programs for
  analysis and synthesis of cam mechanisms, 253–268
  translating roller follower, 254–266
  swinging roller follower, 266–268
Conical cam, 6
Constant diameter cams, 134–137
Constant-velocity motion curve, 11, 12
Crossover shock, 186
Cubic curve No. 1, 18
Cubic curve No. 2, 20
Cubic curve No. 3, 20, 21
Curves, comparison of cam displacement, 30
  constant-velocity motion, 11, 12
  cubic curve No. 1, 18
  cubic curve No. 2, 20
  cubic curve No. 3, 20, 21
Curves, cycloidal motion, 15, 16
  double harmonic motion, 17, 18
  dwell-rise-dwell-return (DRDR), 170–176
  dwell-rise-return (DRR), 176, 177

4-5-6-7 polynomial, 24
  modified cycloid, 30
  modified sinusoidal acceleration, 28, 29
  modified trapezoidal acceleration, 26–28
  parabolic motion, 13, 14
  rise-return (RR), 175
  simple harmonic motion, 14, 15
  3-4 polynomial, 22
  3-4-5 polynomial, 23, 24
Cycloidal motion curve, 15, 16
Cylindrical cam with translating roller follower, 4

Deflection
  dynamic, 182
  increment, 181, 182
Determination of acceleration, 142–145
  for the modified cycloid, 42–45
  in circular-arc straight-line cams, 126–134
  in the crossed slide-crank, 242–244
  in the four-bar linkage, 207–219
  in the inverted crossed slide-crank, 235–242
  in the slider-crank, 216–219
Determination of cam size
  for swinging roller follower, 86–90
  for translating follower, 77–80
Determination of velocities
  in circular-arc straight-line cams, 126–134
  in the four-bar linkage, 206–207
  in the modified cycloid, 42–45
Dimensions, linkage, 146
Displacement diagram, simple, 8
Double-faced cam for double the stroke at half the pressure angle, 200
Double harmonic motion curve, 17, 18
Drawing of cam profile directly on steel plate, 161
Dwell-rise-dwell-return (DRDR) curve, 170–176
Dwell-rise-return (DRR) curve, 176, 177

Dynamic deflection, 182
Dynamic forces, 142

Effective weight, 183–185
  of push rod, 183
  of rocker arm, 184
  of tappet, 184
  of valve unit, 184
Engraving mechanism, 188

Factors
  affecting wear life, 154, 155
  influencing cam forces, 141
  influencing stresses in cams, 141
Fall and rise of follower over only
    72 degree cam shaft rotation, 201,
    202
Finite difference calculations in
    kinematic analysis, 244–249
Flat-faced translating follower cam
    profile determination, 54–56
Flocke method, theory of the, 92–94
Followers, cam, 9, 10
  offset translating roller, 1, 2
  radial translating roller, 1
  swinging flat-faced, 3
  swinging roller, 2
  translating flat-faced, 2
Forces
  dynamic, 142
  spring, 150, 151
4-5-6-7 Polynomial curve, 24

Globidal cams with swinging roller
    followers, 7
Graphical construction of the
    modified cycloid for given
    requirements, 39–41
Graphical method for proportioning a
    circular cam mechanism, 106–116

Increment deflection, 181, 182
Intermittent motion from two
    synchronized cams, 192–194

Kinematic inversion of the spherical
    cam, 7

Layout of a circular-arc straight-line
    cam profile, 126
Linkage dimensions, 146

Machines for cam copying, 169
Manufacturing
  cam profiles composed of circular
    arcs and straight lines, 161, 162
  circular cams, 159, 160
  precision-point cam, 162–168
Matching of constant velocity and
    parabolic motion curves, 32–36
Matching point, significance of the
    position of the, 37–39
Material for cams, 152
Mechanisms
  for changing stroke while machine
    is running, 196
  for making quick change in angular
    positions of feed-cams, 195
  for reducing size of cam, 204, 205
  for varying angular velocity, 203
Method of finite difference calculations
    in kinematic analysis, 244–249
Minimizing vibration, 186
Modified cycloid cam
  curve of, 30
  determination of velocity and
    acceleration for the, 42–45
  graphical construction of the, 39–41
Modified sinusoidal acceleration curve,
    28, 29
Modified trapezoidal acceleration
    curve, 26–28

Offset translating roller follower, 1, 2
  cam profile determination, 46–72
Open track conical cam, 6

Parabolic motion curve, 13, 14
Polydyne cam design, 170–187
Precision-point cam manufacture,
    162–168
Pre-lift, 181
Pressure angle, (def.) 75, 146–150
  importance of, 75, 76
Proportioning
  circular cam mechanism, 108–118

Proportioning (*continued*)
  circular cam with swinging roller
    follower, 117–119
  circular cam with translating roller
    follower, 119

Radial translating roller follower, 1
  cam profile determination of, 48–50,
    57–65
Radius of curvature, 94–105
Ramp, 181
Rise-return (RR) curve, 178

Significance of the position of the
    matching point, 37–39
Simple displacement diagram, 8
Simple harmonic motion curve, 14, 15
Slide motion differential, 196, 197
Slider-crank, 120–125
Sliding cam, 198, 199
Spherical cam with swinging roller
    follower
  and closed track, 7
  and open track, 7
Spring arrangement, 190
Spring forces, 150, 151
Stress calculation, 153–157
Stroke-multiplying mechanism, 200
Swinging arm follower, illustrative
    problem, 137–140
Swinging centric flat-faced follower
    cam profile calculation, 72, 73
Swinging eccentric flat-faced follower
    cam profile calculation, 73, 74
Swinging flat-faced follower, 3
Swinging flat-faced follower cam
    profile determination, 72, 73

Swinging roller follower, 2
  cam profile determination, 50–52,
    68–71
  determination of cam size for, 85–89

Theory of the Flocke method, 92–93
3-4 Polynomial curve, 22
3-4-5 Polynomial curve, 23
Time-displacement diagram, 12
Transferring cam drawing directly to
    steel plate, 160, 161
Translating flat-faced follower, 2
  cam profile calculation of, 71, 72
Translating follower, determination of
    cam size for, 75–85
Trapezoidial acceleration curve, 24, 25
Two movements obtained from one
    cam, 188, 189
Types of cam displacement curves,
    10–30

Varing the cam dwell with two
    adjustable follower rollers, 194,
    195
Velocity determination
  in circular-arc straight-line cams,
    126–134
  in the four-bar linkage, 206,
    210–216
  in the modified cycloid, 42–45
Vibration minimization, 186
Vibrations of cams designed by the
    polydyne method, 187

Wear life, factors affecting, 154, 155
Worm-drive mechanism
  cam-actuated intermittent, 190–192

Milton Keynes UK
Ingram Content Group UK Ltd.
UKHW051852071024
449327UK00025B/1917

9 780367 451509